T0305982

X-Parameters

This is the definitive guide to X-parameters, written by the original inventors and developers of this powerful new paradigm for nonlinear RF and microwave components and systems.

Learn how to use X-parameters to overcome intricate problems in nonlinear RF and microwave engineering, as the general theory behind X-parameters is carefully and intuitively introduced, and then simplified down to specific, practical cases, providing you with useful approximations that will greatly reduce the complexity of measuring, modeling, and designing for nonlinear regimes of operation.

Containing real-world case studies, definitions of standard symbols and notation, detailed derivations within the appendices, and exercises with solutions, this is the definitive stand-alone reference for researchers, engineers, scientists, and students looking to remain on the cutting edge of RF and microwave engineering.

David E. Root is an Agilent Research Fellow at Agilent Technologies. He co-led the Agilent research and technical development of X-parameters through its commercialization. He is a Fellow of the IEEE, and co-editor of *Nonlinear Transistor Model Parameter Extraction Techniques* (2011).

Jan Verspecht is a Master Research Engineer at Agilent Technologies. He invented X-parameters in 1996, and is a Fellow of the IEEE.

Jason Horn is an Expert Design Engineer at Agilent Technologies, who has been heavily involved in the development of X-parameter measurements.

Mihai Marcu is a Senior Consultant at Agilent Technologies, who is deeply involved in the development and application of X-parameters for nonlinear modeling.

The Cambridge RF and Microwave Engineering Series

Series Editor
Steve C. Cripps, Distinguished Research Professor, Cardiff University

Peter Aaen, Jaime Plá and John Wood, *Modeling and Characterization of RF and Microwave Power FETs*

Dominique Schreurs, Máirtín O'Droma, Anthony A. Goacher and Michael Gadringer, *RF Amplifier Behavioral Modeling*

Fan Yang and Yahya Rahmat-Samii, *Electromagnetic Band Gap Structures in Antenna Engineering*

Enrico Rubiola, *Phase Noise and Frequency Stability in Oscillators*

Earl McCune, *Practical Digital Wireless Signals*

Stepan Lucyszyn, *Advanced RF MEMS*

Patrick Roblin, *Nonlinear RF Circuits and the Large-Signal Network Analyzer*

Matthias Rudolph, Christian Fager and David E. Root, *Nonlinear Transistor Model Parameter Extraction Techniques*

John L. B. Walker, *Handbook of RF and Microwave Solid-State Power Amplifiers*

Anh-Vu H. Pham, Morgan J. Chen and Kunia Aihara, *LCP for Microwave Packages and Modules*

Sorin Voinigescu, *High-Frequency Integrated Circuits*

Richard Collier, *Transmission Lines*

Valeria Teppati, Andrea Ferrero and Mohamed Sayed, *Modern RF and Microwave Measurement Techniques*

Nuno Borges Carvalho and Dominique Schreurs, *Microwave and Wireless Measurement Techniques*

David E. Root, Jan Verspecht, Jason Horn and Mihai Marcu, *X-Parameters*

Forthcoming

Richard Carter, *Theory and Design of Microwave Tubes*

Hossein Hashemi and Sanjay Raman, *Silicon mm-Wave Power Amplifiers and Transmitters*

Earl McCune, *Dynamic Power Supply Transmitters*

Isar Mostafanezad, Olga Boric-Lubecke and Jenshan Lin, *Medical and Biological Microwave Sensors*

X-Parameters

Characterization, Modeling, and Design of Nonlinear RF and Microwave Components

DAVID E. ROOT
Agilent Technologies, Inc.

JAN VERSPECHT
Agilent Technologies, Inc.

JASON HORN
Agilent Technologies, Inc.

MIHAI MARCU
Agilent Technologies, Inc.

CAMBRIDGE
UNIVERSITY PRESS

University Printing House, Cambridge CB2 8BS, United Kingdom

One Liberty Plaza, 20th Floor, New York, NY 10006, USA

477 Williamstown Road, Port Melbourne, VIC 3207, Australia

314-321, 3rd Floor, Plot 3, Splendor Forum, Jasola District Centre, New Delhi-110025, India

79 Anson Road, #06-04/06, Singapore 079906

Cambridge University Press is part of the University of Cambridge.

It furthers the University's mission by disseminating knowledge in the pursuit of education, learning and research at the highest international levels of excellence.

www.cambridge.org
Information on this title: www.cambridge.org/9780521193238

First published 2013

A catalogue record for this publication is available from the British Library

Library of Congress Cataloging in Publication data
Root, David E.
X-parameters : characterization, modeling, and design of nonlinear RF and microwave components / David E. Root, Agilent Technologies Inc., Jan Verspecht, Agilent Technologies Inc., Jason Horn, Agilent Technologies Inc., Mihai Marcu, Agilent Technologies Inc.
pages cm – (The Cambridge RF and microwave engineering series)
Includes bibliographical references.
ISBN 978-0-521-19323-8 (Hardback)
1. Microwave circuits–Design and construction–Mathematics. 2. Electric circuits, Nonlinear–Design and construction–Mathematics. 3. Parametric devices–Design and construction–Mathematics. 4. Differential equations. I. Verspecht, Jan. II. Horn, Jason. III. Marcu, Mihai. IV. Title.
TK7876.R66 2013
621.3841′2–dc23 2013013915

ISBN 978-0-521-19323-8 Hardback

For Marilyn, with thanks for her patience, support, and, most of all, her love.
David

In memory of Petrus Verspecht.
Jan

To Jessica, Jonathan, and Elise, my inspiration.
Jason

To Domnica, for the patience shown in the many evenings and weekends that I have spent away from her.
Mihai

"Just as the S-parameters revolutionized linear microwave circuit engineering nearly 60 years ago, the relatively new development of the X-parameters and the mixer-based VNA provides a truly scientific approach to nonlinear RF and microwave circuit design. This book, written by experts, contains a wealth of information about the characterization and modeling of nonlinear components as well as their applications to various types of designs. I can only wish that such capability and textbook had been available when I was a design engineer."

Les Besser
Founder of Compact Software and Besser Associates

"S-parameters revolutionized linear RF and Microwave design in the 1970s and X-parameters are doing the same for non-linear design today. Starting with the familiar foundation of S-parameters, the text guides the reader through the additional non-linear terminology needed to provide a clear and practical view of X-parameters. Many practical examples show how to apply them in real world designs and answers are provided to some of the more subtle concepts of cross-frequency phase and memory effects. In a world where wireless is proliferating, this book will be an invaluable reference for any RF designer to reduce design turns and improve their first-pass designs."

Mark Pierpont
Agilent Technologies

Contents

Preface

The need for a rigorous, yet practical, framework for characterization, modeling, and design of nonlinear electronic components at high frequencies has never been more urgent. The communications revolution is inexorably forcing active devices into more and more strongly nonlinear regimes of operation. This is a consequence of the relentless drive for more efficiency in order to save power, extend battery life, and minimize cooling. The price for efficiency is nonlinearity. Dealing with nonlinearity means that new measurement instrumentation and new modeling and design methodologies are required that go far beyond linear S-parameters. Fortunately, there is an overarching, interoperable paradigm combining all these pieces of the nonlinear puzzle together, seamlessly. The new paradigm is called X-parameters,[1] and that is what this book is about.

The book is intended as a comprehensive introduction to X-parameters. It is aimed at a diverse audience with a wide range of backgrounds. This is quite a challenging undertaking! We are targeting professional microwave engineers, device modeling engineers and scientists, RF and microwave circuit designers, electronic and communications engineers, CAE professionals developing simulator algorithms, and microwave and RF professionals developing new high-speed instrumentation for a wide range of nonlinear characterization applications. The inherent interdisciplinary nature of X-parameters is the prime reason we seek to appeal to this broad audience. The practical solutions based on X-parameters deployed by industry over the past several years depend on contributions in all of these areas.

With this diverse audience in mind, we have chosen a particular sequence with which to introduce the subject. We start with a concise summary of the well-known time-invariant linear theory, namely S-parameters. We choose this context, familiar to many readers, to introduce more advanced concepts that will be needed for the remainder of the book. Chapter 2 introduces X-parameters, based on multi-tone nonlinear spectral maps defined on a harmonic grid, and goes into significant detail about the application and implications of the constraint of time invariance. Chapter 3 simplifies the general discussion to simple practical cases, based on the application of spectral linearization, a useful approximation that reduces complexity, enabling practical applications. Several examples are presented demonstrating the power, utility, and relative simplicity of these

[1] "X-parameters" is a trademark of Agilent Technologies, Inc.

simplest X-parameters. The origins of "conjugate" terms in the spectral linearization are discussed. Chapter 4 is devoted to how X-parameters are measured, and also to how they are computed (generated) from within a circuit simulator. The functional block diagram of the main instrument (the nonlinear vector network analyzer – NVNA) is discussed, and the application of measurements using a pulse generator phase reference to obtain the key X-parameter quantities is reviewed. Chapter 5 extends the treatment of X-parameters to multiple large signals and multiple ports, as is necessary in the treatment of many mixers, the treatment of intermodulation with phase, and the large-signal response of power amplifiers as nonlinear functions of both input power and reflections of electrical signals back into the device due to large mismatch, going beyond the first spectral linearization approximation of Chapter 3. Finally, Chapter 6 extends the treatment of X-parameters to dynamic "memory effects," important phenomena exhibited by practical modern high-speed devices in response to wide-band communication signals, for example. Several appendices are provided for detailed derivations, standard symbol and notational definitions, and further elaboration of some parts of the main text to help serve as a reference for workers in the field.

The book is appropriate as a text for an advanced undergraduate or graduate course in electrical engineering. In fact, we perceive an acute need to make X-parameters a standard part of the electrical engineering curriculum. The book may also be appropriate for applied mathematicians and scientists with an interest in rigorous and practical foundations for applications to a wide range of nonlinear systems well beyond electronics.

The background needed by readers of this book is not much more than first-year calculus, basic circuit theory, and simple Fourier analysis. Rudimentary knowledge of electronic power amplifiers and transistors, S-parameter fundamentals, differential equations, circuit design, and circuit simulation would certainly be helpful.

Acknowledgments

The authors are profoundly grateful to our many dedicated and talented colleagues who collaborated with us to develop and deploy X-parameter technology, products, support, and services. We are grateful to our many customers, academic researchers, and practicing professionals for their thoughtful feedback, stimulating discussions, and creative applications of this technology. We thank Agilent management for their vital support. Finally, we thank the staff at Cambridge University Press for their commitment to this project, their cheerful professionalism, and their patience.

1 S-parameters – a concise review

1.1 Introduction

This chapter presents a concise treatment of S-parameters, meant primarily as an introduction to the more general formalism of X-parameters. The concepts of *time invariance* and *spectral maps* are introduced at this stage to enable an easier generalization to X-parameters in the ensuing chapters. The interpretations of S-parameters as calibrated measurements, intrinsic properties of the device under test (DUT), IP-secure component behavioral models, and composition rules for linear system design are presented. The cascade of two linear S-parameter components is considered as an example to be generalized to the nonlinear case later. The calculation of S-parameters for a transistor from a simple nonlinear device model is used as an example to introduce the concepts of (static) *operating point* and *small-signal conditions*, both of which must be generalized for the treatment of X-parameters.

1.2 S-parameters

Since the 1950s, S-parameters, or scattering parameters, have been among the most important of all the foundations of microwave theory and techniques.

S-parameters are easy to measure at high frequencies with a vector network analyzer (VNA). Well-calibrated S-parameter measurements represent intrinsic properties of the DUT, independent of the VNA system used to characterize it. Calibration procedures [1] remove systematic measurement errors and enable a separation of the overall values into numbers attributable to the device, independent of the measurement system used to characterize it. These DUT properties (gain, loss, reflection coefficient, etc.) are familiar, intuitive, and important [2]. Another key property of S-parameters is that the S-parameters of a composite system are completely determined from knowledge of the S-parameters of the constituent components and their connectivity. S-parameters provide the complete specification of how a linear component responds to an arbitrary signal. Therefore designs of linear systems with S-parameters are predictable with absolute certainty. S-parameters define a complete behavioral description of the linear component at the external terminals, independent of the detailed physics or specifics of the realization of the component. S-parameters can be shared between component vendors and system integrators freely, without the possibility that the component implementation can be reverse engineered, protecting IP and promoting sharing and reuse. Indeed, one may ask the question, "are S-parameters measurements, or do they constitute a model?" The answer is really "both."

S-parameters need not come only from measurements. They can be calculated from physics by solving Maxwell's equations, by linearizing the semiconductor equations, or computed from matrix analysis of linear equivalent circuits. In this way, the many benefits of S-parameters can be realized, starting from a more detailed representation of the component from first principles or from a complicated linear circuit model.

Graphical methods based around the Smith chart were invented to visualize and interpret S-parameters, and graphical design methodologies soon followed for circuit design [2][3]. These days, electronic design automation (EDA) tools provide simulation components – S-parameter blocks – and design capabilities using the familiar S-parameter analysis mode.

One of the great utilities of S-parameters is the interoperability among the measurement, modeling, and design capabilities they provide. One can characterize the component with measured S-parameters, use them as a high-fidelity behavioral model of the component with complete IP protection, and design systems with them in the EDA environment.

1.3 Wave variables

The term "scattering" refers to the relationship between incident and scattered (reflected and transmitted) traveling waves.

By convention, in this text the circuit behavior is described using generalized power waves [2]. There are alternative wave definitions used in the industry. These are briefly reviewed in Appendix A, together with the general notations used in this text.

The wave variables, A and B, corresponding to a specific port of a network, are defined as simple linear combinations of the voltage and current, V and I, at the same port, according to Figure 1.1 and equations (1.1):

$$A = \frac{V + Z_0 I}{2\sqrt{Z_0}},$$
$$B = \frac{V - Z_0 I}{2\sqrt{Z_0}}. \tag{1.1}$$

The reference impedance for the port, Z_0, is, in general, a complex value. For the purpose of simplifying the concepts presented, the reference impedance is restricted to real values in this text.

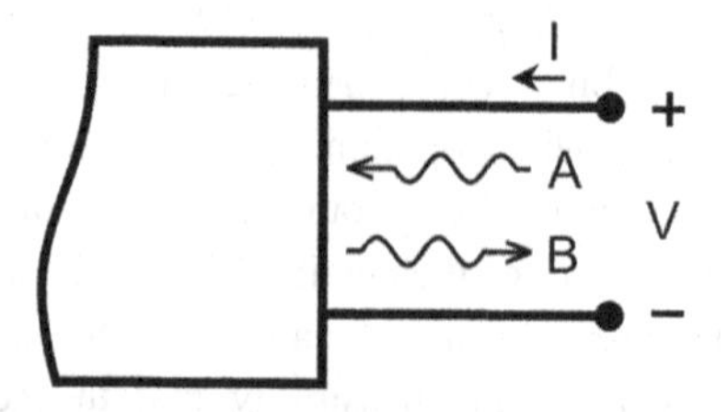

Figure 1.1 Wave definitions.

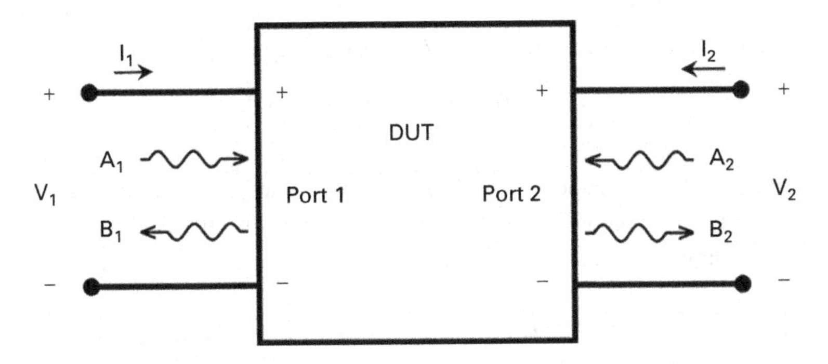

Figure 1.2 Incident and scattered waves of a two-port device.

The currents and voltages can be recovered from the wave variables, according to equations (1.2):

$$V = \sqrt{Z_0}\,(A + B),$$
$$I = \frac{1}{\sqrt{Z_0}}\,(A - B). \tag{1.2}$$

Here, A and B represent the incident and scattered waves, V and I are the port voltage and current, respectively, and Z_0 is the reference impedance for the port. A typical value of Z_0 is 50 Ω by convention, but other choices may be more practical for some applications. A value for Z_0 closer to 1 Ω is more appropriate for S-parameter measurements of power transistors, for example, given that power transistors typically have very small output impedances.

The variables in equations (1.1) and (1.2) are complex numbers representing the RMS-phasor description of sinusoidal signals in the frequency domain (see Appendix A for a more detailed discussion of the notations used). Later this will be generalized to the envelope domain by letting these complex numbers vary in time.

A, B, V, and I can be considered RMS vectors, the components of which indicate the values associated with sinusoidal signals at particular ports labeled by positive integers. Thus A_j is the incident wave RMS phasor at port j and I_k is the current RMS phasor at port k. For now, Z_0 is taken to be a fixed real constant, in particular, 50 Ω.

A graphical representation of the wave description is given in Figure 1.2.

To retrieve the time-dependent sinusoidal voltage signal at the ith port, the complex value of the phasor and also the angular frequency, ω, to which the phasor corresponds, must be known. The voltage is then given by

$$v_i(t) = \mathrm{Re}\{V_i^{(pk)} e^{j\omega t}\}, \tag{1.3}$$

and similarly for the other variables, where $V_i^{(pk)}$ are peak values. Equation (1.3) for the voltage, and a similar equation for the time-dependent current, can be used to define real, time-dependent "wave" quantities using the same linear combinations as in (1.1):

$$a(t) = \frac{1}{2\sqrt{Z_0}}\,\big(v(t) + Z_0 i(t)\big),$$
$$b(t) = \frac{1}{2\sqrt{Z_0}}\,\big(v(t) - Z_0 i(t)\big); \tag{1.4}$$

$$
\begin{aligned}
v(t) &= \sqrt{Z_0}\,\big(a(t)+b(t)\big),\\
i(t) &= \frac{1}{\sqrt{Z_0}}\,\big(a(t)-b(t)\big).
\end{aligned}
\tag{1.5}
$$

It is convenient to keep track of the frequency associated with a particular set of phasors by rewriting (1.1) according to (1.6), and (1.2) according to (1.7), where the port indexing notation is made explicit:

$$
\begin{aligned}
A_i(\omega) &= \frac{1}{2\sqrt{Z_0}}\,\big(V_i(\omega)+Z_0I_i(\omega)\big),\\
B_i(\omega) &= \frac{1}{2\sqrt{Z_0}}\,\big(V_i(\omega)-Z_0I_i(\omega)\big);
\end{aligned}
\tag{1.6}
$$

$$
\begin{aligned}
V_i(\omega) &= \sqrt{Z_0}\,\big(A_i(\omega)+B_i(\omega)\big),\\
I_i(\omega) &= \frac{1}{\sqrt{Z_0}}\big(A_i(\omega)-B_i(\omega)\big).
\end{aligned}
\tag{1.7}
$$

For each angular frequency, ω, (1.6) is a set of two equations defined at each port.

The assumption behind the S-parameter formalism is that the system being described is *linear* and therefore there must be a *linear relationship* between the phasor representation of incident and scattered waves. This is expressed in (1.8) for an N-port network as follows:

$$
B_i(\omega) = \sum_{j=1}^{N} S_{ij}(\omega)A_j(\omega), \qquad \forall i \in \{1, 2, \ldots, N\}.
\tag{1.8}
$$

The set of complex coefficients, $S_{ij}(\omega)$, in (1.8) defines the S-parameter matrix or, simply, the S-parameters at that frequency. Equation (1.8), for the fixed set of complex S-parameters, determines the output phasors for any set of input phasors. The summation is over all port indices, so that incident waves at each port, j, contribute in general to the overall scattered wave at each output port, i. For now we consider all frequencies to be positive ($\omega > 0$). Note that contributions to a scattered wave at frequency ω come only from incident waves at the same frequency. This is not the case for the more general X-parameters, where a stimulus at one frequency can lead to scattered waves at different frequencies.

The set of equations (1.8) represents a model of the network under study. However, this model is valid only if the network has the topological connections shown in Figure 1.2. For example, the model might not accurately represent the behavior of the network when connected as shown in Figure 1.3 because potential losses between the reference pins of the two ports are not individually identified in the set of S-parameters in (1.8).

For the purpose of creating a model for the network, all ports should be referenced to the same pin, as shown in Figure 1.4.

Such connectivity is the natural option for the measurement and modeling process of a three-pin network (like a transistor), but it has to be extended in the general case of an arbitrary network, and it is necessary for all networks, linear and/or nonlinear. This connectivity convention is considered by default (unless otherwise specified) for the remainder of this text.

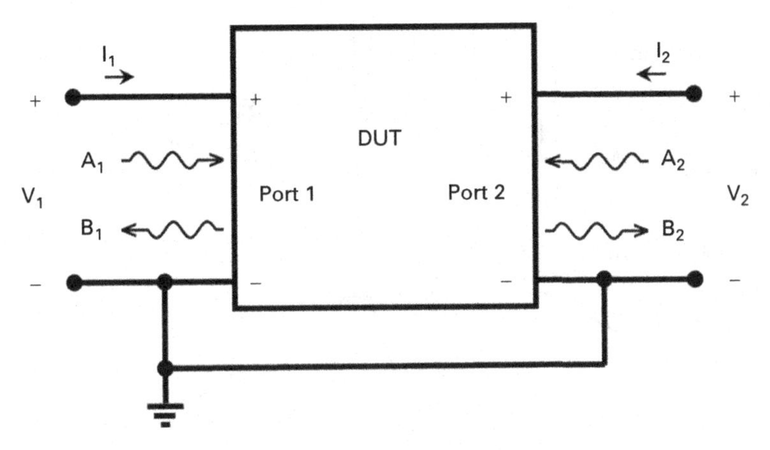

Figure 1.3 Potential losses between reference pins are not individually identified by the model in (1.8).

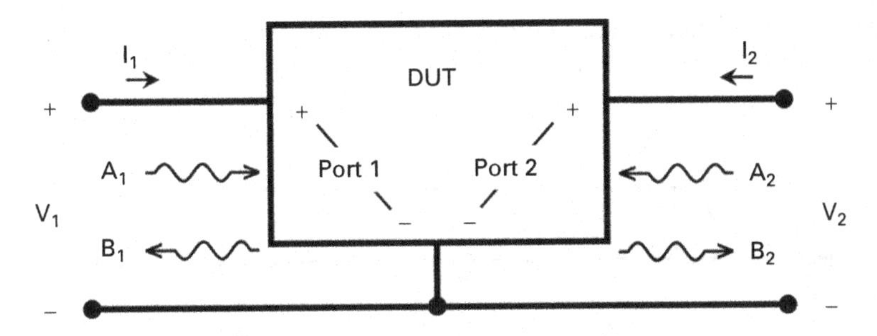

Figure 1.4 All ports should be referenced to the same pin for modeling purposes.

Using the topological connection in Figure 1.4, the set of equations (1.8) represents a complete model of the network under test.

From (1.8) we note that a stimulus (incident wave) at a particular port j will produce a response (scattered wave) at all ports, including the port at which the stimulus is applied.

Equation (1.8) shows that the scattered waves are linear functions of the complex amplitudes (the phasors) of the incident waves. The dependence on frequency of the S-parameter matrix elements, $S_{ij}(\omega)$, can be highly nonlinear, even for a linear device. For example, an ideal band-pass filter response is linear in the incident wave variable, but the filter response is a nonlinear function of the frequency of the incident wave. This is shown in Figure 1.5.

1.4 S-parameter measurement

By setting all incident waves to zero in (1.8), except for A_j, one can deduce the simple relationship between a given S-parameter (S-parameter matrix element) and a particular ratio of scattered to incident waves according to (1.9):

$$S_{ij}(\omega) = \left.\frac{B_i(\omega)}{A_j(\omega)}\right|_{\substack{A_k=0 \\ \forall k \neq j}}. \tag{1.9}$$

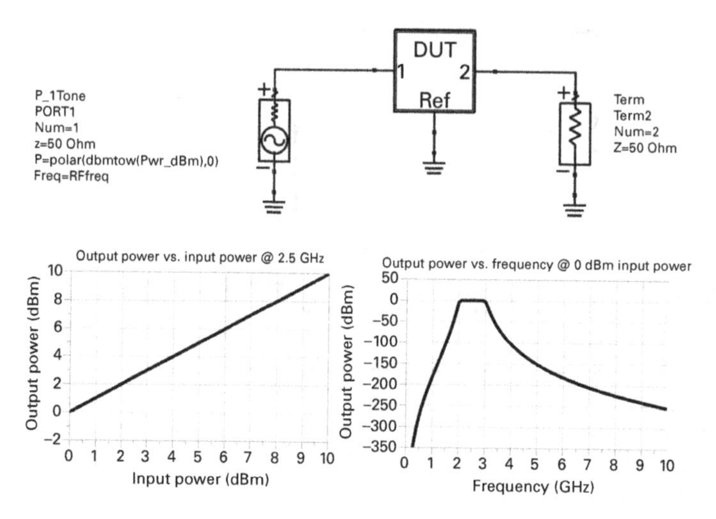

Figure 1.5 A linear network has a linear behavior when plotted versus input power level, but the dependence on frequency is usually not linear.

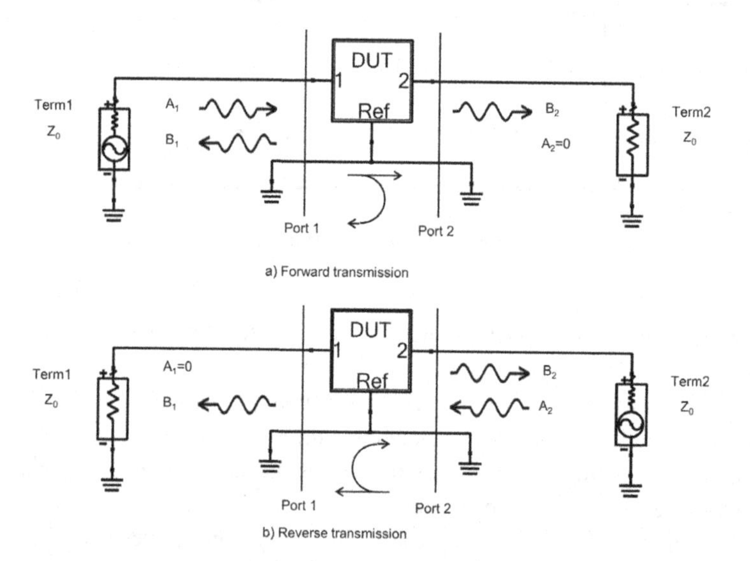

Figure 1.6 S-parameter experiment design: (a) forward transmission; (b) reverse transmission.

Equation (1.9) corresponds to a simple graphical representation shown in Figure 1.6 for the simple case of a two-port component. In Figure 1.6(a), the stimulus is a wave incident at port 1. The fact that A_2 is not present ($A_2 = 0$) is interpreted to mean that the B_2 wave scattered and traveling away from port 2 is not reflected back into the device at port 2. Under this condition, the device is said to be *perfectly matched* at port 2. Two of the four complex S-parameters, specifically S_{11} and S_{21}, can be identified using (1.9) for this case of exciting the device with only A_1. Figure 1.6(b) shows the case where the device is stimulated with a signal, A_2, at port 2, and assumed to be perfectly matched at port 1 ($A_1 = 0$). The remaining S-parameters, S_{12} and S_{22}, can be identified from this ideal experiment.

It is important to note that the ratio on the right-hand side of (1.9) can be computed from independent measurements of incident and scattered waves for actual components corresponding to any non-zero value for the incident wave, A_j. The value of this ratio, however, will generally vary with the magnitude of the incident wave. Therefore, the identification of this ratio with "the S-parameters" of the component is valid for any particular value of incident A_j only if the component behaves linearly, namely according to (1.8). In other words, the values of the incident waves, A_j, need to be in the linear region of operation for this identification to be valid. For nonlinear components, such as transistors biased at a fixed voltage, the scattered waves eventually do not increase as the incident waves become larger in magnitude (this is compression). Therefore, different values of (1.9) result from different values of incident waves. A better definition of S-parameters for a *nonlinear component* is a modification of (1.9), given by (1.10):

$$S_{ij}(\omega) \equiv \lim_{|A_j|\to 0} \left. \frac{B_i(\omega)}{A_j(\omega)} \right|_{\substack{A_k=0 \\ \forall k \neq j}} . \tag{1.10}$$

That is, for a general component, bias at a constant DC stimulus, the S-parameters are related to ratios of output responses to input stimuli in the limit of small input signals. This emphasizes that S-parameters properly apply to nonlinear components only in the *small-signal* limit.

1.5 S-parameters as a spectral map

If there are multiple frequencies present in the input spectrum, one can represent the output spectrum in terms of a matrix giving the contributions to each output frequency from each input frequency.

An example in the case of three input frequencies is given by equation (1.11):

$$\begin{bmatrix} B(\omega_1) \\ B(\omega_2) \\ B(\omega_3) \end{bmatrix} = \begin{bmatrix} S(\omega_1) & 0 & 0 \\ 0 & S(\omega_2) & 0 \\ 0 & 0 & S(\omega_3) \end{bmatrix} \begin{bmatrix} A(\omega_1) \\ A(\omega_2) \\ A(\omega_3) \end{bmatrix} . \tag{1.11}$$

Here we assume a single port, for simplicity, and therefore drop the port indices. It is clear from (1.11) that S-parameters are a diagonal map in frequency space. This means that each output frequency contains contributions only from inputs at that same frequency. Or, in other words, each input frequency never contributes to outputs at any different frequency.

A graphical representation is given in Figure 1.7 for the case of forward transmission through a two-port network with matched terminations at both ports.

The interpretation of Figure 1.7, mathematically represented by (1.11), is that S-parameters define a particularly simple *linear spectral map* relating incident to scattered waves. S-parameters are diagonal in the frequency part of the map, namely they predict a response only at the particular frequencies of the corresponding input stimuli. It will be demonstrated in later chapters that X-parameters provide for richer behavior.

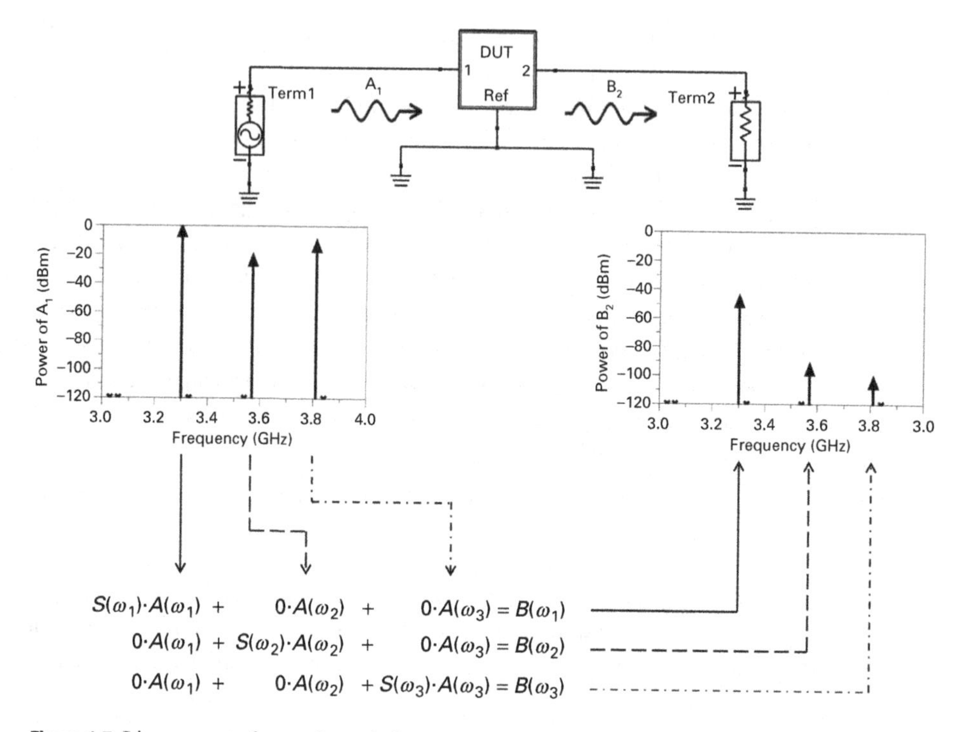

Figure 1.7 Linear spectral map through S-parameters matrix.

For signals with a continuous spectrum, the diagonal nature of the S-parameter spectral map can be written as equation (1.12):

$$B_i(\omega) = \sum_{j=1}^{N} \int S_{ij}(\omega)\delta(\omega - \omega')A_j(\omega')d\omega', \qquad \forall i \in \{1, 2, \ldots, N\}. \tag{1.12}$$

Performing the integral over input frequencies in (1.12) results in equation (1.8), the form usually given for S-parameters.

1.6 Superposition

Any linear theory, such as S-parameters, enables the general response to an arbitrary input signal to be computed by *superposition* of the responses to unit stimuli. Superposition enables great simplifications in analysis and measurement. Superposition is the reason S-parameters can be measured by independent experiments with one sinusoidal stimulus at a time, one stimulus per port per frequency using (1.9). The general response to any set of input signals can be obtained by superposition using (1.8).

An example of superposition is shown in Figure 1.8, with all signals represented in both time and frequency domains.

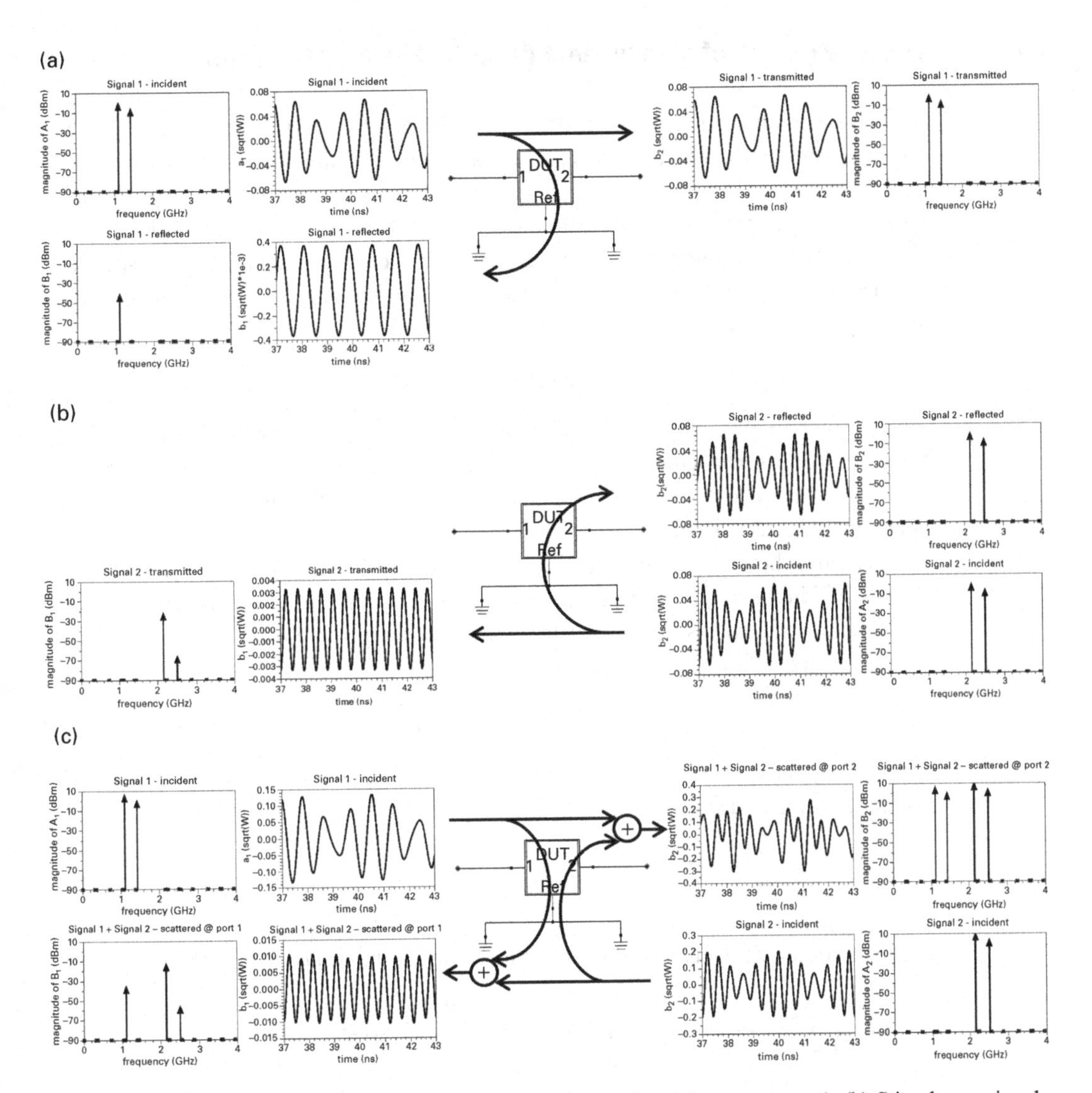

Figure 1.8 Superposition example. (a) Stimulus = signal 1 ⇒ response 1. (b) Stimulus = signal 2 ⇒ response 2. (c) Stimulus = 2 (signal 1) + 3 (signal 2); response = 2 (response 1) + 3 (response 2).

This example uses two signals, each containing two frequency components, as stimuli incident at each port, independently, with the other port perfectly matched. The example shows that the response to a linear combination of the stimuli is the same linear combination of the individual responses.

As always, the caveat is that the component actually behaves linearly over the range of signal levels used to stimulate the device. There is no a-priori way to know whether a component will behave linearly without precise knowledge about its composition or physical measured characteristics.

1.7 Time invariance of components described by S-parameters

A DUT description in terms of S-parameters defined by (1.8) naturally embodies an important principle known as *time invariance.* Time invariance states that if $y(t)$ is the DUT response to an excitation $x(t)$, the DUT response to the time-shifted excitation, $x(t-\tau)$, must be $y(t-\tau)$. This must be true for all time shifts, τ. That is, if the input is shifted in time, the output is shifted by the corresponding amount, but is otherwise identical with the DUT response to the non-shifted input. This is stated mathematically in equation (1.13), where O is the operator taking input to output:

$$\forall\tau \in \mathbb{R} \qquad y(t) = O\big[x(t)\big] \Rightarrow y(t-\tau) = O\big[x(t-\tau)\big]. \tag{1.13}$$

Time invariance is a property of common linear and nonlinear components, such as passive inductors, capacitors, resistors, and diodes, and active devices, such as transistors. Examples of components not time invariant (in the usual sense) are oscillators and other autonomous systems.

The proof follows from elementary properties of the Fourier transform, where a phase shift by $e^{j\omega\tau}$ in the frequency domain corresponds to a time shift of τ in the time domain. The time-domain waves incident at the ports (the stimuli) are $a_k(t)$, and their Fourier transforms are $A_k^{(pk)}(\omega)$, as in equation (1.14):

$$\mathfrak{F}\{a_k(t)\} = A_k^{(pk)}(\omega). \tag{1.14}$$

The time-domain waves scattered from the ports (the response) are $b_i(t)$, and their Fourier transforms are $B_i^{(pk)}(\omega)$, as in equation (1.15):

$$\mathfrak{F}\{b_i(t)\} = B_i^{(pk)}(\omega) = \sum_{k=1}^{N} S_{ik}(\omega)A_k^{(pk)}(\omega) = \sum_{k=1}^{N} S_{ik}(\omega)\, F\{a_k(t)\}. \tag{1.15}$$

If all stimuli are delayed with the same time delay, τ, the response becomes

$$\begin{aligned}\mathfrak{F}^{-1}\left\{\sum_{k=1}^{N} S_{ik}(\omega)\,\mathfrak{F}\{a_k(t-\tau)\}\right\} &= \mathfrak{F}^{-1}\left\{\sum_{k=1}^{N} S_{ik}(\omega)\,\mathfrak{F}\{a_k(t)\}e^{j\omega\tau}\right\}\\ &= \mathfrak{F}^{-1}\left\{B_i^{(pk)}(\omega)e^{j\omega\tau}\right\} = b_i(t-\tau).\end{aligned} \tag{1.16}$$

Equation (1.16) proves that S-parameters are automatically consistent with the principle of time invariance. Therefore, any set of S-parameters describes a time-invariant system.

Unlike the case for S-parameters, a more general (e.g. nonlinear) relationship between incident A waves and scattered B waves is not automatically consistent with the property (1.13) of time invariance. This will be demonstrated in Chapter 2. Therefore, in order to have a consistent representation of a nonlinear time-invariant DUT, the time-invariance property is manifestly incorporated into the mathematical formulation of X-parameters relating input to output waves. A representation of a time-invariant DUT by equations not consistent with (1.13) means the model is fundamentally wrong, and can yield very inaccurate results for some signals, even if the model "fitting" (or identification) appears good at time t.

1.8 Cascadability

Another key property of S-parameters is that a linear circuit or system can be designed with perfect certainty knowing only the S-parameters of the constituent components and their interconnections. The overall S-parameters of the composite design can be calculated by using (1.8) in conjunction with Kirchhoff's voltage law (KVL) and Kirchhoff's current law (KCL), applied at the internal nodes created by connections between two or more components.

This is illustrated for a cascade of two 2-ports in Figure 1.9(a). Each component is characterized by its own S-parameter matrix, $S^{(1)}$ and $S^{(2)}$, respectively. The output port (subscript number 2) of the first component is connected to the input port (subscript number 1) of the second component, creating an internal node.

It is a straightforward exercise to write the two equations that follow from applying KVL and KCL at the internal node. These are given by equations (1.17) and (1.18):

$$B_2^{(1)} = A_1^{(2)}, \tag{1.17}$$

$$A_2^{(1)} = B_1^{(2)}. \tag{1.18}$$

Each 2-port relates two input and two output variables. When cascaded, equations (1.17) and (1.18) can be used to compute the overall S-parameter matrix of the composite. The result is given in equation (1.19):

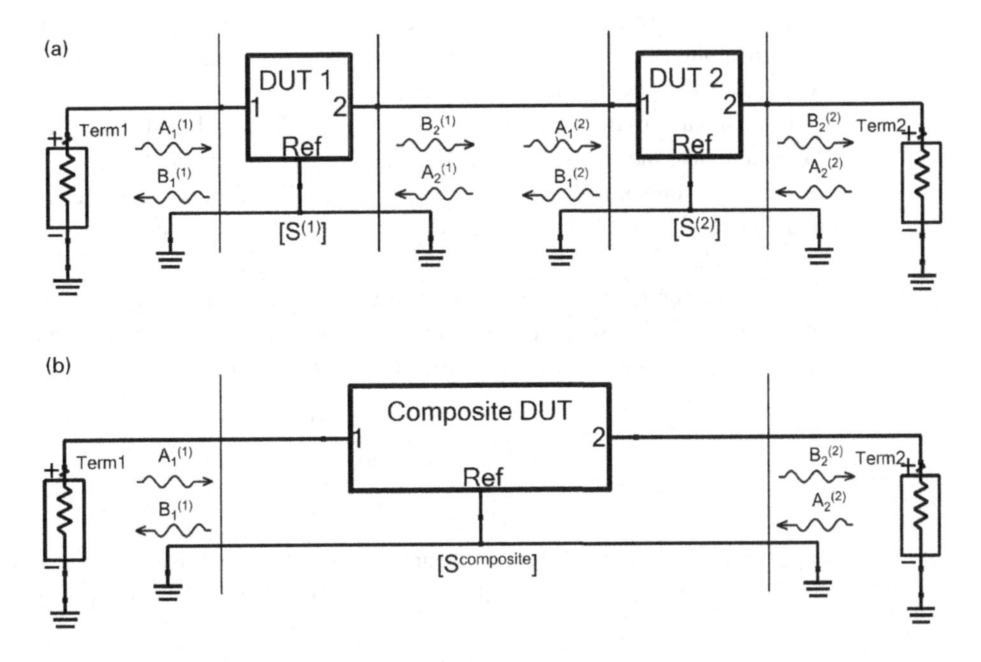

Figure 1.9 The cascade of two 2-ports. (a) Cascade of two DUTs. (b) The equivalent network, yielding the same response as the cascade of two DUTs.

$$S^{composite} = \begin{pmatrix} S_{11}^{(1)} + S_{11}^{(2)} \dfrac{S_{12}^{(1)} S_{21}^{(1)}}{1 - S_{22}^{(1)} S_{11}^{(2)}} & \dfrac{S_{12}^{(1)} S_{12}^{(2)}}{1 - S_{22}^{(1)} S_{11}^{(2)}} \\ \dfrac{S_{21}^{(1)} S_{21}^{(2)}}{1 - S_{22}^{(1)} S_{11}^{(2)}} & S_{22}^{(1)} \dfrac{S_{12}^{(2)} S_{21}^{(2)}}{1 - S_{22}^{(1)} S_{11}^{(2)}} + S_{22}^{(2)} \end{pmatrix}. \tag{1.19}$$

A network characterized by the $S^{composite}$ is equivalent to the cascade of the two DUTs, yielding the same response under the same stimuli. This is shown in Figure 1.9(b).

1.9 DC operating point

For devices such as diodes and transistors, the S-parameter values depend very strongly on the DC bias conditions defining the component operating conditions. For example, a field-effect transistor (FET) biased in the saturation region of operation will have S-parameters appropriate for an active amplifier (where $|S_{21}| \gg 1 \gg |S_{12}|$), whereas when biased "off" (say $V_{DS} = 0$) the S-parameters will represent the characteristics of a passive switch ($S_{21} = S_{12}$ with $|S_{ij}| < 1$).

1.10 S-parameters of a nonlinear device[1]

It is possible to derive the S-parameters of a transistor starting from a simple nonlinear model of the intrinsic device. The process illustrates the concept of linearizing a nonlinear mapping about an operating point. This concept, suitably generalized, is used as a foundation for X-parameters in Chapter 2.

The equivalent circuit of a simple, unilateral FET model is shown in Figure 1.10. More complete model topologies and equations are discussed in [4].

A current source, i_{DS}, represents the nonlinear bias-dependent channel current from drain to source, and a simple nonlinear two-terminal capacitor, q_{GS}, represents the nonlinear charge storage between gate and source terminals.

The large-signal model equations corresponding to this equivalent circuit are given by (1.20) and (1.21), where the stimulus applied to the device is the set of port voltages, v_{GS} and v_{DS}, and the device response is the set of port currents, i_{GS} and i_{DS}:

$$i_G(t) = \frac{dq_{GS}(v_{GS}(t))}{dt}, \tag{1.20}$$

$$i_D(t) = i_{DS}(v_{GS}(t), v_{DS}(t)). \tag{1.21}$$

Equations (1.20) and (1.21) are evaluated for time-varying voltages assumed to have a fixed DC component and a small sinusoidal component at a single RF or microwave

[1] This section can be omitted on first reading.

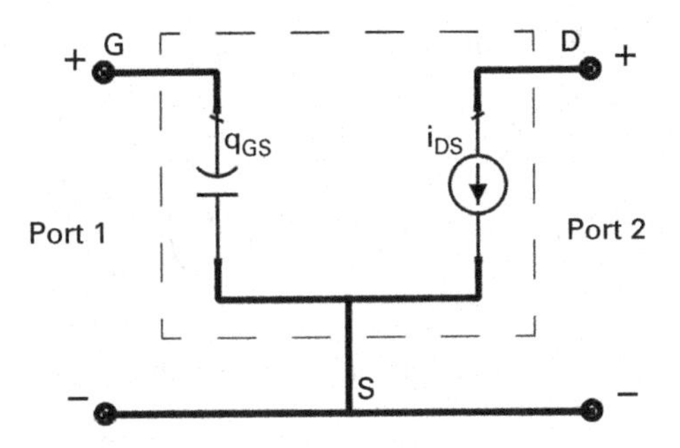

Figure 1.10 Simple nonlinear equivalent circuit model of a FET.

angular frequency, $\omega = 2\pi f$. Port 1 is associated with the gate and port 2 is associated with the drain terminals, referenced to the source.

For a stimulus comprising a combination of DC and one sinusoidal signal, the voltages are formally expressed as shown in (1.22):

$$v_i(t) = v_i^{(DC)} + \delta V_i \cos(\omega t + \phi_i). \tag{1.22}$$

Expressions in (1.23) and (1.24) are obtained by substituting (1.22) for $i = 1{,}2$ into (1.20) and(1.21), and evaluating the result to first order in the real quantities δV_i:

$$\begin{aligned} i_1(t) &= i_1^{(DC)} + \delta i_1(t) \\ &= \frac{d}{dt}\left(q_{GS}\left(v_1^{(DC)}\right) + \left.\frac{dq_{GS}}{dV_1}\right|_{v_1^{(DC)}, v_2^{(DC)}} \delta V_1 \cos(\omega t + \phi_1) \right) \\ &= c_{GS}\left(v_1^{(DC)}\right) \frac{d}{dt}\left(\delta V_1 \cos(\omega t + \phi_1)\right); \end{aligned} \tag{1.23}$$

$$\begin{aligned} i_2(t) &= i_2^{(DC)} + \delta i_2(t) = i_{DS}\left(v_1^{(DC)}, v_2^{(DC)}\right) \\ &\quad + g_m\left(v_1^{(DC)}, v_2^{(DC)}\right) \delta V_1 \cos(\omega t + \phi_1) + g_{DS}\left(v_1^{(DC)}, v_2^{(DC)}\right) \delta V_2 \cos(\omega t + \phi_2). \end{aligned} \tag{1.24}$$

The following definitions of the "linearized equivalent circuit elements" have been used:

$$\begin{aligned} c_{GS}\left(v_1^{(DC)}\right) &= \left.\frac{dq_{GS}}{dv_1}\right|_{v_1^{(DC)}}, \\ g_m\left(v_1^{(DC)}, v_2^{(DC)}\right) &= \left.\frac{\partial i_{DS}}{\partial v_1}\right|_{v_1^{(DC)}, v_2^{(DC)}}, \\ g_{DS}\left(v_1^{(DC)}, v_2^{(DC)}\right) &= \left.\frac{\partial i_{DS}}{\partial v_2}\right|_{v_1^{(DC)}, v_2^{(DC)}}. \end{aligned} \tag{1.25}$$

By equating the zeroth-order terms in δV_i, the operating point conditions are obtained as follows:

$$\begin{aligned} i_1^{(DC)} &= 0, \\ i_2^{(DC)} &= i_{DS}\left(v_1^{(DC)}, v_2^{(DC)}\right). \end{aligned} \tag{1.26}$$

Equating the first-order terms in δV_i leads to

$$\delta i_1(t) = c_{GS}\left(v_1^{(DC)}\right)\frac{d}{dt}(\delta V_1 \cos(\omega t + \phi_1)),$$
$$\delta i_2(t) = g_m\left(v_1^{(DC)}, v_2^{(DC)}\right)\delta V_1 \cos(\omega t + \phi_1) + g_{DS}\left(v_1^{(DC)}, v_2^{(DC)}\right)\delta V_2 \cos(\omega t + \phi_2). \tag{1.27}$$

Equation (1.27) is expressed in the frequency domain by defining complex phasors $\delta I(\omega)$ and $\delta V(\omega)$ according to

$$\begin{aligned}\delta i_i(t) &= \mathrm{Re}\left(\delta I_i(\omega)e^{j\omega t}\right),\\ \delta v_i(t) &= \mathrm{Re}\left(\delta V_i(\omega)e^{j\omega t}\right) = \mathrm{Re}\left(\delta V_i e^{j\phi_i}e^{j\omega t}\right).\end{aligned} \tag{1.28}$$

Equations (1.27) can be rewritten for the complex phasors in matrix notation as

$$\begin{bmatrix}\delta I_1(\omega)\\ \delta I_2(\omega)\end{bmatrix} = Y\left(v_1^{(DC)}, v_2^{(DC)}, \omega\right)\begin{bmatrix}\delta V_1(\omega)\\ \delta V_2(\omega)\end{bmatrix}, \tag{1.29}$$

where

$$Y\left(v_1^{(DC)}, v_2^{(DC)}, \omega\right) = \begin{bmatrix} j\omega c_{GS}\left(v_1^{(DC)}\right) & 0\\ g_m\left(v_1^{(DC)}, v_2^{(DC)}\right) & g_{DS}\left(v_1^{(DC)}, v_2^{(DC)}\right)\end{bmatrix}. \tag{1.30}$$

Equation (1.30) defines the (common source) admittance matrix of the model. The matrix elements are evidently functions of the DC operating point (bias conditions) of the transistor, and also the (angular) frequency of the excitation.

Since the phasors representing the port voltage and currents in (1.29) can be re-expressed as linear combinations of incident and scattered waves using (1.7), it is possible to derive the expression for the S-parameters in terms of the Y-parameters (admittance matrix elements). This results in the well-known conversion formula (1.31) [2]:

$$S = [I - Z_0 Y][I + Z_0 Y]^{-1}. \tag{1.31}$$

Here I is the 2×2 unit matrix, Z_0 is the reference impedance used in the wave definitions in (1.6), Y is the two-port admittance matrix, and S is the corresponding S-parameter matrix. Substituting (1.30) into (1.31) results in an explicit expression, (1.32), for the S-parameters corresponding to this simple model in terms of the linear equivalent circuit element values given in (1.25):

$$S\left(v_1^{(DC)}, v_2^{(DC)}, \omega\right) = \begin{pmatrix} \dfrac{1 - j\omega c_{GS}\left(v_1^{(DC)}\right)Z_0}{1 + j\omega c_{GS}\left(v_1^{(DC)}\right)Z_0} & 0\\ \dfrac{-2g_m\left(v_1^{(DC)}, v_2^{(DC)}\right)Z_0}{\left(1 + g_{DS}\left(v_1^{(DC)}, v_2^{(DC)}\right)Z_0\right)\left(1 + j\omega c_{GS}\left(v_1^{(DC)}\right)Z_0\right)} & \dfrac{1 - g_{DS}\left(v_1^{(DC)}, v_2^{(DC)}\right)Z_0}{1 + g_{DS}\left(v_1^{(DC)}, v_2^{(DC)}\right)Z_0}\end{pmatrix}. \tag{1.32}$$

In summary, the S-parameters of a nonlinear component can be derived or computed by linearizing the full nonlinear characteristics of the circuit equations around a static (DC) operating point defined by the voltage or current bias conditions. The S-parameters define a linear relationship between the incident and scattered waves at a fixed DC operating point of the device and fixed frequency for the incident waves. The S-parameters are an accurate description of how the device responds to signals, provided the signal amplitude is sufficiently small that the DC operating point is not significantly affected by the signal. This will almost always be the case for signals of sufficiently small amplitude.

1.11 Additional benefits of S-parameters

1.11.1 S-parameters are applicable to distributed components at high frequencies

S-parameters can accommodate an arbitrary frequency dependence in the linear spectral mapping. S-parameters therefore apply when describing distributed components for which lumped approximations are not very accurate or efficient. This is especially true for high-frequency microwave components when the typical wavelengths of the stimulus approach and become smaller than the physical size of the component. The simplest example is the case of linear transmission lines. Another common example is the case of an active device, for which measured S-parameters of a transistor can be much more accurate than those computable from the linearized lumped nonlinear model. This is especially true as the frequency approaches and exceeds the device cutoff frequency, f_T, beyond which a distributed representation is generally required.

1.11.2 S-parameters are easy to measure at high frequencies

S-parameters contain no more information than the familiar Y- and Z-parameters of elementary linear circuit theory, yet they have great practical advantages. S-parameters are much easier to measure at high frequencies. Y- and Z-parameters require short- and open-circuit boundary conditions, respectively, on the components for a direct measurement. Short- and open-circuit conditions are hard to achieve at microwave frequencies, and so are impractical. Moreover, such conditions presented to a power transistor can create oscillations that can destroy the component.

It will be demonstrated that X-parameters combine the accuracy and ease of measurement of a frequency-domain approach based on wave variables with the ability to handle nonlinearities that go beyond the linear relationship assumed by equation (1.8).

1.11.3 Interpretation of two-port S-parameters

S-parameters relate to familiar quantities, such as gain, return loss, and output match. They provide insight into the component behavior. Table 1.1 lists the four complex S-parameters and their corresponding interpretation for generic linear 2-ports [2]. The third column expresses common amplifier quantities in terms of the corresponding S-parameters.

Table 1.1 S-parameters of a generic 2-port and the corresponding figures of merit of a two-port amplifier

S-parameter	Generic 2-port (with input and output ports properly terminated)	Amplifier figure of merit
S_{11}	input reflection coefficient	return loss: dB$\lvert S_{11}\rvert$
S_{12}	reverse transmission coefficient	isolation: dB$\lvert S_{12}\rvert$
S_{21}	forward transmission coefficient	gain: dB$\lvert S_{21}\rvert$
S_{22}	output reflection coefficient	output match: dB$\lvert S_{22}\rvert$

1.11.4 Hierarchical behavioral design with S-parameters

S-parameters can be measured at any level of the electronic technology hierarchy, from the transistor (in small-signal conditions) to the linear circuit, or all the way to the complete linear system. S-parameters provide a complete and accurate behavioral representation of the (linear) component at that level of the design hierarchy for efficient design at the next. S-parameters can dramatically reduce the complexity of a multi-component linear design by eliminating all internal nodes through the application of linear algebra, as discussed in Section 1.8. S-parameters protect the intellectual property (IP) of the design by providing only the behavioral relationship at the external terminals, with no information about the internal realization of the functionality in terms of circuit elements arranged in a particular topology (e.g. the schematic). S-parameters are therefore a *black-box* behavioral approach, requiring no a-priori information about the component, except that it is linear.

S-parameters are hierarchical. A system of two connected linear components can be represented by the overall S-parameters of the composite and used as a behavioral model at a higher level of the design hierarchy. The composite S-parameters can be obtained from direct measurement of the cascaded system, or by composition of the S-parameters of the constituent components according to the procedure of Section 1.8. Any subset of interacting linear components can be represented by an S-parameter block and inserted into a larger design. The process can be repeated iteratively from one level to the next level.

1.12 Limitations of S-parameters

Interestingly, S-parameters are still commonly used for nonlinear devices, such as transistors and amplifiers. The problem, often forgotten or taken for granted, is that S-parameters only describe properly the behavior of a nonlinear component in response to *small-signal* stimuli, for which the device behavior can be approximated as linear around a fixed DC, or static, operating point. That is, only when the nonlinear device is assumed to depend linearly on all RF components of the incident signals is the S-parameter paradigm valid. S-parameters contain no information about how a non-linear component generates distortion, such as that manifested by energy generated at harmonics and intermodulation frequencies by the component in response to excitation by one or more tones (sinusoidal signals). S-parameters are inadequate to describe nonlinear devices excited by large signals, despite many ad hoc attempts. Attempts to generalize S-parameters to the nonlinear realm have led to a wide variety of problems, including measurements that appear non-repeatable, and the inability to make predictable design inferences from such measurements or simulations. Techniques such "hot S-parameter" measurements and modeling nonlinear components using "large-signal S-parameter analysis" are examples of incomplete, insufficient, and ultimately incorrect approaches. Under large-signal conditions, totally new phenomena, for which there are no analogs in S-parameter theory, appear and must be taken into account.

For example, when a nonlinear component, such as a transistor or power amplifier, is stimulated simultaneously by two sinusoidal signals at different frequencies, the output spectrum is not consistent with the superposition principle discussed in Section 1.6. Rather, the response of a nonlinear system to two or more excitations generally contains more non-zero frequency components than were present in the input signal (see Figure 1.11). These terms are generally referred to as *intermodulation*

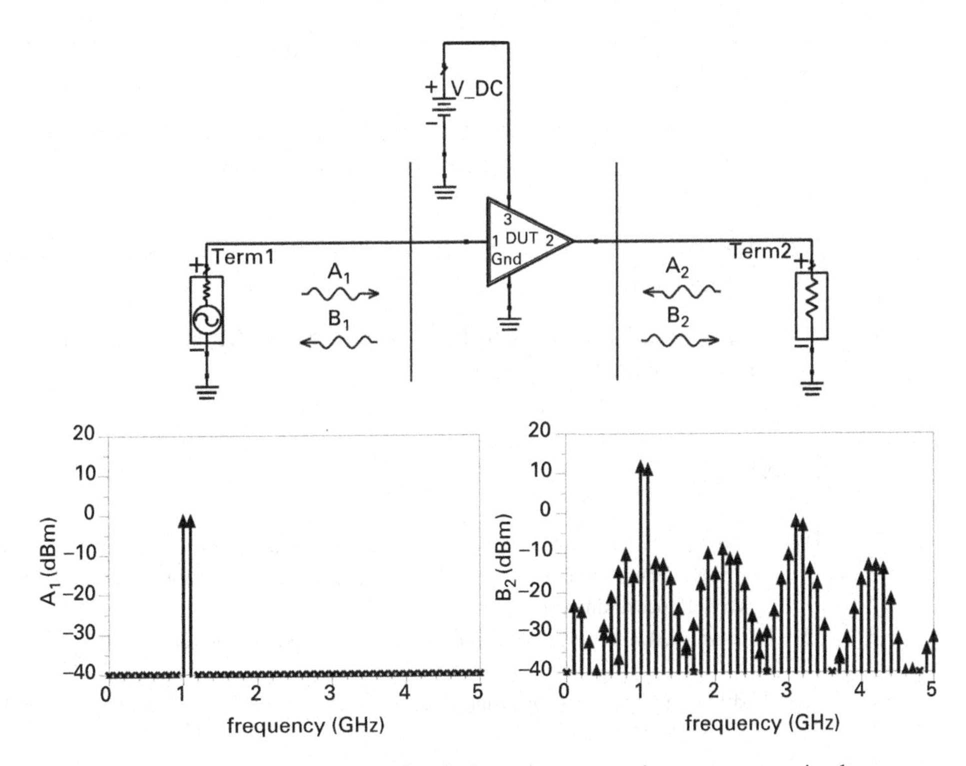

Figure 1.11 Intermodulation spectrum of typical transistor output in response to a simultaneous excitation by two sinusoidal signals at different frequencies. Superposition is not obeyed in general.

distortion.[2] It is clear that this phenomenon cannot be described by S-parameters. The more general framework of X-parameters is required to address this type of behavior.

1.13 Summary

S-parameters are linear time-invariant spectral maps defined in the frequency domain. They represent intrinsic properties of the DUT, enabling the hierarchical design of linear circuits and systems given only the S-parameters of the constituent functional blocks and their topological arrangement in the design. S-parameters can be defined, measured, and calculated for nonlinear components, but they are valid descriptions only under small-signal conditions. Phenomena such as harmonic and intermodulation distortion, generated by nonlinear components, require a more comprehensive framework for their consistent description and application.

Exercises

1.1 Prove that (1.17) and (1.18) are equivalent to the circuit laws. *Hint*: Start by expressing KVL and KCL in the time domain, use (1.4) to write the laws in wave variables, and then transform them to the frequency domain for the corresponding phasors.

1.2 Derive (1.19). *Hint*: Eliminate the four internal variables appearing in (1.17) and (1.18). The matrix producing $B_1^{(1)}$ and $B_2^{(2)}$ from $A_1^{(1)}$ and $A_2^{(2)}$ is the required result.

1.3 Derive (1.31) from (1.1) and the definition of the Y-matrix (reviewed in Appendix A).

1.4 Derive (1.32), the explicit form of the simple FET model S-parameter matrix, from the model of Figure 1.10.

References

[1] D. Rytting, *Calibration and Error Correction Techniques for Network Analysis*, IEEE Microwave Theory and Techniques Short Course, 2007; available at http://ieeexplore.ieee.org/servlet/opac?mdnumber=EW1062.

[2] G. Gonzalez, *Microwave Transistor Amplifiers*, 2nd edn. Englewood Cliffs, NJ: Prentice Hall, 1984.

[3] "S-parameter techniques for faster, more accurate network design," Hewlett-Packard application note AN 95–1, 1968.

[2] In fact, there are even cases where "passive connectors" cause intermodulation distortion [5].

[4] D. E. Root, J. Xu, J. Horn, and M. Iwamoto, "The large-signal model: theoretical foundations, practical considerations, and recent trends," in *Nonlinear Transistor Model Parameter Extraction Techniques*, M. Rudolf, C. Fager, and D. E. Root, Eds. Cambridge: Cambridge University Press, 2012, chap. 5.

[5] J. J. Henrie, A. J. Christianson, and W. J. Chappell, "Linear-nonlinear interaction and passive intermodulation distortion," *IEEE Trans. Microw. Theory Tech.*, vol. **58**, no. 5, pp. 1230–1237, May 2010.

Additional reading

K. Kurokawa, "Power waves and the scattering matrix," *IEEE Trans. Microw. Theory Tech.*, vol. **13**, no. 2, pp. 194–202, Mar. 1965.

2 X-parameters – fundamental concepts

2.1 Overview

The chapter begins with a brief review of nonlinear DUT behavior that cannot be described by linear S-parameters. In a simple but important case, it is demonstrated that a DUT description in terms of a nonlinear spectral mapping of signals defined on a harmonic frequency grid overcomes this limitation. It is shown that cascadability of nonlinear components described in this way follows from the circuit laws with no further approximation. Practical considerations of such a description are discussed. The principle of time invariance is invoked to obtain the final form for "single-tone" X-parameters. Key applications are presented.

2.2 Nonlinear behavior and nonlinear spectral mapping

It is useful to start the discussion about nonlinear behavior by observing the output of an amplifier when a relatively small sinusoidal signal, of frequency f_1, is applied at its input, port 1:

$$a_1(t) = \mathrm{Re}\{A_1 \cdot e^{j \cdot \omega_1 \cdot t}\}, \qquad A_1 \in \mathbb{C}. \tag{2.1}$$

The output contains an amplified replica of the sinusoidal signal and a number of additional sinusoidal signals at frequencies that are harmonics of f_1. These waves are displayed in a graphical form in Figure 2.1 for two power levels, -50 dBm and -20 dBm, of the incident wave $a_1(t)$ (that contains energy only at frequency ω_1).

The harmonic frequencies generated by the DUT at the output port have negligible levels if the input signal is small (as is the case at -50 dBm in Figure 2.1). As usual in engineering applications, the level below which the harmonics are negligible varies from one application to another. Once their contribution is negligible, the behavior of the amplifier is accurately described by the S-parameters, as discussed in Chapter 1, provided also the ratio between response and stimulus is independent of the stimulus power. Compression of the fundamental frequency response is a nonlinear process that also precludes the applicability of S-parameters.

Levels of scattered waves at harmonic frequencies rise and become non-negligible as the level of the input signal increases. The wave scattered from the output port, $b_2(t)$,

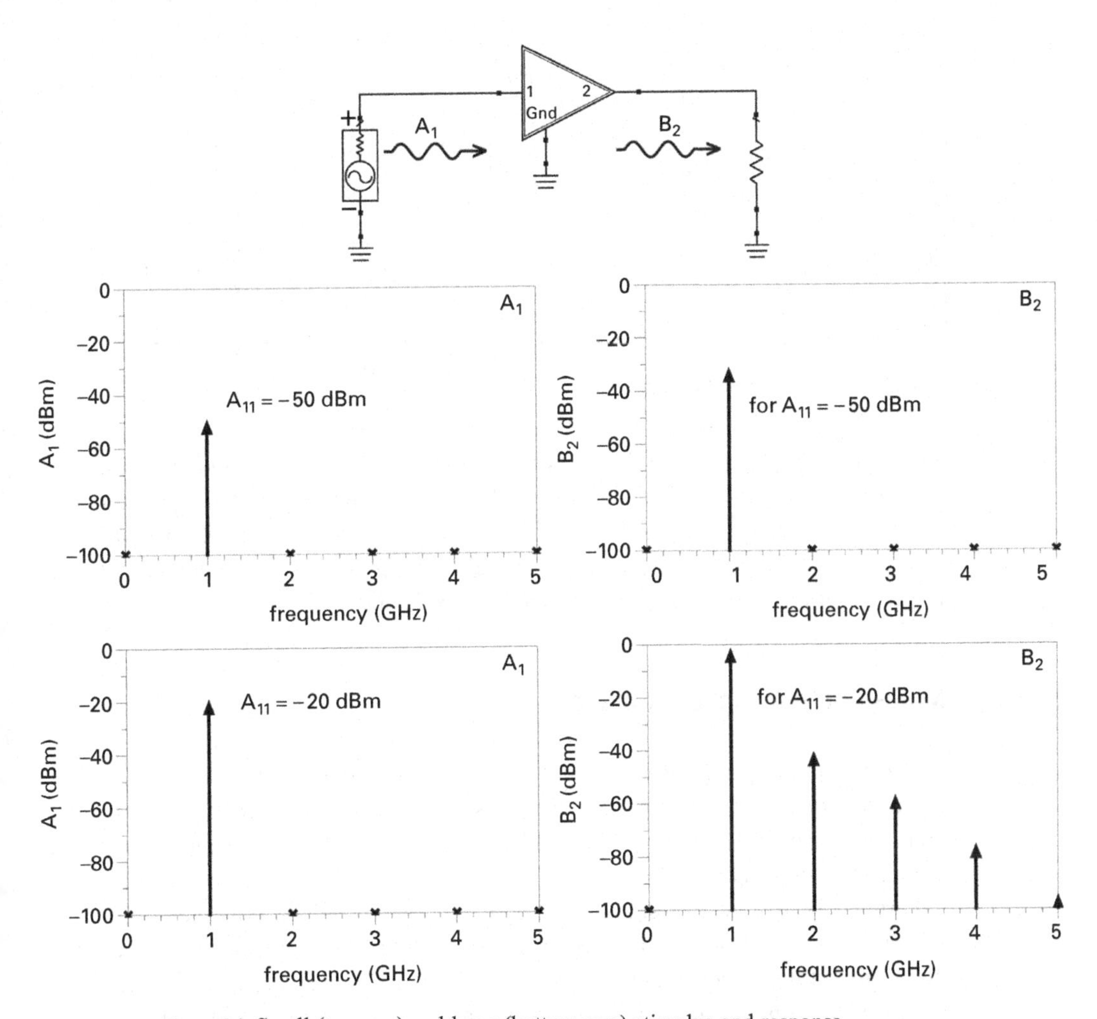

Figure 2.1 Small (top row) and large (bottom row) stimulus and response.

then contains significant energy at harmonic frequencies. Generally, the energy is distributed in the frequency domain on a grid, ω_k, defined by the stimulus, ω_1:

$$\omega_k = k\omega_1, \qquad k \in \mathbb{N}. \tag{2.2}$$

Two indices are required to specify the location (port) and frequency component of the variables in the frequency domain: the first index identifies the port and the second index identifies the component on the frequency grid.

The complex amplitude (hence the magnitude and the phase) of the response at each harmonic at any of the ports depends on the magnitude and on the phase (and hence on the complex amplitude) of the stimulus, $A_{1,1}$. This dependence is formally represented in equation (2.3), which has been generalized to an arbitrary number of ports, N:

$$B_{p,k} = F_{p,k}(A_{1,1}), \qquad \begin{array}{l} p = 1, 2, \ldots, N, \\ k = 1, 2, \ldots, K. \end{array} \tag{2.3}$$

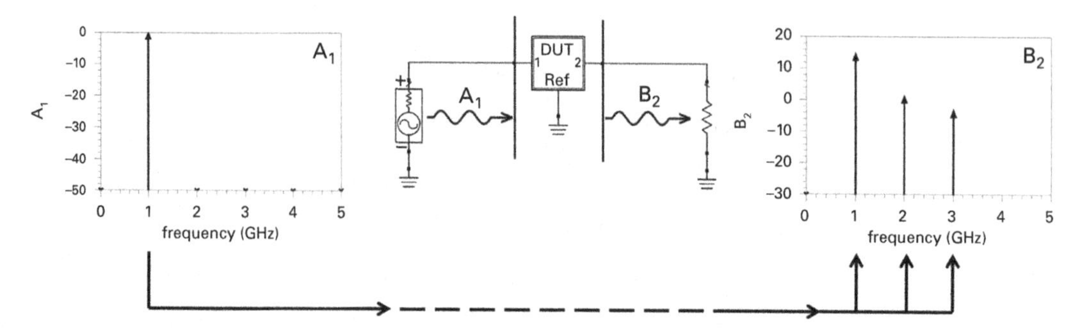

Figure 2.2 Nonlinear spectral map.

The notation $F_{p,k}$ in equation (2.3) specifies the scattered wave behavior of the amplifier at port p and harmonic index k as a particular nonlinear function of the input stimulus. There are $N \times K$ such functions if we consider N ports and K harmonics per port.

Equation (2.3) represents a nonlinear spectral mapping of the stimulus, $A_{1,1}$, to the responses of the system, $B_{p,k}$; this is depicted graphically in Figure 2.2.

2.3 Multi-harmonic spectral maps

The generation of scattered waves with energies at harmonic multiples of the incident stimulus frequency raises important questions for the cascading of multiple DUTs. The output waves of DUT1 become the incident waves for DUT2 in the simple cascade of 2-ports shown in Figure 2.3.

Fortunately, for incident signals with spectral components on a harmonic grid, the scattered waves will also have spectral components precisely on the same frequency grid. This is true for components such as diodes, transistors, and power amplifiers. For now, we only consider these types of devices that do not produce outputs in the absence of input stimuli (that is, unlike an oscillator).[1]

This means that the general problem, for this class of excitations and DUTs, reduces to characterizing the functional dependence of the scattered waves on all the spectral components of the incident signals. That is, the input–output functional relationship that completely characterizes the DUT steady-state nonlinear behavior can be formulated as shown in equation (2.4):

$$B_{p,k} = F_{p,k}(A_{1,1}, A_{1,2}, \ldots, A_{1,K}, A_{2,1}, A_{2,2}, \ldots, A_{2,K}, \ \ldots, A_{N,1}, A_{N,2}, \ldots, A_{N,K}). \tag{2.4}$$

For an N-port device, for given bias conditions at all DUT ports, there are $N \times K$ complex-valued functions, $F_{p,k}$, each of which is defined on $N \times K$ complex amplitudes representing the magnitudes and phases of the spectral components incident at each port.

[1] Phenomena such as sub-harmonic oscillations are neglected for now.

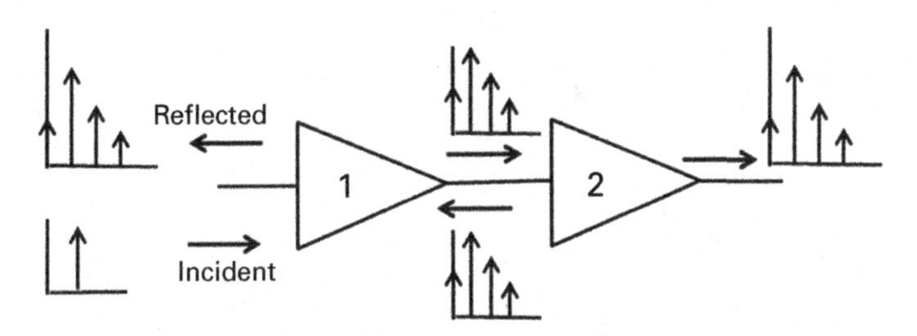

Figure 2.3 Cascaded 2-ports showing incident and scattered waves.

Considering that the DC-bias stimuli at each of the N ports, DCS_p, $p = 1,2,\ldots,N$, may vary, these additional independent variables must also be considered, as shown in (2.5). The DC response, DCR_p, at each of the N ports of the DUT also depends on all of these arguments and is identified separately in (2.5):

$$\begin{aligned} DCR_p &= DCF_p(DCS_1,\ldots,DCS_N,A_{1,1},\ldots,A_{1,K},A_{2,1},\ldots,A_{2,K},\ldots,A_{N,1},A_{N,1},\ldots,A_{N,K}),\\ B_{p,k} &= F_{p,k}(DCS_1,\ldots,DCS_N,A_{1,1},\ldots,A_{1,K},A_{2,1},\ldots,A_{2,K},\ldots,A_{N,1},A_{N,1},\ldots,A_{N,K}). \end{aligned} \tag{2.5}$$

Equation (2.5) can be written in the more compact form shown in (2.6), where the generic notations $q=\overline{1,N}$ and $h=\overline{1,K}$ are used to show that q takes all integer values between 1 and N, and k takes all integer values between 1 and K:

$$\begin{aligned} DCR_p &= DCF_p\left(\left\{DCS_q, q=\overline{1,N}\right\},\ \left\{A_{r,h}, r=\overline{1,N}, h=\overline{1,K}\right\}\right),\\ B_{p,k} &= F_{p,k}\left(\left\{DCS_q, q=\overline{1,N}\right\},\ \left\{A_{r,h}, r=\overline{1,N}, h=\overline{1,K}\right\}\right). \end{aligned} \tag{2.6}$$

Though explicit and unambiguous, the notation of (2.6) is still cumbersome. Therefore, unless specifically needed, the index ranges will be dropped, reducing (2.6) to the simpler form (2.7):

$$\begin{aligned} DCR_p &= DCF_p\left(\left\{DCS_q\right\},\ \left\{A_{r,h}\right\}\right),\\ B_{p,k} &= F_{p,k}\left(\left\{DCS_q\right\},\ \left\{A_{r,h}\right\}\right). \end{aligned} \tag{2.7}$$

The arguments of the functions in (2.7) must be understood to include *all* the DC stimuli and *all* the RF incident waves at *all* ports, as explicit in (2.6). There are distinct functions $DCF_p(.)$ and $F_{p,k}(.)$ corresponding to each value of the indexes p and k. The DC stimuli, DCS_p, may be either voltages or currents. The DUT response at DC at each of its ports, DCR_p, will be current or voltage, whichever complements the stimulus at the respective port.

For a DC-voltage bias at port p, the functions $DCF_p(.)$ are denoted $FI_p(.)$, and the response is the current, I_p, as shown in (2.8):

$$I_p = FI_p(.). \tag{2.8}$$

For a DC-current bias at port p, the functions $DCF_p(.)$ are denoted $FV_p(.)$, and the response is the current, V_p, as shown in (2.9):

$$V_p = FV_p(.). \tag{2.9}$$

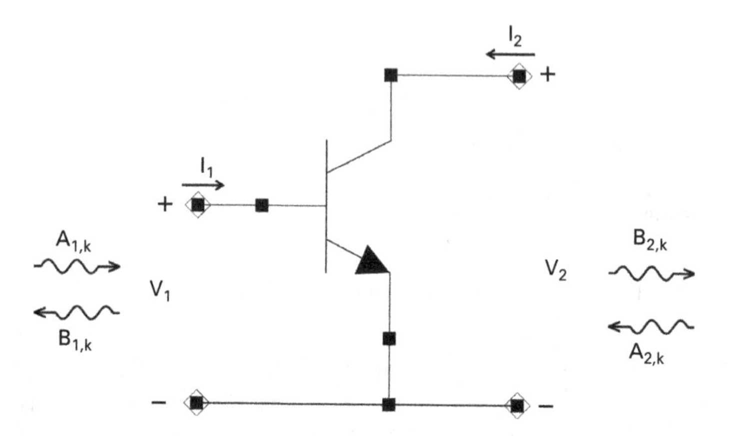

Figure 2.4 Characterization of nonlinear 2-port in A–B / I–V space on harmonic grid.

As a result, the functions $DCF_p(.)$ (which could be either $FV_p(.)$ or $FI_p(.)$) are real valued, and both sets of functions $DCF_p(.)$ and $F_{p,k}(.)$ are defined on $N \times K$ complex amplitudes and N real-valued bias conditions. Equations (2.7), in mixed DC-bias and wave variables, provide a complete representation for the N-port, equivalent to using spectral components of voltages and currents.

There is an additional constraint on the physically admissible functions defining relations (2.6). This constraint follows directly from the DUT property of time invariance. This constraint will be applied after simplifying the general expression (2.6) for practical applications.

As an example, consider the characterization of a transistor as a nonlinear 2-port in the A–B / I–V space (the generalized power waves and DC current–voltage), as depicted in Figure 2.4.

The set of equations describing the steady-state behavior of the transistor, considering only three harmonics have significant energy levels, is shown in (2.10):

$$\begin{aligned} V_1 &= FV_1\left(I_1, V_2, A_{1,1}, A_{1,2}, A_{1,3}, A_{2,1}, A_{2,2}, A_{2,3}\right), \\ B_{1,k} &= F_{1,k}\left(I_1, V_2, A_{1,1}, A_{1,2}, A_{1,3}, A_{2,1}, A_{2,2}, A_{2,3}\right), \\ I_2 &= FI_2\left(I_1, V_2, A_{1,1}, A_{1,2}, A_{1,3}, A_{2,1}, A_{2,2}, A_{2,3}\right), \\ B_{2,k} &= F_{2,k}\left(I_1, V_2, A_{1,1}, A_{1,2}, A_{1,3}, A_{2,1}, A_{2,2}, A_{2,3}\right). \end{aligned} \tag{2.10}$$

Note that, for port 1, current is the independent variable and voltage is the dependent variable, while, for port 2, the roles are reversed: voltage is the independent variable and current is the dependent variable. This is consistent with the most commonly used biasing scheme for a bipolar junction transistor (BJT), the device used in this example. Mathematically, however, either choice for the independent and dependent DC-bias variables can be selected at each port, independently, to describe accurately the behavior of the device, regardless of the actual biasing scheme used in the application.

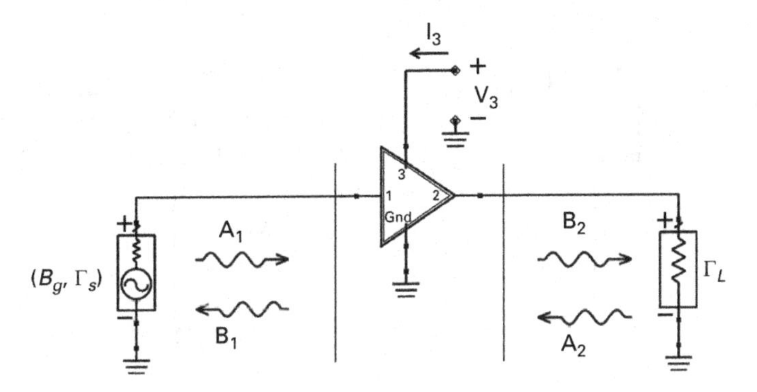

Figure 2.5 Load and source mismatch.

2.4 Load- and source-mismatch effects

The set of equations (2.7) is used to characterize the steady-state behavior of a nonlinear system under source- and load-mismatch conditions. This case is shown in Figure 2.5, where both Γ_S and Γ_L may have any arbitrary value. The wave generated and scattered by the source is marked by B_g in this figure.

The boundary conditions for source and load show the waves in complex-vector notation, as in the example for the A_1 wave in (2.11):

$$A_1 = [A_{1,1}, A_{1,2}, A_{1,3}]. \tag{2.11}$$

The system of equations completely describing such behavior under mismatch conditions is composed by equations (2.7) plus the source and load conditions, as shown in (2.12), where only three harmonics are considered:

$$\begin{aligned}
A_{1,k} &= B_{g,k} + \Gamma_{S,k} B_{1,k}, \\
A_{2,k} &= \Gamma_L B_{2,k}, \\
I_3 &= FI_3\left(V_3, A_{1,1}, A_{1,2}, A_{1,3}, A_{2,1}, A_{2,2}, A_{2,3}\right), \\
B_{1,k} &= F_{1,k}\left(V_3, A_{1,1}, A_{1,2}, A_{1,3}, A_{2,1}, A_{2,2}, A_{2,3}\right), \qquad k = 1,2, \\
B_{2,k} &= F_{2,k}\left(V_3, A_{1,1}, A_{1,2}, A_{1,3}, A_{2,1}, A_{2,2}, A_{2,3}\right), \qquad k = 1,2.
\end{aligned} \tag{2.12}$$

The solution to the system of equations (2.12) is the steady-state behavior of the system under the specified multi-harmonic source and load mismatch.

2.5 Cascading DUTs

The set of equations (2.7) is also used to find the steady-state behavior of a cascade of nonlinear systems. A generic example is shown in Figure 2.6. When two components are cascaded, an internal node is created at which the circuit laws KVL and KCL must be satisfied.

(a)

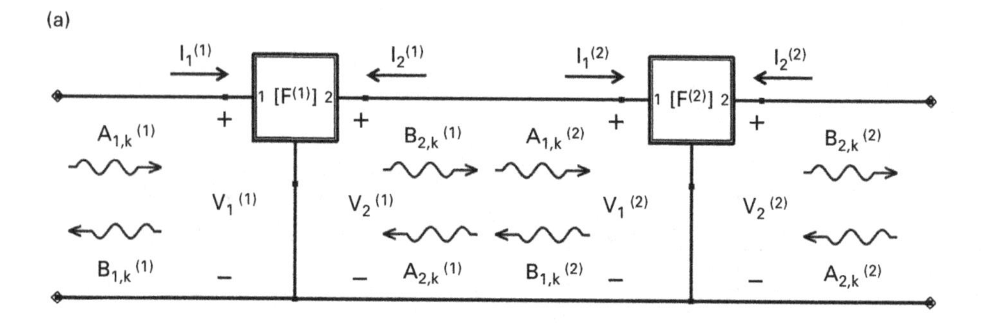

(b)

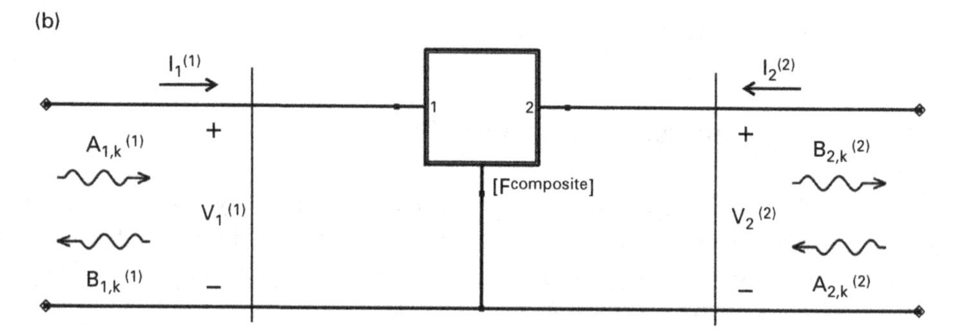

Figure 2.6 Cascade of two nonlinear systems. (a) Cascade of two DUTs. (b) The equivalent network, yielding the same response as the cascade of two DUTs.

It is a straightforward exercise to write the equations that follow from applying KVL and KCL at the internal node for wave variables in the frequency domain, and DC voltage and current. These relationships are given in (2.13),

$$\begin{aligned} V_2^{(1)} &= V_1^{(2)}, \\ I_2^{(1)} &= -I_1^{(2)}, \\ B_{2,k}^{(1)} &= A_{1,k}^{(2)}, \\ A_{2,k}^{(1)} &= B_{1,k}^{(2)}. \end{aligned} \tag{2.13}$$

which define a set of $2 \times (K + 1)$ equations:

- two real-numbered equations, one for DC voltage and one for DC current;
- $2 \times K$ complex equations, two per harmonic index.

The set of equations (2.13) is augmented by

- the boundary conditions, at the source and at the load, and
- the model (2.6) for each of the two nonlinear elements in the cascade.

The boundary conditions are determined by the external interactions at the source and load. The source boundary conditions, considered known, are shown in (2.14):

$$\begin{aligned} V_1^{(1)} &= V_{source}, \\ I_1^{(1)} &= I_{source}, \\ A_{1,k}^{(1)} &= A_{source}, \\ B_{1,k}^{(1)} &= B_{source}. \end{aligned} \tag{2.14}$$

The load boundary conditions, considered known, are shown in (2.15):

$$\begin{aligned} V_2^{(2)} &= V_{load}, \\ I_2^{(2)} &= I_{load}, \\ A_{2,k}^{(2)} &= A_{load}, \\ B_{2,k}^{(2)} &= B_{load}. \end{aligned} \tag{2.15}$$

The behavior of the first nonlinear element in the cascade is modeled in (2.16):

$$\begin{aligned} DCR_p^{(1)} &= DCF_p^{(1)}\left(\left\{DCS_q^{(1)}\right\}, \left\{A_{q,k}^{(1)}\right\}\right), \\ B_{p,k}^{(1)} &= F_{p,k}^{(1)}\left(\left\{DCS_q^{(1)}\right\}, \left\{A_{q,k}^{(1)}\right\}\right). \end{aligned} \tag{2.16}$$

The behavior of the second nonlinear element in the cascade is modeled in (2.17):

$$\begin{aligned} DCR_p^{(2)} &= DCF_p^{(2)}\left(\left\{DCS_q^{(2)}\right\}, \left\{A_{q,k}^{(2)}\right\}\right), \\ B_{p,k}^{(2)} &= F_{p,k}^{(2)}\left(\left\{DCS_q^{(2)}\right\}, \left\{A_{q,k}^{(2)}\right\}\right). \end{aligned} \tag{2.17}$$

The solution to the system of nonlinear equations (formed by (2.13), (2.14), (2.15), (2.16), and (2.17)) is the steady-state solution of the cascade at the internal node. This contains the waves and the DC voltages and currents at the internal node, identified in (2.13).

The entire system of equations is very similar to the one formulated for the linear operation. Unlike the case of Chapter 1, where the device was linear in all spectral components (except DC), elimination of the internal variables at the internal node defined by a cascade now requires the solution of nonlinear equations. If each wave variable has several harmonic components, there are multiple nonlinear equations in many variables, expressed by (2.7) in the general case, to be solved self-consistently.

2.6 Example: cascading two RF power amplifiers with independent bias

The mathematical formalism used to describe the nonlinear behavior should be flexible enough to allow formulation of the cascade equations when the DC and RF are present at different physical ports. A typical example is shown in Figure 2.7, where the harmonic indices are suppressed and the A and B waves are considered as vectors at the given numbered ports.

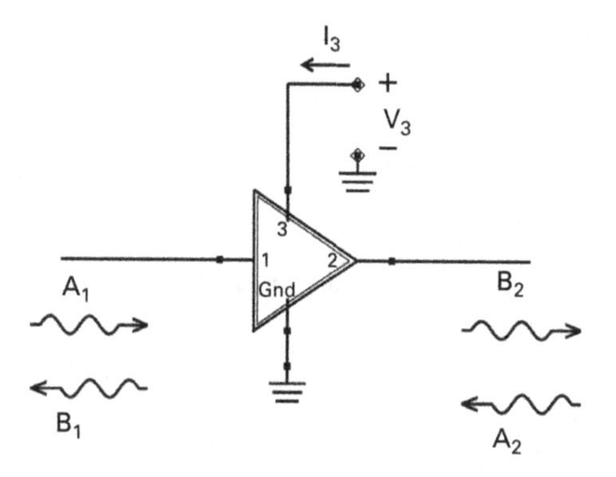

Figure 2.7 Two-port (w.r.t. RF) component.

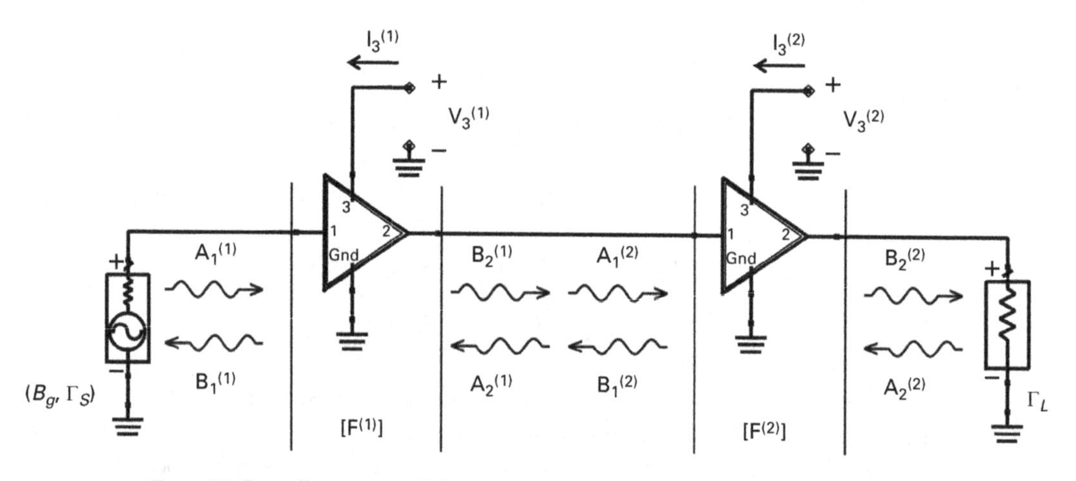

Figure 2.8 Cascading two amplifiers.

The situation of Figure 2.8 is obtained when cascading two such components, which depicts the common case of a cascade of two amplifiers working on a passive load.

The techniques described in both Sections 2.4 and 2.5 are used to solve for the steady-state behavior of the cascade under mismatched source and load conditions.

The internal node is now described by (2.18):

$$\begin{aligned} B_2^{(1)} &= F_2^{(1)}\left(V_3^{(1)}, A_1^{(1)}, A_2^{(1)}\right) = A_1^{(2)}, \\ B_1^{(2)} &= F_1^{(2)}\left(V_3^{(2)}, A_1^{(2)}, A_2^{(2)}\right) = A_2^{(1)}. \end{aligned} \tag{2.18}$$

The source and load conditions are described by (2.19):

$$\begin{aligned} A_1^{(1)} &= B_g + B_1^{(1)}\Gamma_S, \\ A_2^{(2)} &= B_2^{(2)}\Gamma_L, \end{aligned} \tag{2.19}$$

where B_g is the wave generated by the source when terminated on its own matched load, Γ_S^*.

For a complete description of the cascade, the DC equations need to be added, as shown in (2.20):

$$
\begin{aligned}
I_3^{(1)} &= DCF_3\left(V_3^{(1)}, A_1^{(1)}, A_2^{(1)}\right), \\
I_3^{(2)} &= DCF_3\left(V_3^{(2)}, A_1^{(2)}, A_2^{(2)}\right).
\end{aligned}
\tag{2.20}
$$

For the particular example shown in Figure 2.8, solving (2.18) and (2.19) gives the complete RF solution for the cascade, independent of (2.20), given that there is no interaction through the bias network between the two amplifiers. Equations (2.20) can then be used to determine the current through each amplifier at the corresponding individual operating conditions.

In the general case, interaction through the power distribution network may also be present, and then (2.18), (2.19), and (2.20) must be solved simultaneously, together with the other equations describing the behavior of the network connecting the bias ports of the two amplifiers.

The only assumption about cascading nonlinear blocks defined this way is that enough harmonics are used to characterize the nonlinear mappings (2.4), and that enough samples of the values of the dependent and independent variables are obtained (if measurement-based) to ensure accuracy over the ranges of signals likely to be used in an application. Of course, here we are still also assuming that the signals have spectral components on a harmonic grid and that each component is properly characterized on the same grid. This restriction will be relaxed in a later chapter.

It is important to understand that no approximation is made in formulating the system of equations (2.18) and (2.19). A complete solution of wave values (voltages/currents) at all interfaces in the cascade is determined for arbitrary values of the amplitude and phase of all harmonics in the stimulus.

2.7 Relationship to harmonic balance

Solving for cascaded nonlinear blocks as described in Section 2.5 is exactly analogous to the process a harmonic balance simulator uses to solve the nonlinear algebraic circuit equations for designs with multiple components in the frequency domain. The only difference is that here we are defining the component's nonlinear relationships in *A–B* space for the wave variables, whereas most simulators work in *I–V* space.

Harmonic balance analysis, introduced in the mid 1980s, was a great advance for the steady-state circuit simulation of RF and microwave nonlinear analog circuits [1]. It provided more efficient solutions compared to conventional time-domain methods (e.g. SPICE) for most practical nonlinear microwave circuits containing time constants spanning many orders of magnitude. A big difference, however, is that for most harmonic balance simulators the model description of the nonlinear components – transistors and diodes – is still formulated as time-domain nonlinear ordinary differential equations. The harmonic balance algorithm therefore has to transform these model equations into nonlinear algebraic equations in the frequency domain, and back again, for every iteration of the nonlinear solver.

Modeling and measurement methods have lagged far behind the simulation algorithms. One can say that by measuring or otherwise prescribing the functions (2.4), one is writing the model equations directly in the mathematical language native to the simulation algorithm that most efficiently solves the problem. The X-parameter approach is precisely this for large-signal multi-tone steady-state (and later modulated) signals.

2.8 Cross-frequency phase

Equation (2.3) includes a phase relationship between signals at different frequencies. It is important to clarify the definition of this phase relationship, called cross-frequency phase. This is discussed in Section 2.8.2, but the concept of commensurate signals needs to be introduced first.

2.8.1 Commensurate signals

Sinusoidal signals in a set are commensurate if all signals in the set are located on a frequency grid $f_k = kf_{fund}, k \in \mathbb{N}$, defined by one frequency, f_{fund}, called the fundamental. This concept will be revisited in Chapter 5 for the case of multiple fundamentals.

A set of commensurate sinusoidal signals allows the definition of the cross-frequency phase between any pair of two signals in the set.

For a nonlinear amplifier under the stimulus of a single large tone, all responses are a combination of the harmonics of the stimulus. All harmonics are sinusoids located on a frequency grid defined by the stimulus, as shown in (2.2); hence they are commensurate.

2.8.2 Definition of cross-frequency phase

In general, any combination of commensurate signals is periodic, with the period set by the fundamental frequency of the set, in this case f_{fund}.

The cross-frequency phase of any signal in a set of commensurate signals is the phase of that signal as measured at a moment in time when the signal at the fundamental frequency has a phase of 0°.

The cross-frequency phase is not defined for non-commensurate signals.

One important aspect is that the signals in the set do not have to coexist in the same node of the network. They just need to coexist in time so that they can be measured simultaneously.

For example, let us consider the measurement process of the steady-state behavior of an amplifier under a single-tone stimulus. The input and output waves exist at different

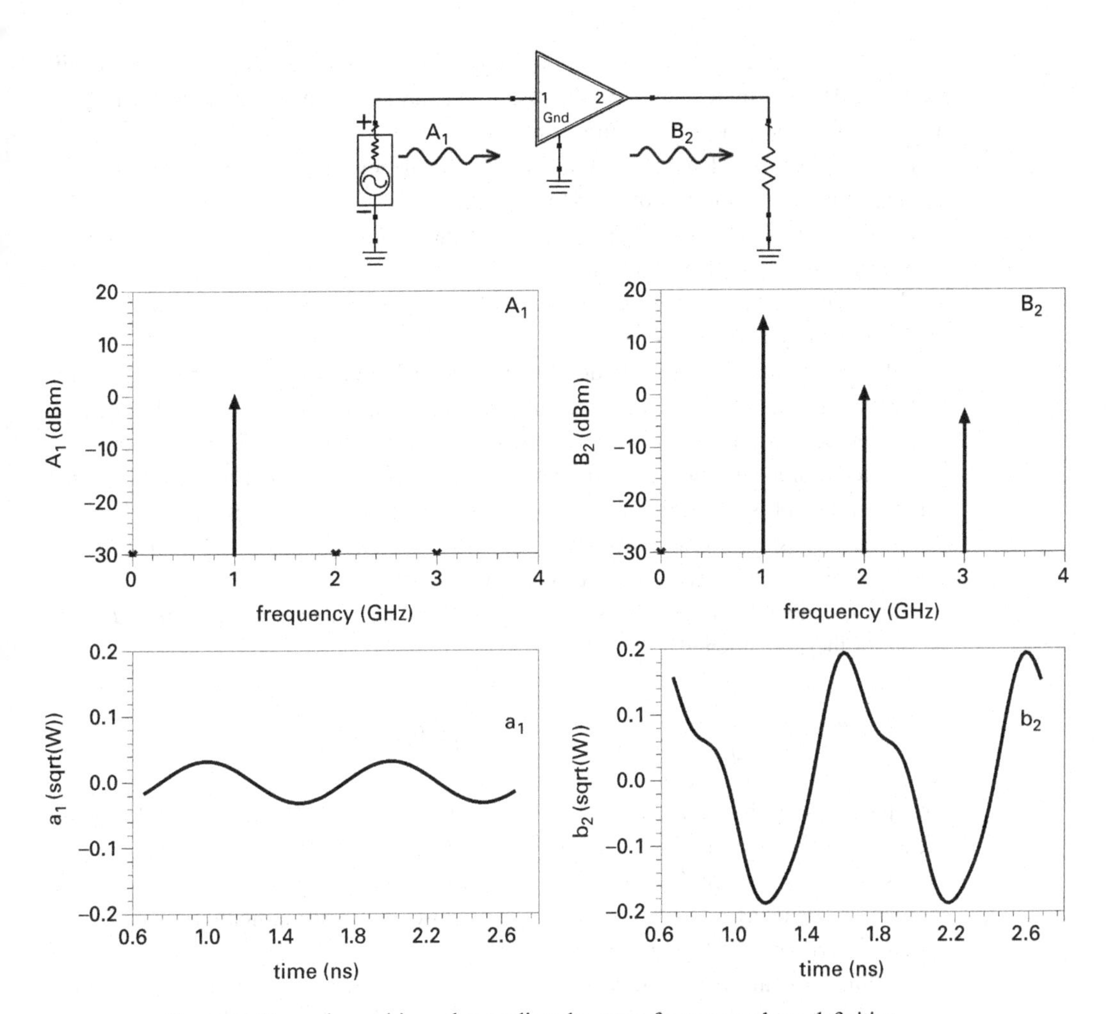

Figure 2.9 Example used in understanding the cross-frequency phase definition.

nodes of the network (input and output), but they coexist at the same time. All these waves form a set of commensurate signals, with the fundamental frequency set by the stimulus signal (which exists at the input port).

Once the waves are generated, the cross-frequency measurement process is not related to the DUT itself (the amplifier), but simply captures the relationship between the signals themselves. This process is not influenced in any way by the origin of the waves.

The DUT itself and its behavior have no influence on how the cross-frequency phase is measured. The cross-frequency phase is a very important aspect in the characterization of the DUT behavior.

The output scattered wave, B_2, is used as an example in this discussion, but the concepts are applicable to all waves scattered from the ports of a network under the excitation of a single-tone stimulus.

Figure 2.9 shows a simple example of an amplifier used as the system under test for a better understanding of the cross-frequency phase definition. The stimulus, A_1, and the response at port 2, B_2, are shown in both frequency and time domain. The time-domain view is helpful in understanding the cross-frequency phase relationship as used in the analysis of the behavior of nonlinear systems.

The stimulus, A_1, defines the fundamental frequency of the set of commensurate signals contained in the response of the amplifier.

Without any loss of generality, only three harmonics are used in this discussion, considering all higher harmonics have negligible power levels.

The waves A and B considered in Figure 2.9 are measured in either the time domain or the frequency domain.

If the measurement is carried out in the time domain, the waveforms are captured for one complete period of the fundamental frequency, starting at an arbitrary moment in time, marked as t_m in Figure 2.10, considered the time reference. Figure 2.10 displays two periods for a better understanding of the overall phenomenon. The period captured for measurement is highlighted. Due to the periodic nature of the response, a Fourier-series decomposition of the time-domain signals yields the frequency-domain representation, complete with amplitude and phase information.

The coefficients of the Fourier series are calculated as shown in (2.20), where T is the period of the fundamental:

$$B_{2,k}^{(m)} = \frac{2}{T} \int_{t_m}^{t_m+T} b_{2,k}^{(m)}(t)\, e^{-jk\frac{2\pi}{T}(t-t_m)}\, dt = \frac{2}{T} \int_{0}^{T} b_{2,k}^{(m)}(t+t_m)\, e^{-jk\frac{2\pi}{T}t}\, dt \tag{2.20}$$

The phase of each individual spectral line, $B_{2,k}^{(m)}$, reported by (2.20), is the phase measured at the moment t_m, the beginning of the time interval over which the integration is performed for the Fourier analysis, as also shown in Figure 2.10. In general, the moment t_m can have any arbitrary value because the measurement moment is not synchronized with the waveform itself.

The time-domain waveform is related to the frequency-domain spectrum through the Fourier series, as shown in (2.21):

$$b_2(t) = \mathrm{Re}\left(\sum_{k=1}^{3} B_{2,k}^{(m)}\, e^{j\omega_k \cdot t} \right). \tag{2.21}$$

In (2.21), all $B_{2,k}^{(m)}$ are peak values. The spectral components $B_{2,k}^{(m)}$ are obtained directly, with both magnitude and phase, if the waves A and B considered in Figure 2.9 are measured in the frequency domain at the moment in time t_m. The process of measuring these waves is discussed in detail in Chapter 4.

The superscript m in (2.21) indicates that $B_{2,k}^{(m)}$ is the measured wave component at an arbitrary moment in time, the moment when the measurement is taken. This superscript is used to distinguish the variables actually measured from the variables that will be used to characterize the intrinsic properties of the system under test, which will not take the superscript m.

In the case shown in Figure 2.10(a), the time reference (the beginning of the time interval used for the integration period in the Fourier analysis) is located at an arbitrary moment in time, t_m. The Fourier series in (2.21) can be reformatted as shown in (2.22) by explicitly showing the phase of each frequency component:

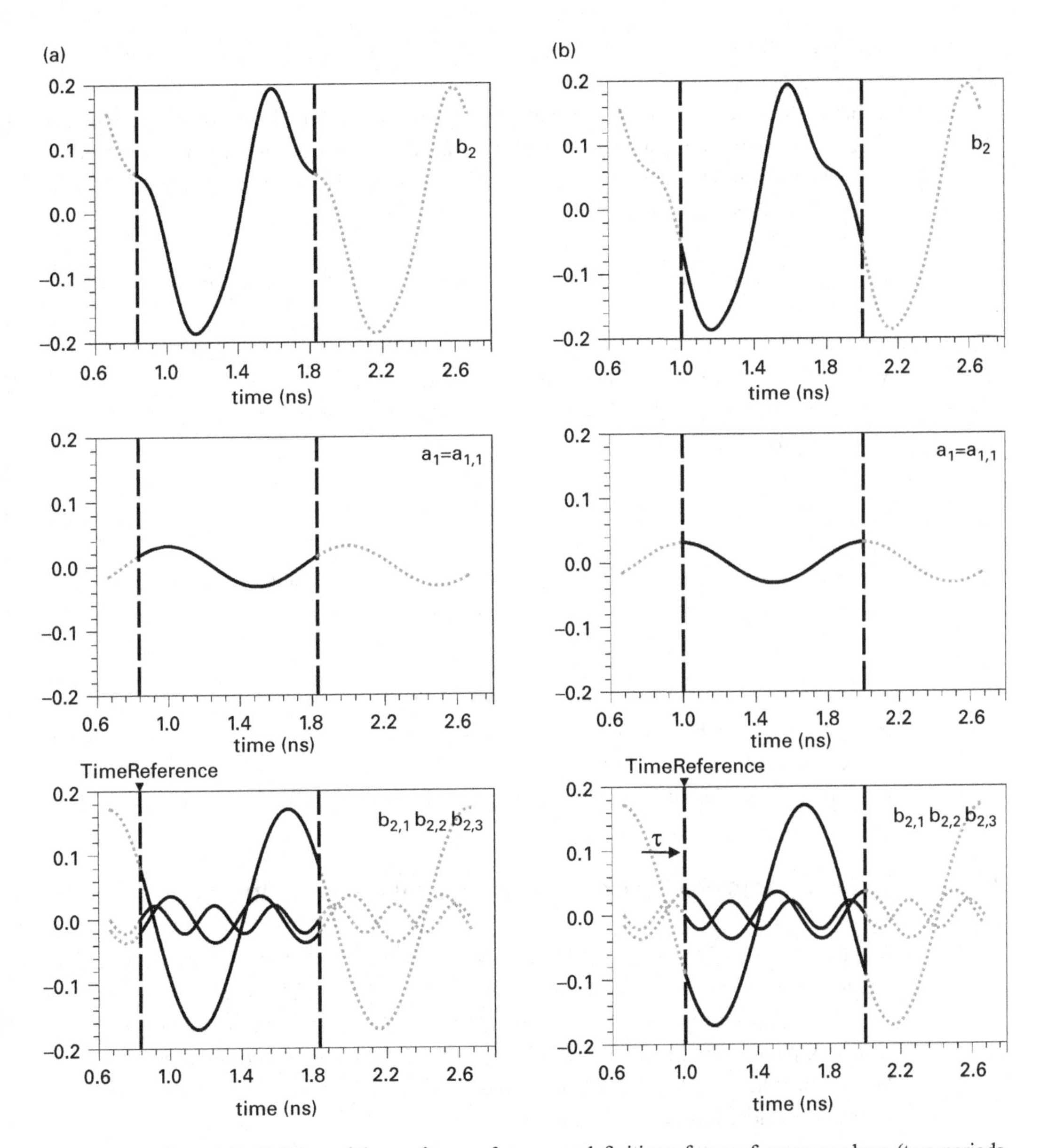

Figure 2.10 Shifting of time reference for proper definition of cross-frequency phase (two periods are displayed, the measured period is highlighted). (a) Arbitrary time reference. (b) Time shift such that fundamental $a_{1,1}$ has phase 0°.

$$b(t) = \mathrm{Re}\left(\left|B_{2,1}^{(m)}\right| e^{j\cdot\phi_{2,1}^{(m)}} e^{j\cdot\omega_1\cdot t} + \left|B_{2,2}^{(m)}\right| e^{j\cdot\phi_{2,2}^{(m)}}.e^{j\omega_2\cdot t} + \left|B_{2,3}^{(m)}\right| e^{j\phi_{2,3}^{(m)}} e^{j\cdot\omega_3\cdot t}\right). \quad (2.22)$$

The phases may have any values because the reference is an arbitrary moment in time.

The time reference can be shifted to another position in time, which is specific and more convenient: the moment in time when the fundamental has a phase of 0°.

This time shift can be achieved mathematically, without any additional measurements, as shown below.

In this treatment of the topic of nonlinear behavior, a cosine-based phase definition is used. As a result, the new time reference, delayed by τ, is the moment in time when the fundamental sinusoid has its peak value. This situation is shown in Figure 2.10(b).

The Fourier series becomes

$$b(t-\tau)=\text{Re}\left(\left|B_{2,1}^{(m)}\right| e^{j\left(\phi_{2,1}^{(m)}-\omega_1\tau\right)} e^{j\omega_1 t}+\left|B_{2,2}^{(m)}\right| e^{j\left(\phi_{2,2}^{(m)}-\omega_2\tau\right)} e^{j\omega_2 t}+\left|B_{2,3}^{(m)}\right| e^{j\left(\phi_{2,3}^{(m)}-\omega_3\tau\right)} e^{j\omega_3 t}\right), \tag{2.23}$$

where the time-shift property of the Fourier series was used.

The time shift, τ, is determined by the condition of the fundamental phase to be 0° after the time shift, as shown in equation (2.24):

$$\phi_{2,1}^{(m)} - \omega_1\tau = 0 \quad \Rightarrow \quad \tau = \frac{\phi_{2,1}^{(m)}}{\omega_1}. \tag{2.24}$$

The phases of all other harmonics after the time shift can now be calculated, and, after using (2.24) and (2.2), the new Fourier coefficients are shown in equation (2.25):

$$B_{2,k} = \left|B_{2,k}^{(m)}\right| e^{j\left(\phi_{2,k}^{(m)}-k\phi_{2,1}^{(m)}\right)} = \left|B_{2,k}\right| e^{j\phi_{2,k}}. \tag{2.25}$$

The phases of the Fourier-series coefficients in equation (2.25) are the cross-frequency phases with respect to the fundamental. These values can now be used to determine the cross-frequency relative phase between any two signals in the set of commensurate signals.

2.9 Basic X-parameters for multi-harmonic multi-port stimulus

The functions $F_{p,k}$ have been used until now to describe the behavior of the system. This method requires knowledge of the functions for all possible combinations of values for their multiple arguments (the DC stimuli and the complex multiple incident RF waves at all ports).

The X-parameters have been developed such that they are determined on a minimal but comprehensive set of conditions, and they can still describe the intrinsic behavior of the system under any other conditions of use.

In the general case, multiple harmonics are applied simultaneously to all the ports of the DUT. They are all commensurate signals on the same harmonic grid.

The DUT responds to this multi-harmonic stimulus with scattered waves at its ports, on all harmonics, $B_{p,k}$. This response depends on all incident waves applied by the stimulus, in their complex form.

To clarify these ideas, a two-port DUT is considered here together with a stimulus containing only the first two harmonics, $A_{1,1}$, $A_{1,2}$, $A_{2,1}$, and $A_{2,2}$, simultaneously incident on each of the two ports, as shown in (2.26):

$$B_{p,k} = F_{p,k}(A_{1,1}, A_{1,2}, A_{2,1}, A_{2,2}). \tag{2.26}$$

2.9.1 Time invariance and related properties of $F_{p,k}(.)$ functions

As already discussed in Chapter 1, we consider the case of time-invariant systems. The time-invariance condition imposes some useful properties upon the nonlinear functions $F_{p,k}$.

Considering the response at port 2 of the DUT, $b_2(t)$, due to the stimulus applied at port 1, $a_1(t)$, the behavior of the amplifier is described by an operator O, as shown in (2.27):

$$b_2(t) = O\left[a_1(t)\right]. \tag{2.27}$$

If the stimulus is delayed by a time τ, the response must be delayed by the same time duration τ, as shown in (2.28):

$$b_2(t-\tau) = O\left[a_1(t-\tau)\right]. \tag{2.28}$$

The wave $b_2(t)$ is the summation of the waves at port 2 on all harmonic frequencies, which means that all harmonics would be delayed by the same amount of time, τ:

$$b_2(t-\tau) = \sum_{k=1}^{K} b_{2,k}(t-\tau). \tag{2.29}$$

In the frequency domain, the delayed version of the response can be expressed as shown in (2.30):

$$B_{p,k}\, e^{-j\omega_k\tau} = F_{p,k}\left(A_{1,1}\, e^{-j\omega_1\tau}, A_{1,2}\, e^{-j\omega_2\tau}, A_{2,1}\, e^{-j\omega_1\tau}, A_{2,2}\, e^{-j\omega_2\tau}\right). \tag{2.30}$$

Combining (2.30) and (2.26), and using the frequency-grid definition in (2.2), the functions $F_{p,k}$ must satisfy the relationship shown in (2.31):

$$\begin{aligned} &F_{p,k}\left(A_{1,1}\, e^{-j\omega_1\tau}, A_{1,2}(e^{-j\omega_1\tau})^2, A_{2,1}\, e^{-j\omega_1\tau}, A_{2,2}(e^{-j\omega_1\tau})^2\right) \\ &\quad = F_{p,k}(A_{1,1}, A_{1,2}, A_{2,1}, A_{2,2})e^{-j\omega_k\tau}. \end{aligned} \tag{2.31}$$

This leads us to the conclusion that the nonlinear operators $F_{p,k}$ must satisfy the general property expressed in (2.32):

$$\begin{aligned} &F_{p,k}\left(A_{1,1}\, e^{j\phi}, A_{1,2}\left(e^{j\phi}\right)^2, A_{2,1}\, e^{j\phi}, A_{2,2}\left(e^{j\phi}\right)^2\right) \\ &\quad = F_{p,k}(A_{1,1}, A_{1,2}, A_{2,1}, A_{2,2})\left(e^{j\phi}\right)^k, \qquad \forall\phi \in \mathbb{R}. \end{aligned} \tag{2.32}$$

This property can now be used to separate the dependence on the magnitude and phase of the fundamental frequency, as shown in (2.33):

$$\begin{aligned} &F_{p,k}\left(A_{1,1}\, e^{-j\phi_{A1,1}}, A_{1,2}\left(e^{-j\phi_{A1,1}}\right)^2, A_{2,1}\, e^{-j\phi_{A1,1}}, A_{2,2}\left(e^{-j\phi_{A1,1}}\right)^2\right) \\ &\quad = F_{p,k}(A_{1,1}, A_{1,2}, A_{2,1}, A_{2,2})\left(e^{-j\phi_{A1,1}}\right)^k, \qquad \forall\phi \in \mathbb{R}. \end{aligned} \tag{2.33}$$

The notation P is introduced by (2.34):

$$P = e^{j\cdot phase\left(A_{1,1}\right)}. \tag{2.34}$$

Using the notation introduced in (2.34) and considering $A_{1,1}\,e^{-j\phi_{A1,1}} = |A_{1,1}|$, equation (2.33) can be reformulated as shown in (2.35):

$$F_{p,k}(A_{1,1}, A_{1,2}, A_{2,1}, A_{2,2}) = F_{p,k}\left(|A_{1,1}|, A_{1,2}P^{-2}, A_{2,1}P^{-1}, A_{2,2}P^{-2}\right)P^{k}. \quad (2.35)$$

Equation (2.35) and its implications are very significant. First, the property of taking the dependence of the *phase* ($A_{1,1}$) outside the arguments of the $F_{p,k}$ functions enforces the time invariance of the model formulation. This also shows that not any complex function can represent a time-invariant system. Only functions for which (2.35) is possible may be considered in the nonlinear system modeling process.

A second significance of (2.35) is that the set of functions $F_{p,k}$ depends on one fewer real argument than might be assumed by counting the arguments of (2.26). This reduces the dimensionality of the state space over which the system needs to be characterized. This becomes a very important practical benefit, saving characterization time and measured data file size, and improving the ultimate speed of simulation when the model is used in design.

2.9.2 Definition of X-parameters and X-parameter behavioral model

The X-parameters for the multi-harmonic stimulus can now be defined as shown in (2.36):

$$X_{p,k}^{(FB)}\left(|A_{1,1}|, A_{1,2}P^{-2}, A_{2,1}P^{-1}, A_{2,2}P^{-2}\right) = \frac{F_{p,k}(A_{1,1}, A_{1,2}, A_{2,1}, A_{2,2})}{P^{k}}. \quad (2.36)$$

The set of X-parameters can now be used to formulate the equations of a behavioral model that can calculate the response of the system to any steady-state stimulus containing energy at any of the three harmonics initially considered. The model equations are shown in (2.37), where the number of harmonics has been extended to K:

$$B_p(\omega) = \sum_{k=1}^{K} X_{p,k}^{(FB)}\left(|A_{1,1}|, \left\{A_{\substack{q,k \\ (q,k)\neq(1,1)}} P^{-k}\right\}\right)P^{k}\delta(\omega_k). \quad (2.37)$$

The X-parameters are not dependent on the phase of the fundamental due to the multiplication of each harmonic complex-valued argument with the coefficient P^{-k}, which time shifts each harmonic to the moment when the fundamental has a phase of 0°, effectively removing the dependence on the phase of the fundamental.

Although the X-parameters do not depend on the phase of the fundamental, the X-parameter behavioral model in (2.37) does capture the dependence of the response, $B_p(\omega)$, on the phase of $A_{1,1}$ through the multiplication of $X_{p,k}^{(FB)}$ with the factor P^k.

The DC behavior must be added to the set of equations shown in (2.37) to obtain a complete description of the nonlinear behavior of the DUT. This is shown in (2.38):

$$\begin{aligned} DCR_p &= X_p^{(FDCR)}\left(\{DCS_q\},\ |A_{1,1}|,\ \left\{A_{\substack{q,k \\ (q,k)\neq(1,1)}} P^{-k}\right\}\right), \\ B_{p,k} &= X_{p,k}^{(FB)}\left(\{DCS_q\},\ |A_{1,1}|,\ \left\{A_{\substack{q,k \\ (q,k)\neq(1,1)}} P^{-k}\right\}\right)P^{k}. \end{aligned} \quad (2.38)$$

The parameters $X_p^{(FDCR)}$ in (2.38) are defined directly by the functions $DCF_p(.)$ in (2.6), with the forms shown in (2.39) for voltage bias and in (2.40) for current bias, as per (2.8) and (2.9):

$$X_p^{(FDCR)}(.) = X_p^{(FI)}(.) = FI_p(.), \tag{2.39}$$

$$X_p^{(FDCR)}(.) = X_p^{(FV)}(.) = FV_p(.). \tag{2.40}$$

The system of equations in (2.38) is a behavioral model that completely characterizes the nonlinear behavior of the DUT. This model is known as the X-parameter model. We use the term "model" with no implication of any approximation of the system description beyond the postulate of time invariance.

The model is valid for all power levels and phases of the K harmonics. Note that the X-parameter values depend only on the amplitude of the fundamental, but they depend on both the amplitude and phase of the harmonics.

The set of basic X-parameters is thus composed of the terms identified in (2.41):

$$stimulus \begin{cases} DCS_p \\ A_{p,k} \end{cases}; \qquad X\text{-}parameters \begin{cases} X_p^{(FDCR)} \\ X_{p,k}^{(FB)} \end{cases} \qquad p = \overline{1,N},\ k = \overline{1,K}. \tag{2.41}$$

2.9.3 Example: a set of X-parameters

For example, the general model in (2.38) can be explicitly written as shown in (2.42) for the following specific case:

- a two-port DUT,
- with three-harmonic stimulus applied (incident) at both port 1 and port 2,
- with only two harmonics containing significant energy in the response (scattered from ports 1 and 2),
- with voltage bias on port 1, and
- with current bias on port 2.

$$\begin{aligned}
I_1 &= X_1^{(FI)}\left(VDC_1, IDC_2, |A_{1,1}|, A_{1,2}P^{-2}, A_{1,3}P^{-3}, A_{2,1}P^{-1}, A_{2,2}P^{-2}, A_{2,3}P^{-3}\right),\\
B_{1,1} &= X_{1,1}^{(FB)}\left(VDC_1, IDC_2, |A_{1,1}|, A_{1,2}P^{-2}, A_{1,3}P^{-3}, A_{2,1}P^{-1}, A_{2,2}P^{-2}, A_{2,3}P^{-3}\right)P^1,\\
B_{1,2} &= X_{1,2}^{(FB)}\left(VDC_1, IDC_2, |A_{1,1}|, A_{1,2}P^{-2}, A_{1,3}P^{-3}, A_{2,1}P^{-1}, A_{2,2}P^{-2}, A_{2,3}P^{-3}\right)P^2,\\
V_2 &= X_2^{(FV)}\left(VDC_1, IDC_2, |A_{1,1}|, A_{1,2}P^{-2}, A_{1,3}P^{-3}, A_{2,1}P^{-1}, A_{2,2}P^{-2}, A_{2,3}P^{-3}\right),\\
B_{2,1} &= X_{2,1}^{(FB)}\left(VDC_1, IDC_2, |A_{1,1}|, A_{1,2}P^{-2}, A_{1,3}P^{-3}, A_{2,1}P^{-1}, A_{2,2}P^{-2}, A_{2,3}P^{-3}\right)P^1,\\
B_{2,2} &= X_{2,2}^{(FB)}\left(VDC_1, IDC_2, |A_{1,1}|, A_{1,2}P^{-2}, A_{1,3}P^{-3}, A_{2,1}P^{-1}, A_{2,2}P^{-2}, A_{2,3}P^{-3}\right)P^2.
\end{aligned} \tag{2.42}$$

The set of basic X-parameters for this example is shown in (2.43), where the arguments (the independent variables) are not shown for convenience of notation.

$$stimulus \begin{cases} VDC_1 \\ A_{1,1} \\ A_{1,2} \\ A_{1,3} \\ IDC_2 \\ A_{2,1} \\ A_{2,2} \\ A_{2,3} \end{cases} ; \quad X\text{-}parameters \begin{cases} X_1^{(FI)} \\ X_{1,1}^{(FB)} \\ X_{1,2}^{(FB)} \\ X_2^{(FV)} \\ X_{2,1}^{(FB)} \\ X_{2,2}^{(FB)} \end{cases} . \tag{2.43}$$

2.10 Physical meaning of the basic X-parameters

The basic X-parameters for the case of multi-harmonic stimulus of a multi-port network are defined by the mathematical relationships shown in (2.36), and the X-parameter behavioral model (which calculates the nonlinear behavior of the system based on the X-parameters) is shown in (2.38). It is very useful to understand the physical meaning of these mathematical formulations.

2.10.1 Reference stimulus and response

Insight into this physical interpretation can be gained by observing that the response of the system is identical to the values of the basic X-parameters if the coefficient P has a value of 1, as shown in (2.44):

$$\text{for } P = 1 \Rightarrow \begin{cases} DCR_p = X_p^{(FDCR)}\left(\{DCS_q\}, |A_{1,1}|, \left\{ \underset{(q,k) \neq (1,1)}{A_{q,k}} P^{-k} \right\}\right) \\ B_{p,k} = X_{p,k}^{(FB)}\left(\{DCS_q\}, |A_{1,1}|, \left\{ \underset{(q,k) \neq (1,1)}{A_{q,k}} P^{-k} \right\}\right). \end{cases} \tag{2.44}$$

Considering the definition (2.34) of the factor P, condition $P = 1$ practically means that the phase of the fundamental, *phase* $(A_{1,1})$, must be 0°. The stimulus for this particular case is called the reference stimulus, and it is shown in (2.45):

$$\{DCS_q\}, |A_{1,1}|, \left\{ \underset{(q,k) \neq (1,1)}{A_{q,k}} P^{-k} \right\}. \tag{2.45}$$

The response of the system to the reference stimulus is called the reference response.

Any test stimulus with an arbitrary value of *phase*$(A_{1,1})$ but with the same phase relationships between the harmonics is a delayed version of the reference stimulus and results in a delayed version of the reference response due to the time-invariance condition discussed in Section 2.9.1.

The above property can be used to define the concept of equivalent stimuli. Two stimuli are said to be equivalent if they are delayed versions of each other.

2.10.2 Physical interpretation

From (2.44) it follows that the physical interpretation of the basic X-parameters is that they are the frequency-domain components of the reference response generated by the reference stimulus equivalent to the test stimulus.

It also follows that the X-parameter behavioral model in (2.38) has an easy physical interpretation. The steady-state response of the system to an arbitrary test stimulus can be calculated by applying a corresponding time shift to the reference response generated by the reference stimulus equivalent to the test stimulus.

2.11 Using the X-parameter behavioral model

Let's have a closer look at how the response of the system is calculated by the X-parameter behavioral model.

Once the phase of the fundamental, $phase(A_{1,1})$, in the test stimulus is known, the reference stimulus is obtained by shifting the test stimulus back in time with the duration τ corresponding to $phase(A_{1,1})$, as determined in (2.46):

$$\tau = \frac{phase(A_{1,1})}{\omega_1}. \tag{2.46}$$

This time shift is equivalent to a phase shift on each harmonic, k, of the test stimulus, as determined in (2.47):

$$phase\_shift\_on\_harmonic\_k = \omega_k \tau. \tag{2.47}$$

This phase shift is implemented in the model as a multiplication with the coefficient P^{-k}, as shown in (2.48), where the negative sign for the exponent means a shift back in time:

$$\{DCS_q\},\ |A_{1,1}|,\ \left\{ \underset{(q,k)\neq(1,1)}{A_{q,k} \qquad P^{-k}} \right\}. \tag{2.48}$$

The set of signals in (2.48) comprises the reference stimulus equivalent to the test stimulus.

Considering the X-parameters of the system under test are known, they are the frequency-domain components of the reference response (the response to the reference stimulus (2.48)), as shown in (2.49):

$$\left\{ \begin{array}{l} X_p^{(FDCR)} \\ X_{p,k}^{(FB)} \end{array} \right. . \tag{2.49}$$

The response to the test stimulus is thus obtained by time shifting the reference response (2.49) forward in time with the same value τ, thus bringing it back in time alignment with the test stimulus. This is equivalent to phase shifting each harmonic of the reference response, $X_{p,k}^{(F)}$, with a corresponding phase $\omega_k \tau$. The phase shift is implemented as a multiplication with the coefficient P^k, as shown in (2.50), where the positive sign of the exponent means a shift forward in time:

$$\begin{cases} X_p^{(FDCR)} \\ X_{p,k}^{(FB)} P^k \end{cases} . \tag{2.50}$$

The entire sequence of steps described in (2.46) to (2.50) is represented mathematically by the X-parameter behavioral model in (2.38). Also, this sequence of steps represents the conceptual methodology for measuring X-parameters, which is explained in detail in Chapter 4.

The X-parameters are thus a description of the intrinsic steady-state behavior of the system under test, and they can be used to calculate the steady-state response to an arbitrary stimulus of the same harmonic content as the reference stimulus that generated the set of X-parameters.

2.11.1 Example: amplifier with source and load mismatch

To clarify these ideas, let us consider the example of an amplifier under a single-tone stimulus working under mismatched conditions at both ports (source and load mismatch), as shown in Figure 2.11.

Although a single tone is applied by the signal source at port 1 only, the stimulus comprises incident waves at both ports 1 and 2. The incident wave at port 2, A_2, occurs due to the load mismatch, which reflects energy on all harmonics back towards the amplifier output.

Note that wave A_1, incident on the input port, is not a pure sinusoidal signal, although the source itself applies a single large tone. This is due to the nonlinear behavior of the amplifier itself combined with the source mismatch. The amplifier internally generates harmonics which are scattered from both ports 1 and 2. Some of the harmonic energy scattered from port 1 is reflected back towards the amplifier due to the source mismatch, thus generating the distortion of the wave A_1 observed in Figure 2.11.

The behavior of the amplifier is observed (or measured) at the arbitrary moment in time, t_m, as depicted in Figure 2.11(a).

If the measurement takes place in the frequency domain, this means that both stimulus and response spectra are measured at the moment t_m. If the measurement takes place in the time domain, this means both stimulus and response waveforms are measured (recorded) starting at the moment t_m for a duration equal to a time period of the stimulus.

The amplifier is characterized by its X-parameters, which are the frequency-domain representation of the response to the reference stimulus. The Fourier decomposition of the test stimulus reveals the harmonic content, as shown in Figure 2.11(b). The reference stimulus is obtained by time shifting the test stimulus until the fundamental has a 0° phase at the initial moment of the observation interval, as shown in Figure 2.11(c).

The response to the reference stimulus is known from the X-parameter values.

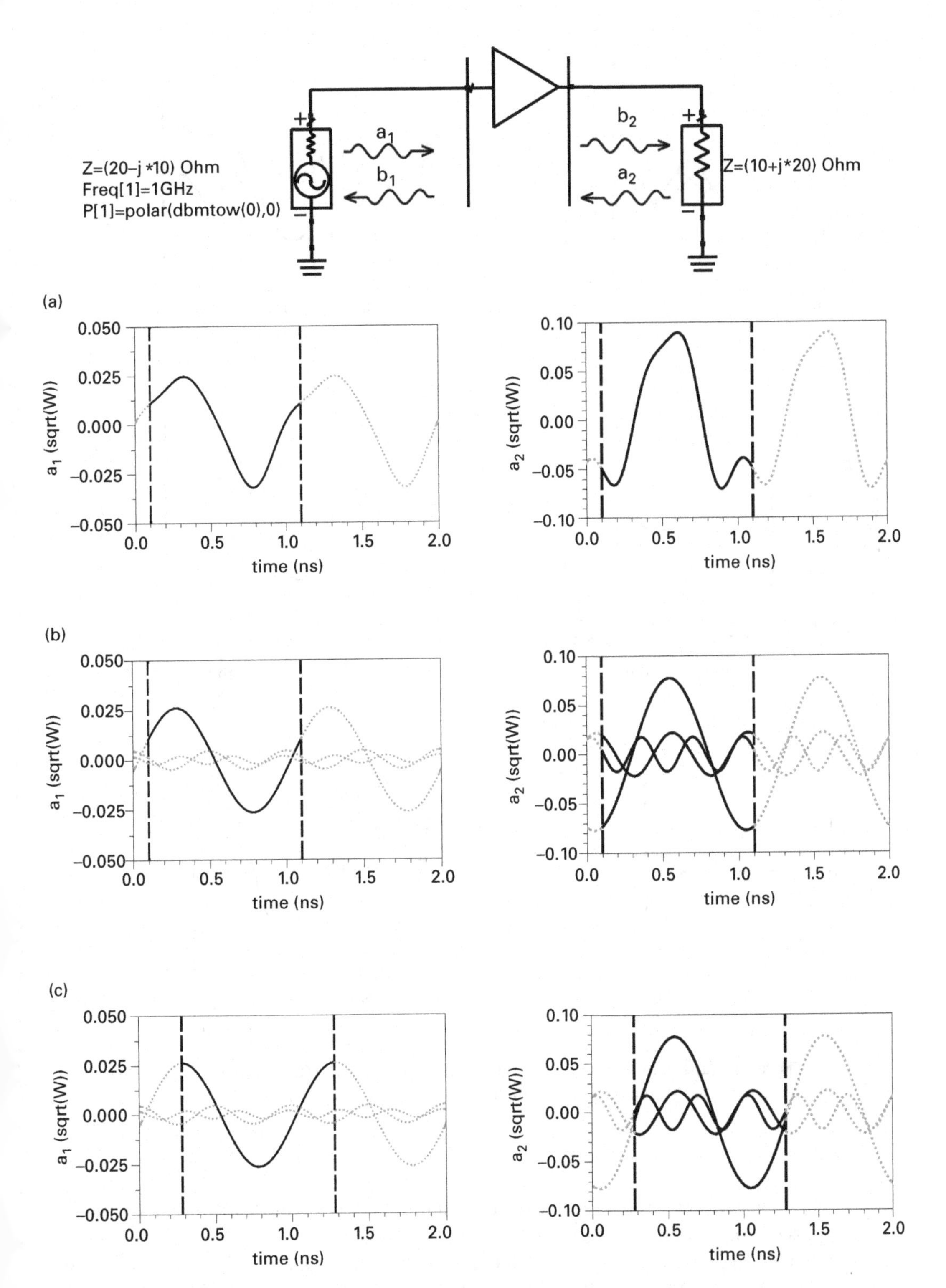

Figure 2.11 (a) Test stimulus, complete waveforms. (b) Test stimulus, Fourier decomposition. (c) Reference stimulus, Fourier decomposition.

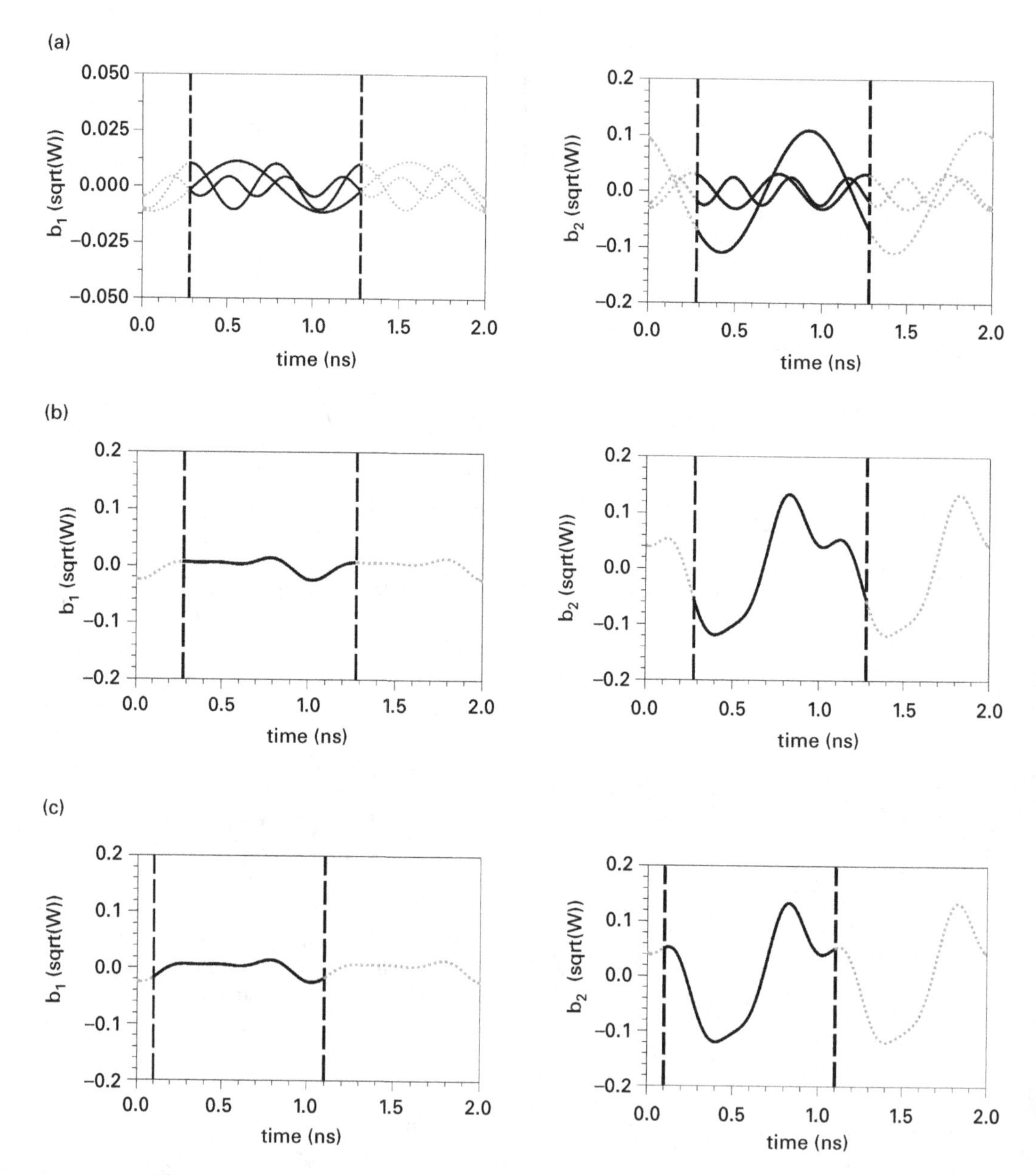

Figure 2.12 (a) Reference response, Fourier decomposition. (b) Reference response, complete waveforms. (c) Response to test stimulus, complete waveforms.

The reference response is shown in Figure 2.12(a), using its Fourier decomposition. It is obtained by converting to the time domain the spectral components represented by the X-parameters of the amplifier. The full waveforms of the reference response, B_1 and B_2, are shown in Figure 2.12(b), after combining all spectral components.

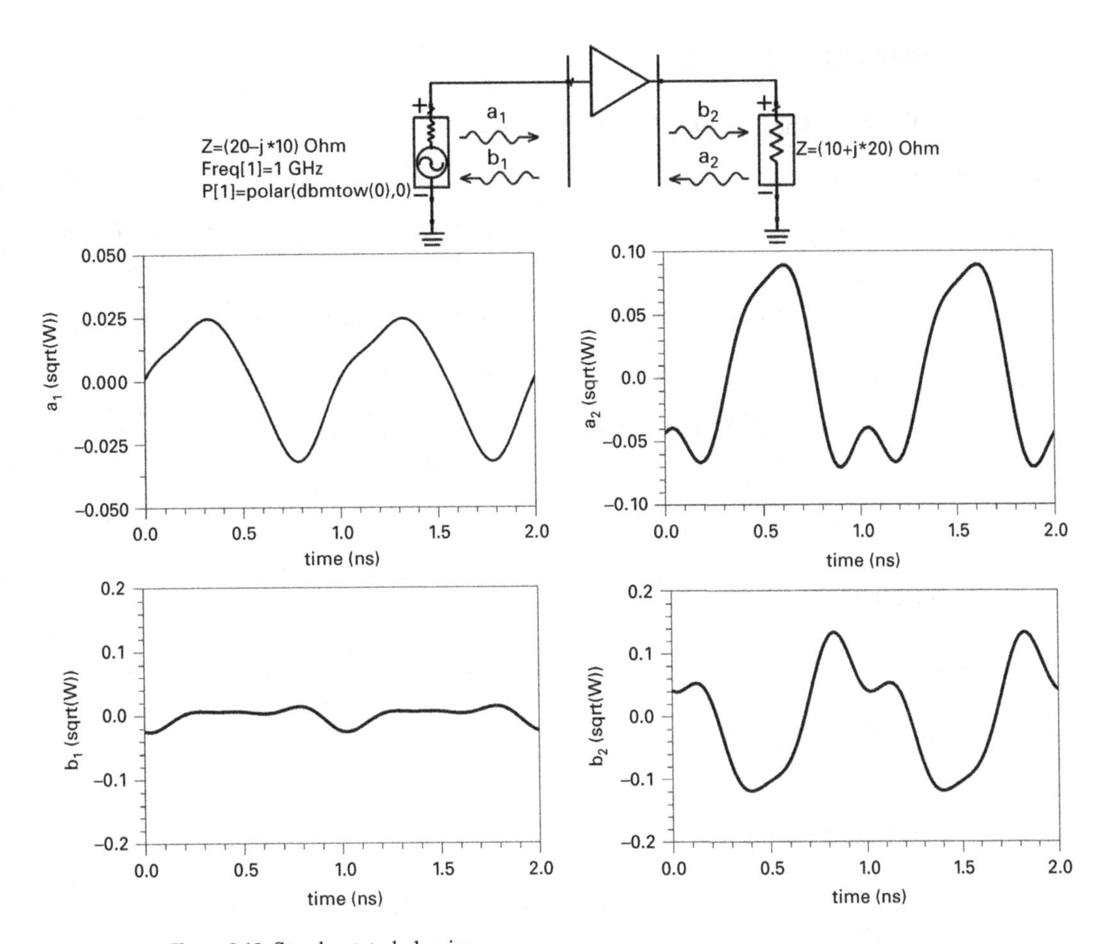

Figure 2.13 Steady-state behavior.

Finally, the response to the test stimulus is obtained by shifting the reference response back in time until the initial moment in time of the fundamental period corresponds to the measurement moment, t_m, as shown in Figure 2.12(c).

The steady-state behavior of the amplifier, complete with both stimulus and response, is shown in Figure 2.13.

2.12 Summary

Nonlinear components exhibit behavior, such as compression and harmonic generation, which is not describable by S-parameters. For the important practical case of signals defined on a harmonic grid, nonlinear spectral mappings defined on incident-wave variables and DC-bias conditions can describe the steady-state DUT behavior completely. Nonlinear system behavior can be predicted without further approximation by cascading the nonlinear description of the constitutive components. One-tone X-parameters are derived by applying the principle of time invariance to the nonlinear mapping.

Exercises

2.1 A nonlinear spectral map defined by (2.51),

$$B_{2,1} = GA_{1,1} - \gamma A_{1,1}^3, \tag{2.51}$$

is proposed to model a time-invariant DUT. Prove that (2.51) is not time invariant.

2.2 Prove that an alternative model, given by (2.52), defines a time-invariant map:

$$B_{2,1} = GA_{1,1} - \gamma |A_{1,1}|^2 A_{1,1}. \tag{2.52}$$

2.3 Reduce (2.52) to the form given by (2.53) and find an explicit expression for $X_{2,1}(.)$:

$$B_{2,1} = X_{2,1}(|A_{1,1}|)P^1. \tag{2.53}$$

References

[1] J. Verspecht, "Describing functions can better model hard nonlinearities in the frequency domain than the Volterra theory," Ph.D. thesis annex, Vrije Univ. Brussel, Belgium, Nov. 1995; available at http://www.janverspecht.com/skynet/Work/annex.pdf.

Additional reading

K. Kundert, *The Designer's Guide to SPICE and Spectre®*. Norwell, MA: Kluwer Academic Publishers, 1995.

D. E. Root, J. Horn, T. Nielsen, *et al.*, "X-parameters: the emerging paradigm for interoperable characterization, modeling, and design of nonlinear microwave and RF components and systems," in *IEEE Wamicon2011 Tutorial*, Clearwater, FL, Apr. 2011.

D. E. Root, J. Verspecht, D. Sharrit, J. Wood, and A. Cognata, "Broad-band, poly-harmonic distortion (PHD) behavioral models from fast automated simulations and large-signal vectorial network measurements," *IEEE Trans. Microw. Theory Tech.*, vol. **53**, no. 11, pp. 3656–3664, Nov. 2005.

J. Verspecht and D. E. Root, "Poly-harmonic distortion modeling," *IEEE Microwave*, vol. **7**, no. 3, pp. 44–57, June 2006.

3 Spectral linearization approximation

3.1 Simplification of basic X-parameters for small mismatch

The considerations used for the definition of X-parameters apply to the steady-state behavior of time-invariant nonlinear components with incident (and hence also scattered) waves on a harmonic frequency grid. The generality of the formalism comes at the cost of considerable complexity. Each spectral map is a nonlinear function of every applied DC-bias condition and all the magnitudes and phases of each spectral component of every signal at every port. Sampling such behavior in all variables for many ports and harmonics would be prohibitive in terms of data acquisition time, data file size, and model simulation speed.

Fortunately, in most cases of practical interest, only a few large-signal components need to be considered with complete generality, while most spectral components can be considered small and dealt with by methods of perturbation theory. Specifically, we can apply the following methodology.

- Identify the few large tones that drive the main nonlinear behavior of the system.
- Identify the nonlinear spectral map defined when only these few large tones drive the system (without any of the other tones identified as small-level signals). This nonlinear spectral map represents a specific steady state the system arrives at due to the large-signal stimulus.
- Linearize the spectral maps around this specific steady state defined by the few important large tones.
- Consider all the rest of the signals (the small signals) to be treated linearly based on the linearized map.

This methodology approximates the complicated multi-variate nonlinear maps with much simpler nonlinear maps, defined on the much smaller number of selected large tones only, and simple linear maps accounting for the contributions of the many tones with small amplitudes. These approximations result in a dramatic reduction of complexity while providing an excellent description of many important practical cases.

The simplest such approximation is to linearize around a DC operating point and assume *all* RF spectral components are treated linearly. This is just the S-parameter approximation.

In this section, it is assumed that there is one large RF signal that must be treated generally and that the remaining spectral components can be treated as small. This is often the case for power amplifiers, at least those "nearly matched" to 50 Ω, where the main stimulus signal is a narrow-band modulation around a fixed carrier incident at the input port. The DUT response includes all the harmonics. If the DUT is not perfectly matched at the output due to an imperfect load impedance, there will be reflections at the fundamental and the harmonics, back into the DUT. These mismatches will be considered small perturbations to simplify the description. Similarly, scattered waves at the harmonic frequencies at the input could be reflected back into the DUT if it is not perfectly matched at the input due to an imperfect source impedance. These incident signals can also be treated as small in many applications. The validity of the approximation depends on whether these signals are small enough such that their contribution appears only linearly. In later sections, this assumption will be relaxed, in particular to deal with arbitrary load dependence at the output port.

The practical example described above of an amplifier operating under mismatched conditions is the same case studied in Section 2.10. It will be demonstrated how the complex multi-variate nonlinear map can be simplified when only one incident tone is large, the fundamental applied at the input. All other tones in the stimulus (all harmonics incident on the ports) have small amplitudes due to the small source and load mismatches.

Before proceeding with linearizing the spectral map around the nonlinear steady state, some properties of the functions establishing the nonlinear spectral map need to be clarified, specifically the analyticity of these functions. This is important because the linearization requires the differentiation of these complex functions of complex variables.

3.1.1 Non-analytic maps

Physically, the large sinusoidal incident signal at the fundamental causes the device to vary periodically in time. The simultaneous presence of small incident signals at other frequencies will produce output signals at all sum and difference frequencies of the large and small tones. Since all incident tones are on a harmonic grid, the "self-mixing" of each small tone will produce an output at each harmonic. As will be seen in this section, the complex amplitude of the spectral response at each harmonic will be a function of the complex amplitude and the *complex conjugate of the amplitude* of the small tone.

The physical situation is equivalent to a mixer with a large LO signal and small RF tone. To clarify the ideas, consider the case of the ideal mixer in Figure 3.1.

Without any loss of generality, the RF and LO input signals are sinusoidal and can be written as in (3.1):

$$\begin{aligned} RF_{in}(t) &= \mathrm{Re}(A_{RF}e^{j\omega_0 t}) = \frac{A_{RF}e^{j\omega_0 t} + A_{RF}^* e^{-j\omega_0 t}}{2}, \\ LO_{in}(t) &= \mathrm{Re}(A_{LO}e^{j\omega_1 t}) = \frac{A_{LO}e^{j\omega_1 t} + A_{LO}^* e^{-j\omega_1 t}}{2}. \end{aligned} \tag{3.1}$$

The output (the IF) is the product of the time-domain signals, as shown in (3.2):

$$IF_{out}(t) = LO_{in}(t) \cdot RF_{in}(t). \tag{3.2}$$

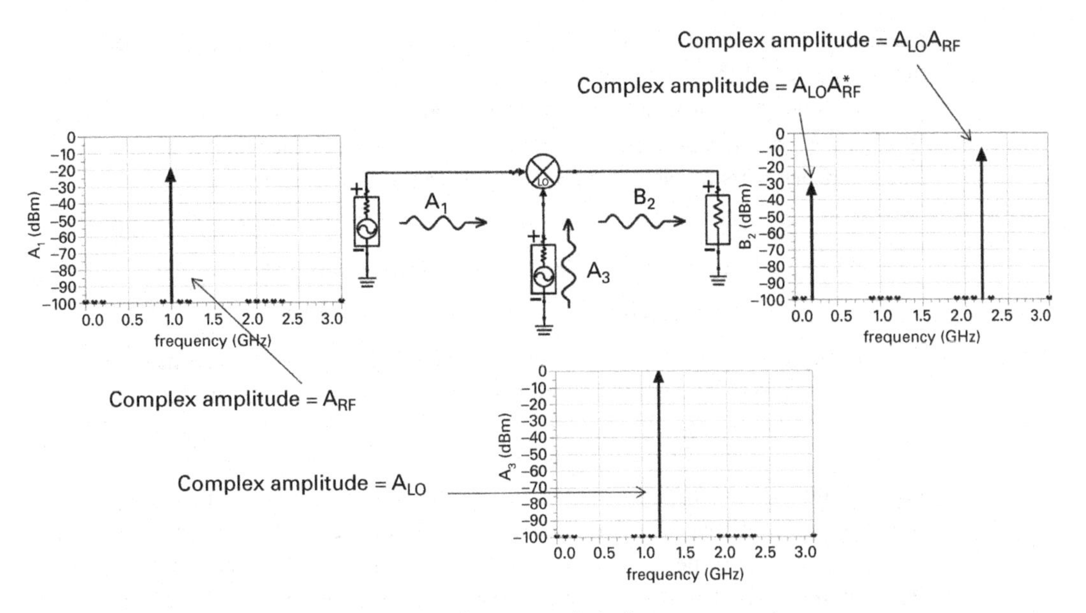

Figure 3.1 Ideal mixer demonstrating non-analytic frequency mappings.

After simple algebraic transformations using (3.1), it can be shown that (3.2) can be expressed as shown in (3.3):

$$IF_{out}(t) = \frac{B_{IF+}e^{j\omega_+ t} + B^*_{IF+}e^{-j\omega_+ t}}{2} + \frac{B_{IF-}e^{j\omega_- t} + B^*_{IF-}e^{-j\omega_- t}}{2}. \tag{3.3}$$

The complex amplitudes B_{IF+} and B_{IF-}, together with the corresponding output frequencies, are shown in (3.4):

$$\begin{aligned} B_{IF+} &= \frac{A_{LO}A_{RF}}{2}, \\ B_{IF-} &= \frac{A_{LO}A^*_{RF}}{2}, \\ \omega_+ &= \omega_1 + \omega_0, \\ \omega_- &= \omega_1 - \omega_0. \end{aligned} \tag{3.4}$$

Expression (3.3) shows that the output contains two frequencies, ω_+ and ω_-, as emphasized after converting (3.3) into (3.5):

$$IF_{out}(t) = \mathrm{Re}\left(B_{IF+}e^{j\omega_+ t}\right) + \mathrm{Re}\left(B_{IF-}e^{j\omega_- t}\right). \tag{3.5}$$

This well-known behavior of the ideal mixer is displayed graphically in Figure 3.1, where the complex amplitudes of the input and output waves are shown.

So, in the frequency domain, we can write the mapping in (3.6):

$$B_{IF-} = A_{LO}A^*_{RF} = F(A_{LO}, A_{RF}). \tag{3.6}$$

The important conclusion is that if we associate the complex variables A_{LO} and A_{RF} with the (positive) single-sided spectrum of incident signals, the complex map $F(.)$, in (3.6), associated with the (positive) single-sided output spectrum, is *non-analytic* because of the appearance of the conjugate term, A_{RF}^*.

This is important when we linearize these maps in the perturbation theory to follow. We therefore consider single-sided spectral maps that depend independently on the real and imaginary parts of each of their complex arguments, as shown in (3.7):

$$\begin{aligned} B_{IF-} &= A_{LO}A_{RF}^* \\ &= (\mathrm{Re}(A_{LO}) + j\mathrm{Im}(A_{LO})) \cdot (\mathrm{Re}(A_{RF}) - j\mathrm{Im}(A_{RF})) \\ &= F(\mathrm{Re}(A_{LO}), \mathrm{Im}(A_{LO}), \mathrm{Re}(A_{RF}), \mathrm{Im}(A_{RF})). \end{aligned} \tag{3.7}$$

Equivalently, the independent variables of the spectral maps may be considered functions of their arguments and their conjugates independently, as shown in (3.8):

$$B_{IF-} = A_{LO}A_{RF}^* = F(A_{LO}, A_{LO}^*, A_{RF}, A_{RF}^*). \tag{3.8}$$

There is one more important observation. The argument of non-analyticity of the nonlinear spectral mapping functions $F(.)$ holds true even for the case of the RF and LO having the same frequency. When $\omega_1 = \omega_0$, $\omega_- = 0$ rad/s and the mixer has a DC response. Based on the above mathematical analysis, the amplitude of the DC response still depends on the complex-conjugate term A_{RF}^*, as shown in (3.6). This proves that, in the most general sense, the nonlinear mapping functions are non-analytic for the DC response also when considered as functions of the complex amplitudes A_{LO} and A_{RF}.

An alternative description can be given in terms of double-sided spectral maps that are analytic in each complex variable.

3.1.2 Large-signal operating point

Let us consider the response of the multi-port system to a multi-harmonic stimulus, as described by the X-parameter model derived in Section 2.9.2 and shown in (2.38). For convenience, this model is shown again here, in (3.9):

$$\begin{aligned} DCR_p &= X_p^{(FDCR)}\left(\{DCS_q\},\ |A_{1,1}|,\ \left\{\underset{(q,k)\neq(1,1)}{A_{q,k}} \quad P^{-k}\right\}\right), \\ B_{p,k} &= X_{p,k}^{(FB)}\left(\{DCS_q\},\ |A_{1,1}|,\ \left\{\underset{(q,k)\neq(1,1)}{A_{q,k}} \quad P^{-k}\right\}\right)P^k. \end{aligned} \tag{3.9}$$

The simplification described in the preceding section refers to the case where all signals are small, except one. For convenience, and without any loss of generality, we assume this signal is applied at port 1, on the fundamental frequency, so it has the complex envelope $A_{1,1}$.

At the limit, all the small signals have zero power, which means that the response of the system in the absence of the small signals depends only on the large-signal magnitude, as shown in equation (3.10). It is easy to recognize that this is the X-parameter model for a multi-port system under one large-tone stimulus, as shown in (3.10):

$$\begin{aligned} DCR_p^{(LS)} &= X_p^{(FDCR)}\left(\{DCS_q\},\ |A_{1,1}|,\ 0,\ 0,\ \ldots,0\right), \\ B_{p,k}^{(LS)} &= X_{p,k}^{(FB)}\left(\{DCS_q\},\ |A_{1,1}|,\ 0,\ 0,\ \ldots,0\right)P^k. \end{aligned} \tag{3.10}$$

The superscript LS in (3.10) identifies this as the response to the large-signal stimulus only. This is the large-signal steady state of the system. The small signals will create a perturbation of the system around this state. Due to its significant role in the perturbation theory discussed next, we need to identify this large-signal steady state.

The large-signal steady state, comprising the totality of the large-signal stimulus and the large-signal response, is the large-signal operating point, usually referred to as the LSOP. The LSOP is thus the state identified in (3.10) and formally shown in (3.11), in which *RFS* stands for RF stimulus:

$$LSOP = \begin{cases} DCS^{(LSOP)} &= \{DCS_q\}, \\ RFS^{(LSOP)} &= A_{1,1}, \\ DCR_p^{(LSOP)} &= X_p^{(FDCR)}\left(\{DCS_q\},\ |A_{1,1}|\right), \\ B_{p,k}^{(LSOP)} &= X_{p,k}^{(FB)}\left(\{DCS_q\},\ |A_{1,1}|\right)P^k. \end{cases} \tag{3.11}$$

As discussed in Section 2.10, a time shift does not change the state of the system. In other words, equivalent stimuli generate equivalent responses, which are all delayed versions of each other and thus represent the same basic large-signal steady state of the system.

In a more general sense, multiple large signals may be applied simultaneously, without any of the small signals. For example, the fundamental may be applied at both input and output simultaneously, as is the case for power-amplifier devices that work on a large load mismatch, so there is a large wave reflected by the load which becomes incident on the output port. The combination of all the large signals creates a large-signal steady state around which the small signals (the other harmonics which are not large) create a perturbation. The LSOP in this case is defined in the same way: the totality of the large-signal stimulus and the corresponding response. The only difference between this case and the case of a single large tone is the higher complexity of both the stimulus and the response.

Formally, the LSOP for the general multi-large-tone stimulus can be identified as shown in (3.13), where $refRFS^{(LSOP)}$ is the reference stimulus equivalent to $RFS^{(LSOP)}$ (i.e. the time-delayed version of $RFS^{(LSOP)}$ that has $phase(A_{1,1}) = 0$):

$$refRFS^{(LSOP)} = \{\text{all large } A_{q,k}P^{-k}\}, \tag{3.12}$$

$$LSOP = \begin{cases} DCS^{(LSOP)} &= \{DCS_q\}, \\ RFS^{(LSOP)} &= \{\text{all large } A_{q,k}\}, \\ DCR_p^{(LSOP)} &= X_p^{(FDCR)}\left(DCS^{(LSOP)}, refRFS^{(LSOP)}\right), \\ B_{p,k}^{(LSOP)} &= X_{p,k}^{(FB)}\left(DCS^{(LSOP)}, refRFS^{(LSOP)}\right)P^k. \end{cases} \tag{3.13}$$

There is one more convention to be set before proceeding further. This relates to some simplified notations that we use sometimes from here on. As shown in (3.13), the formal identification of the LSOP may be quite cumbersome. In order to simplify the formulae, the following conventions are used where needed.

- The notation *LSOPS* is used for the large-signal stimulus (which is the stimulus that is part of the LSOP), which is formally identified in (3.14):

$$LSOPS = \begin{cases} DCS^{(LSOP)} = \{DCS_q\}, \\ RFS^{(LSOP)} = \{\text{all large } A_{q,k}\}. \end{cases} \tag{3.14}$$

- The notation *LSOPR* is used for the large-signal response (which is the response that is part of the LSOP), which is formally identified in (3.15):

$$LSOPR = \begin{cases} DCR_p^{(LSOP)} = X_p^{(FDCR)}\left(DCS^{(LSOP)}, refRFS^{(LSOP)}\right), \\ B_{p,k}^{(LSOP)} = X_{p,k}^{(FB)}\left(DCS^{(LSOP)}, refRFS^{(LSOP)}\right)P^k. \end{cases} \tag{3.15}$$

- The LSOP is the totality of *LSOPS* and *LSOPR*, as shown in (3.16):

$$LSOP = \{LSOPS, LSOPR\}. \tag{3.16}$$

Note that the LSOP is set by the stimulus (i.e. by the *LSOPS*), but it contains both the stimulus and the response (i.e. both *LSOPS* and *LSOPR*).

It is sometimes useful to refer directly to the specific reference stimulus that identifies the LSOP, complete with DC and RF stimulus. This is identified as shown in (3.17):

$$refLSOPS = \left\{DCS^{(LSOP)}, refRFS^{(LSOP)}\right\}. \tag{3.17}$$

Another useful observation is that the LSOP contains both the DC operating point and the RF operating point. Sometimes it is useful to identify separately these elements of the LSOP. The notations shown in (3.18) are used when needed:

$$\begin{aligned} DCOP &= \begin{cases} DCS^{(LSOP)} = DCS_p, \\ DCR_p^{(LSOP)} = X_p^{(FDCR)}(refLSOPS); \end{cases} \\ RFOP &= \begin{cases} RFS^{(LSOP)} = \{\text{all large } A_{q,k}\}, \\ B_{p,k}^{(LSOP)} = X_{p,k}^{(FB)}(refLSOPS)P^k; \end{cases} \\ LSOP &= \{DCOP, RFOP\}. \end{aligned} \tag{3.18}$$

The case of a single large tone and multiple small tones is used in the following application of the perturbation theory, due to its simpler form, but the derivation is easily extendable to an arbitrary number of large and small tones.

3.2 Adding small-signal stimuli (linearized nonlinear spectral mapping)

The objective in this section is to find a more convenient way of formulating the model shown in (3.9) by taking advantage of the knowledge that some of the signals and the stimulus are small. This leads to the formulation of a linearized spectral map around the LSOP set by the large-signal stimulus.

The linearization process leads to the introduction of additional X-parameter terms. Although the number of terms in the set of X-parameters increases significantly, the final model is simpler and easier to formulate and use in a simulation environment. In addition, the new terms introduced in the linearization process provide significantly more insight into the nature of the nonlinear behavior and how the system evolves from one large-signal state to another one.

The LSOP is the steady state achieved by the system when no small-signal perturbations are applied.

When small perturbations are applied on harmonic frequencies around the LSOP stimulus, the system will respond with variations of the scattered RF waves and its DC responses around the values at the LSOP.

The variations of the scattered waves are captured by new X-parameter terms that are referred to as RF-interaction terms.

The variations of the DC responses are captured by new X-parameter terms that are referred to as RF-to-DC-interaction terms.

They will be studied individually.

3.2.1 Small-signal interactions: the RF terms

The case of a single large-tone stimulus is presented here to allow simplification of the mathematical expressions. This does not represent any loss of generality. The results can easily be extended to the more general case of a multi-harmonic stimulus.

The LSOP stimulus is shown in (3.19):

$$LSOPS = \left\{\{DCS_q\}, A_{1,1}\right\}. \tag{3.19}$$

The corresponding reference stimulus is shown in (3.20):

$$refLSOPS = \left\{\{DCS_q\}, |A_{1,1}|\right\}. \tag{3.20}$$

The scattered waves emerging from the system operating at the LSOP are shown in (3.21):

$$B_{p,k} = F_{p.k}\left(LSOPS, \left\{\begin{matrix} A_{q,k} \\ (q,k) \neq (1,1) \end{matrix}\right\}\right). \tag{3.21}$$

The application of the perturbation theory evaluates the complex function $F_{p.k}(.)$ of the complex variables $A_{q,k}P^{-k}$, while the *refLSOPS* remains unchanged. The value of this function is estimated around the fixed *refLSOPS* using a series expansion based on the partial derivatives. The calculation of the partial derivatives requires the $F_{p.k}(.)$ function to be analytical with respect to all independent variables.

The analysis of a mixer, presented in Section 3.1.1, showed that these nonlinear mapping functions, $F_{p.k}(.)$, are in fact non-analytic in the general case. As a result, they must be considered complex functions depending on both sets of complex variables $A_{q,k}P^{-k}$ and $(A_{q,k}P^{-k})^*$ in order to be analytic functions and thus allowing the series expansion based on the partial derivatives.

The scattered waves can then be represented around the LSOP as shown in (3.22):

$$
\begin{aligned}
B_{p,k} &= F_{p,k}\left(refLSOPS, \left\{ \underset{(q,k)\neq(1,1)}{A_{q,k} \qquad P^{-k}} \right\} \right) P^k \\
&\cong F_{p,k}(refLSOPS)P^k \\
&+ \sum_{\substack{q=1 \\ l=1 \\ (q,l)\neq(1,1)}}^{\substack{q=N \\ l=K}} \left[\left.\frac{\partial F_{p,k}}{\partial(A_{q,l}P^{-l})}\right|_{refLSOPS} A_{q,l}P^{k-l} + \left.\frac{\partial F_{p,k}}{\partial((A_{q,l}\cdot P^{-l})^*)}\right|_{refLSOPS} A^*_{q,l}P^{k+l} \right].
\end{aligned}
\tag{3.22}
$$

The partial derivatives in (3.22), of the $F_{p,k}(.)$ nonlinear mapping function, are used to define two sets of RF-interaction X-parameters of type S and of type T, as shown in (3.23) and (3.24), respectively:

$$
X^{(S)}_{p,k;q,l} \equiv \left.\frac{\partial F_{p,k}}{\partial(A_{q,l}P^{-l})}\right|_{refLSOPS} = \left.\frac{\partial F_{p,k}}{\partial A_{q,l}}\right|_{refLSOPS} P^l, \tag{3.23}
$$

$$
X^{(T)}_{p,k;q,l} \equiv \left.\frac{\partial F_{p,k}}{\partial(A_{q,l}P^{-l})^*}\right|_{refLSOPS} = \left.\frac{\partial F_{p,k}}{\partial A^*_{q,l}}\right|_{refLSOPS} P^{-l}. \tag{3.24}
$$

In addition, the first term in (3.22) can easily be expressed using the $X^{(FB)}_{p.k}$, based on the definition in equation (2.36).

With the above definitions and observations, (3.22) can be rewritten as shown in (3.25), where the dependence of all X-parameters on the LSOP is also recognized:

$$
\begin{aligned}
B_{p,k} &\cong X^{(FB)}_{p.k}(refLSOPS)P^k \\
&+ \sum_{\substack{q=1 \\ l=1 \\ (q,l)\neq(1,1)}}^{\substack{q=N \\ l=K}} \left[X^{(S)}_{p,k;q,l}(refLSOPS)A_{q,l}P^{k-l} + X^{(T)}_{p,k;q,l}(refLSOPS)A^*_{q,l}P^{k+l} \right].
\end{aligned}
\tag{3.25}
$$

3.2.2 Small-signal interactions: the DC terms

The system response to the small RF perturbations also includes small variations of the DC-response components.

For DC-voltage-biased ports, the response is the DC current. Considering the definition (2.39) of the $X^{(FI)}_p$ term, the current response for an arbitrary port p can be written as shown in (3.26), where all RF signals are small except for $A_{1,1}$:

$$
\begin{aligned}
I_p &= X^{(FI)}_p\left(refLSOPS, A_{1,2}P^{-2}, A_{1,3}P^{-3}, A_{2,1}P^{-1}, A_{2,2}P^{-2}, A_{2,3}P^{-3}\right) \\
&= FI_p\left(refLSOPS, A_{1,2}P^{-2}, A_{1,3}P^{-3}, A_{2,1}P^{-1}, A_{2,2}P^{-2}, A_{2,3}P^{-3}\right).
\end{aligned}
\tag{3.26}
$$

The variation of the DC current I_p as the system is exercised around the LSOP by the small RF perturbation can be evaluated by a series expansion of (3.26) using partial derivatives.

In the same discussion presented in Section 3.1.1 for the mixer, it was proven that the functions $FI_v(.)$ are non-analytic with respect to the variables $A_{q,k}$, so they have to be considered functions of both $A_{q,k}$ and their complex conjugates $A_{q,k}^*$ in order to ensure analyticity and make possible a series expansion based on partial derivatives.

We consider a single-tone large-signal excitation for a simplified mathematical formulation. This does not represent any loss of generality, and the results can be easily extended to the more general case of a multi-harmonic stimulus.

The LSOP stimulus and corresponding reference stimulus for this case are shown in (3.27):

$$\begin{aligned} LSOPS &= \left\{\{DCS_q\}, A_{1,1}\right\}, \\ refLSOPS &= \left\{\{DCS_q\},\ |A_{1,1}|\right\}. \end{aligned} \tag{3.27}$$

The current response of the system operating at the LSOP is shown in (3.28):

$$I_p = FI_p\left(refLSOPS,\ \left\{A_{\substack{q,l \\ (q,l)\neq(1,1)}}\right\}\right). \tag{3.28}$$

When the small-signal perturbation is superimposed over the LSOP stimulus, the current will vary around its LSOP value, and its total value can be expressed as shown in (3.29):

$$\begin{aligned} I_p &= FI_p\left(refLSOPS,\ \left\{A_{\substack{q,l \\ (q,l)\neq(1,1)}} P^{-l}\right\}\right) \\ &\cong FI_p(refLSOPS) \\ &\quad + \sum_{\substack{q=1 \\ l=1 \\ (q,l)\neq(1,1)}}^{\substack{q=N \\ l=K}} \left[\left.\frac{\partial FI_p}{\partial\left(A_{q,l}P^{-l}\right)}\right|_{refLSOPS} A_{q,l}P^{-l} + \left.\frac{\partial FI_p}{\partial\left(\left(A_{q,l}P^{-l}\right)^*\right)}\right|_{refLSOPS} \left(A_{q,l}P^{-l}\right)^* \right] \\ &\cong FI_p(refLSOPS) \\ &\quad + \sum_{\substack{q=1 \\ l=1 \\ (q,l)\neq(1,1)}}^{\substack{q=N \\ l=K}} \left(\left.\frac{\partial FI_p}{\partial A_{q,l}}\right|_{refLSOPS} A_{q,l} + \left.\frac{\partial FI_p}{\partial A_{q,l}^*}\right|_{refLSOPS} A_{q,l}^* \right). \end{aligned} \tag{3.29}$$

One important property of the mapping functions $FI_p(.)$ is that they always evaluate to a real-numbered value because the result of the evaluation is a DC current.

The first term in (3.29), $FI_p(refLSOPS)$, is a real number because it represents the DC current at the LSOP (without any small-signal perturbations).

It then follows that the sum of all the terms must also be a real number. However, it is observed that each individual small-signal perturbation on each of the harmonics can be applied individually, by itself, independent of all the other small signals on any other harmonic frequencies. It is thus required that the sum of the two terms for each of the harmonics must be a real number, as shown in (3.30):

$$\left(\left. \frac{\partial FI_p}{\partial A_{q,l}} \right|_{refLSOPS} A_{q,l} + \left. \frac{\partial FI_p}{\partial A^*_{q,l}} \right|_{refLSOPS} A^*_{q,l} \right) \in \mathbb{R} \qquad (3.30)$$

$$\forall\, p, q = \overline{1, N}, \quad 1 = \overline{1, K}.$$

It can be mathematically proven that, for (3.30) to be true for any value $A_{q,l} \in \mathbb{C}$, it is necessary and sufficient for (3.31) to be true:

$$\frac{\partial FI_p}{\partial A_{q,l}} = \left(\frac{\partial FI_p}{\partial A^*_{q,l}} \right)^* . \qquad (3.31)$$

Expression (3.30) can be rewritten as in (3.32) when using (3.31):

$$2 \cdot \mathrm{Re}\left(\left. \frac{\partial FI_p}{\partial A_{q,l}} \right|_{refLSOPS} A_{q,l} \right) \in \mathbb{R}. \qquad (3.32)$$

A new X-parameter term is defined as shown in (3.33):

$$X^{(Y)}_{p;q,l}(refLSOPS) = 2 \left. \frac{\partial FI_p}{\partial A_{q,l}} \right|_{refLSOPS} . \qquad (3.33)$$

The total DC response of the DUT to the superposition of the small RF perturbation over the large RF signal can thus be calculated as shown in (3.34):

$$\begin{aligned} I_p &= FI_p\left(refLSOPS, \left\{ A_{\substack{q,l \\ (q,l) \neq (1,1)}} \quad P^{-l} \right\} \right) \\ &\cong X^{(FI)}_p(refLSOPS) + \sum_{\substack{q=1 \\ l=1 \\ (q,l) \neq (1,1)}}^{\substack{q=N \\ l=K}} \mathrm{Re}\left(X^{(Y)}_{p;q,l}(refLSOPS) A_{q,l} \right). \end{aligned} \qquad (3.34)$$

For the case of current-bias ports, the DUT response is the port voltage. Following a similar process, a new X-parameter term, $X^{(Z)}_{p;q,l}$, is defined as shown in (3.35):

$$X^{(Z)}_{p;q,l}(refLSOPS) = 2 \left. \frac{\partial FV_p}{\partial A_{q,l}} \right|_{refLSOPS} . \qquad (3.35)$$

The $FV_p(.)$ functions must also satisfy condition (3.36) because the DC voltage is also a real-numbered value:

$$\frac{\partial FV_p}{\partial A_{q,l}} = \left(\frac{\partial FV_p}{\partial A^*_{q,l}} \right)^* . \qquad (3.36)$$

It follows that the total voltage response can be calculated as shown in (3.37):

$$V_p = FV_p\left(refLSOPS, \left\{ \underset{(q,l)\neq(1,1)}{A_{q,l}} \quad P^{-l} \right\}\right)$$
$$\cong X_p^{(FV)}(refLSOP) + \sum_{\substack{q=1\\ l=1\\ (q,l)\neq(1,1)}}^{\substack{q=N\\ l=K}} \mathrm{Re}\left(X_{p;q,l}^{(Z)}(refLSOPS)A_{q,l}\right). \tag{3.37}$$

3.3 Physical meaning of the small-signal interaction terms

For practical use of the X-parameter model formulated by (3.25), (3.34), and (3.37), it is important to understand the physical meaning of the small-signal interaction terms.

The distinct X-parameters defining the spectral mappings linearized around a single-tone LSOP can be interpreted as follows.

The terms $X_{p,k}^{(FB)}$ represent the set of mappings from a single-tone RF input to the scattered output waves at port p and harmonic frequency index k for a system perfectly matched at the output port and perfectly matched at each harmonic at all ports.

There are two sets of functions, $X_{p,k}^{(S)}$ and $X_{p,k}^{(T)}$, that determine the sensitivity to mismatch of the system at ports p and harmonic index k. It is noted that, unlike linear S-parameters, there are independent contributions to the scattered waves proportional to both $A_{q,l}$ and $A_{q,l}^*$. The terms $X_{p,k}^{(S)}$ are related to the so-called "hot S-parameters," but contain generalizations including harmonics. It will be seen later the importance of the terms $X_{p,k}^{(T)}$ as the magnitude of the input drive, $|A_{1,1}|$, becomes large.

The terms $X_{p;q,l}^{(Y)}$ give the sensitivity of the DC-bias currents at port p to RF mismatches at port q at the harmonic index l. Similarly, the terms $X_{p;q,l}^{(Z)}$ give the sensitivity of the DC-bias voltages at port p to mismatches at port q at the harmonic index l. These terms are essential for proper modeling of the match-dependent power-added efficiency (PAE) under drive, for example.

We note that, unlike linear S-parameters, X-parameters characterize port-to-port and frequency-to-frequency interactions. That is, a mismatch at a given harmonic at one port can affect the response at another port at each harmonic frequency as well as the DC-bias condition.

Let us analyze how the scattered waves can be estimated by the model around the LSOP using (3.25). When the coefficient P has a value of 1, meaning that the phase of the fundamental, *phase*($A_{1,1}$), is 0° (the system is in the reference-LSOP state), the waves can be expressed as the sum of various X-parameter terms. The small-signal interaction terms are scaled by the small-signal stimuli or their complex-conjugate value. This is shown in (3.38):

$$P=1\Rightarrow.$$

$$B_{p,k}\cong X^{(FB)}_{p.k}(refLSOPS)+\sum_{\substack{q=1\\l=1\\(q,l)\neq(1,1)}}^{\substack{q=N\\l=K}}\left[X^{(S)}_{p,k;q,l}(refLSOPS)A_{q,l}+X^{(T)}_{p,k;q,l}(refLSOPS)A^{*}_{q,l}\right].$$

(3.38)

The resulting wave, $B_{p,k}$, is composed of the LSOP term, the $X^{(FB)}_{p.k}$, plus contributions due to the small-signal stimulus on each individual frequency component, $A_{q,l}$. Each of these small-signal contributions are independent of each other and they depend linearly only on the stimulus applied on the individual frequency. They can thus be separated by applying the small-signal stimuli one at a time, on each frequency and port.

For example, let us consider that all small-signal stimuli are zero, except for the one applied at port q on harmonic l, $A_{q,l}$. The wave scattered at port p on harmonic k, $B_{p,k}$, is then calculated as shown in (3.39):

$$\left.\begin{array}{ll} A_{r,m}=0, & \forall(r,m)\neq(q,l)\\ A_{q,l}\neq 0 & \end{array}\right\}\Rightarrow$$

$$B_{p,k}\cong X^{(FB)}_{p.k}(refLSOPS)+X^{(S)}_{p,k;q,l}(refLSOPS)A_{q,l}+X^{(T)}_{p,k;q,l}(refLSOPS)A^{*}_{q,l}.$$

(3.39)

The small-signal response (i.e. the part of the response that occurs due to the small-signal stimulus) is identified as $b_{p,k}$, and it is the difference between the total response, $B_{p,k}$, and the large-signal response, $X^{(FB)}_{p.k}(refLSOPS)$, as shown in (3.40):

$$\left.\begin{array}{ll} A_{r,m}=0, & \forall(r,m)\neq(q,l)\\ A_{q,l}\neq 0 & \end{array}\right\}\Rightarrow$$

$$b_{p,k}\cong B_{p,k}-X^{(FB)}_{p.k}(refLSOPS)=X^{(S)}_{p,k;q,l}(refLSOPS)A_{q,l}+X^{(T)}_{p,k;q,l}(refLSOPS)A^{*}_{q,l}.$$

(3.40)

From the formulation shown in (3.40), it follows that the $X^{(S)}_{p,k;q,l}$ term is an intrinsic property of the system under test and that it represents the small-signal transfer function from harmonic l applied at port q to harmonic k occurring at port p when the system operates at its reference LSOP (*refLSOP*).

In a similar manner, from the formulation shown in (3.40), it follows that the $X^{(T)}_{p,k;q,l}$ term is an intrinsic property of the system under test and that it represents the small-signal transfer function with respect to the conjugate of the small signal from harmonic l applied at port q to harmonic k occurring at port p when the system operates at its reference LSOP (*refLSOP*).

As an example, let us consider an amplifier, as shown in Figure 3.2. The stimulus A_1 contains only large-signal energy on the fundamental frequency $A_{1,1}$, and the stimulus A_2 contains only small-signal energy on the second harmonic $A_{2,2}$. This case represents a typical situation of an amplifier under a stimulus coming from a very good source (with negligible spectral content) and with a mismatch on the second harmonic.

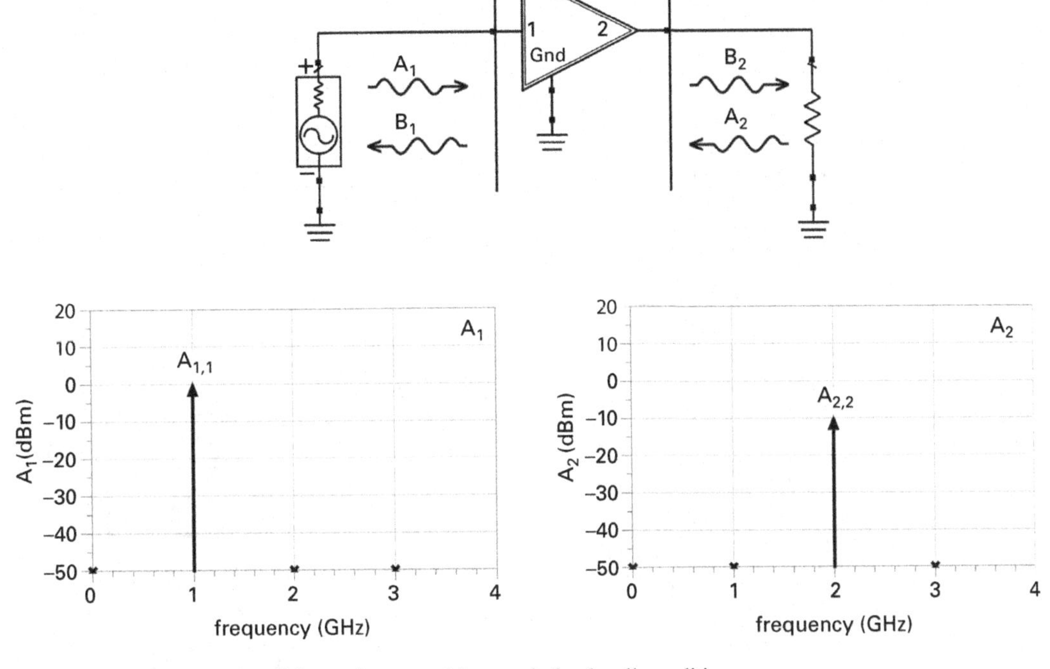

Figure 3.2 Amplifier under second-harmonic load-pull conditions.

In many practical situations, this represents the setup for the second harmonic load pull of an amplifier.

The total response of the amplifier under these conditions is calculated using X-parameters. As an example, the wave scattered from port 2 (the output of the amplifier) on the fundamental frequency, $B_{2,1}$, is observed as a function of the phase of the small-signal stimulus, $A_{2,2}$.

The actual response of the amplifier is shown in Figure 3.3. As the phase of $A_{2,2}$ varies, the $A_{2,2}$ vector describes a circle because its amplitude remains constant during this experiment.

The vector $B_{2,1}$ does not describe a circle as the phase of $A_{2,2}$ varies, as shown in Figure 3.3. A zoomed-in view of the tip of the vector is shown on the lower left side of the figure to provide a better understanding of the physical phenomenon.

It is important to understand how the X-parameter model reproduces this effect. The calculation of $B_{2,1}$ using the X-parameter behavioral model is shown in (3.41):

$$B_{2,1} \cong X_{2,1}^{(FB)}(refLSOPS) + X_{2,1;2,2}^{(S)}(refLSOPS)A_{2,2} + X_{2,1;2,2}^{(T)}(refLSOPS)A_{2,2}^{*}. \quad (3.41)$$

According to this model, the total response is the sum of three vectors: *FX*, *SX*, and *TX*, as shown in Figure 3.4, where the three vectors are identified in (3.42):

$$\begin{aligned} FX &= X_{2,1}^{(FB)}(refLSOPS), \\ SX &= X_{2,1;2,2}^{(S)}(refLSOPS)A_{2,2}, \\ TX &= X_{2,1;2,2}^{(T)}(refLSOPS)A_{2,2}^{*}. \end{aligned} \quad (3.42)$$

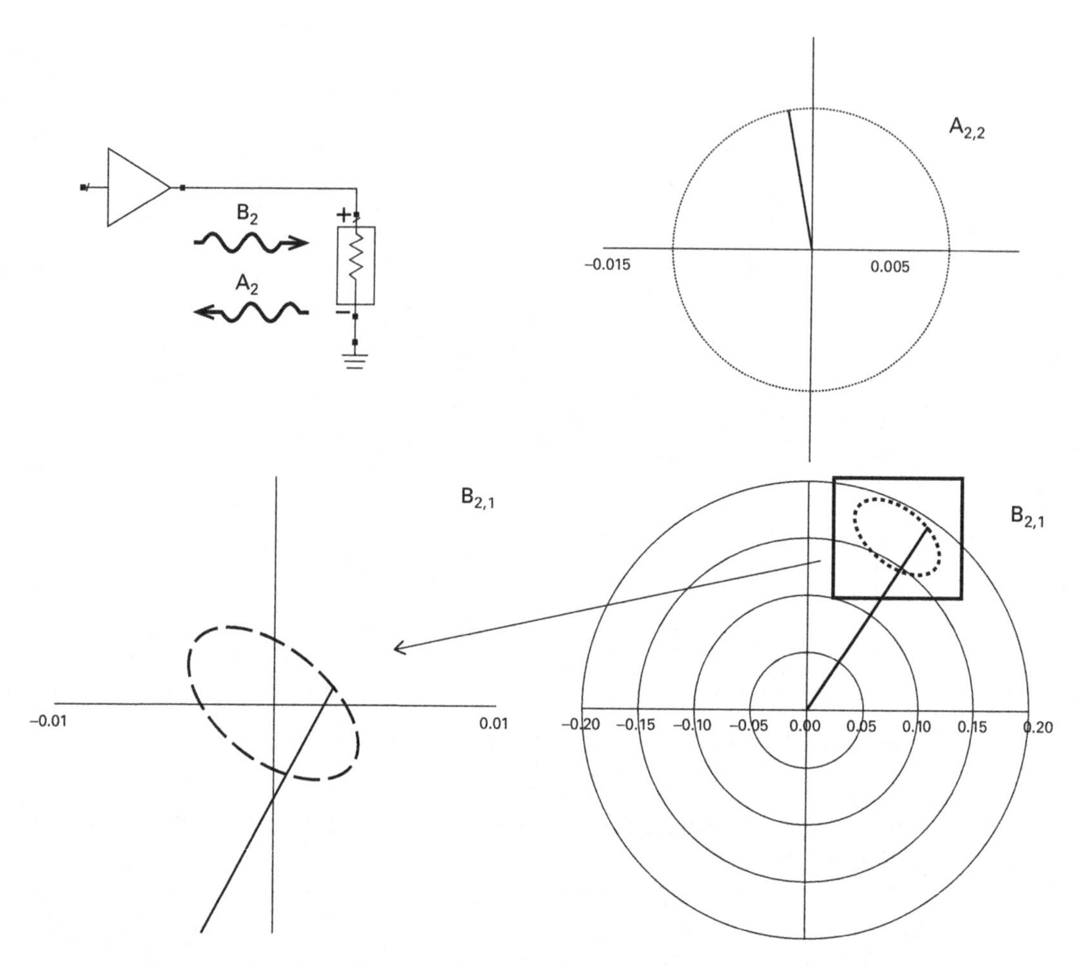

Figure 3.3 Actual $B_{2,1}$ response of the amplifier versus phase of $A_{2,2}$.

As the phase of the $A_{2,2}$ stimulus varies, the *SX* and *TX* vectors rotate in opposite directions, as shown in Figure 3.5. This is because the $A^*_{2,2}$ rotates in the opposite direction to $A_{2,2}$. The end result is that $X^{(S)}A_{2,2} + X^{(T)}A^*_{2,2}$ describes a contour, as shown in Figure 3.4 and Figure 3.5. The center of this contour is defined by the *FX* term, which does not vary when the phase of $A_{2,2}$ changes because it represents the LSOP response.

Figure 3.5 illustrates how the two components rotate to generate the contour centered on the tip of the LSOP vector.

For more complex situations, when multiple small signals are applied at different ports, the locus of the tip of the $B_{2,1}$ vector may have various shapes. These shapes are calculated by the corresponding X-parameter behavioral model by summing multiple vectors proportional to the individual small-signal stimuli and their complex conjugates.

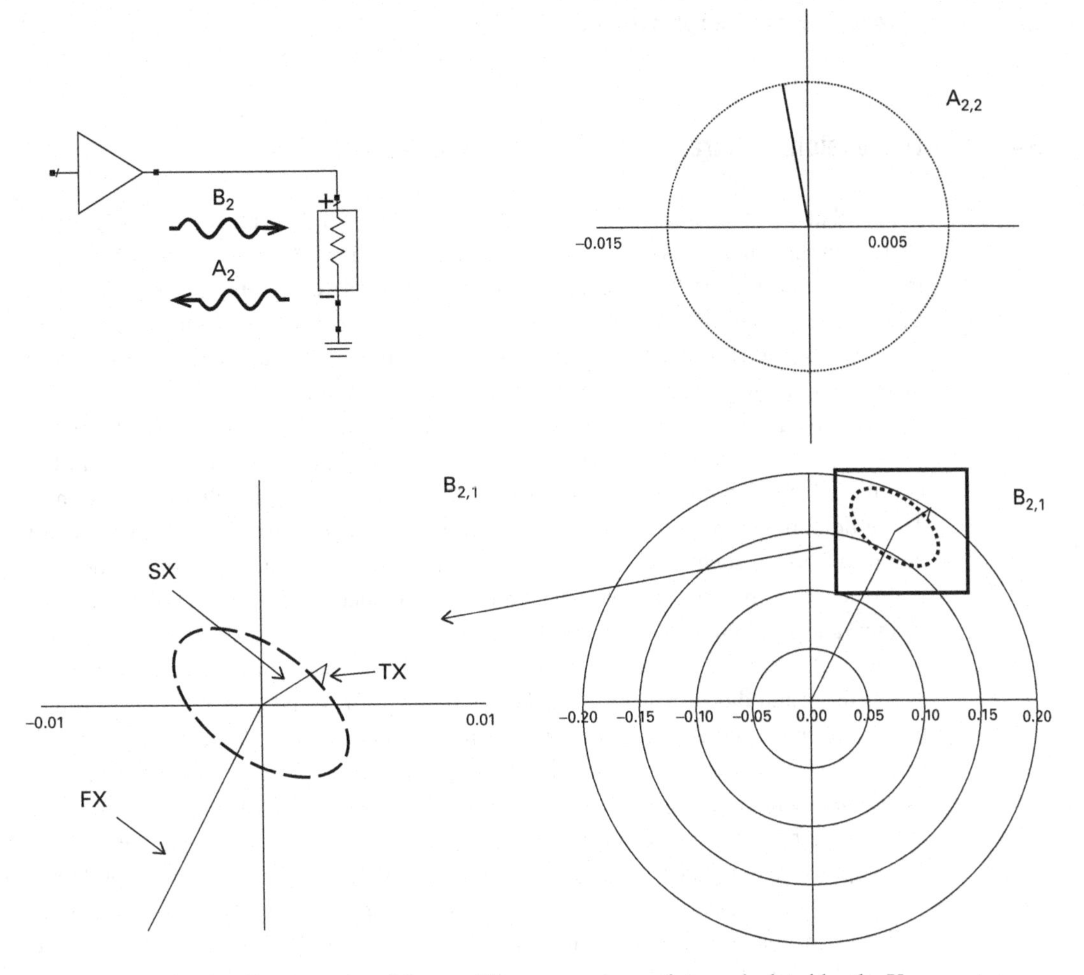

Figure 3.4 $B_{2,1}$ response of the amplifier versus phase of $A_{2,2}$ calculated by the X-parameter behavioral model.

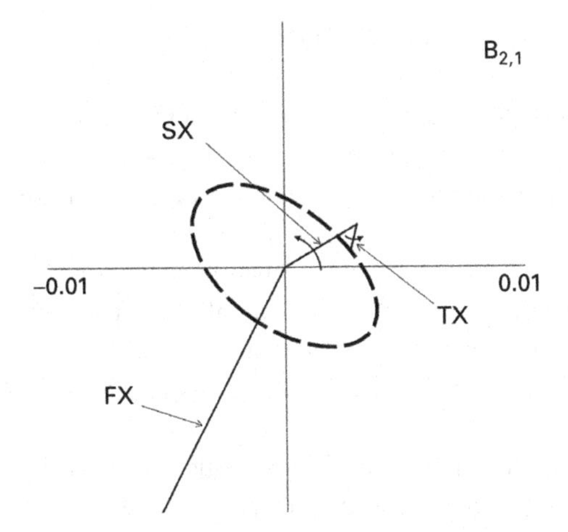

Figure 3.5 The summation of the three vectors yielding the total $B_{2,1}$ response calculated by the X-parameter behavioral model. The circular arrows show the direction in which the vectors rotate as the phase of $A_{2,2}$ increases.

3.4 Discussion: X-parameters and the spectral Jacobian

We see that the spectral linearization in (3.25), (3.34), and (3.37) reduces the complexity of the nonlinear functions defining the scattered waves and DC responses to functions of many fewer variables than the general (harmonic grid) case, like the one shown in Section 2.9.3. Only DC-bias variables and the magnitude of the fundamental incident wave at the fundamental frequency appear as arguments to nonlinear functions. All other spectral components appear linearly in the equations.

The X-parameter functions (3.23), (3.24), (3.33), and (3.35) are related to partial derivatives of the corresponding nonlinear functions, $F(.)$, evaluated at the LSOP established by the bias variables and the single large-tone stimulus and response. These definitions provide a means of computing the X-parameters from any standard nonlinear model, such as the example of the FET model in Chapter 1. Any model that is defined in terms of conventional nonlinear equivalent circuit elements, with well-defined and differentiable constitutive relations for currents and charges (or voltages and fluxes), can be converted into X-parameters. Since most component models, such as transistor and diode models, are defined in terms of I–V and Q–V relations, one can transform these equations into the spectral domain and associate specific X-parameter terms with the spectral Jacobian in A–B space (wave variables). Since nonlinear simulators compute the spectral Jacobian (in I–V space) from knowledge only of these constitutive relations and their partial derivatives, it is evident that X-parameters can be computed in simulators by solving only the single-tone harmonic-balance problem. That is, knowledge of the one-tone nonlinear map and the spectral Jacobian is sufficient to compute the X-parameters that provide the DUT multi-linear response to many RF signals, provided all but the one large tone can be considered to contribute linearly to the output.

3.5 X-parameters as a superset of S-parameters

The X-parameter model describes the intrinsic behavior of a nonlinear system. Because the nonlinear behavior is dependent on the level of the large-signal stimulus, it is common practice to extract the X-parameters as a function of this variable (the level of the large-signal stimulus).

At sufficiently low levels, the stimulus becomes small enough to be considered a small signal and the system behaves linearly. It is thus natural that X-parameters become S-parameters at the limit situation of small-signal stimulus.

We consider the example of an amplifier, as shown in Figure 3.6, to gain a better understanding of this aspect.

The X-parameters of this amplifier are extracted for a single-tone large-signal stimulus applied at its input (port 1) with a power sweep for $A_{1,1}$ from -40 dBm to $+10$ dBm and considering five harmonics.

The large-signal reference response for the first three harmonics, as captured by the X-parameters, is shown in Figure 3.7.

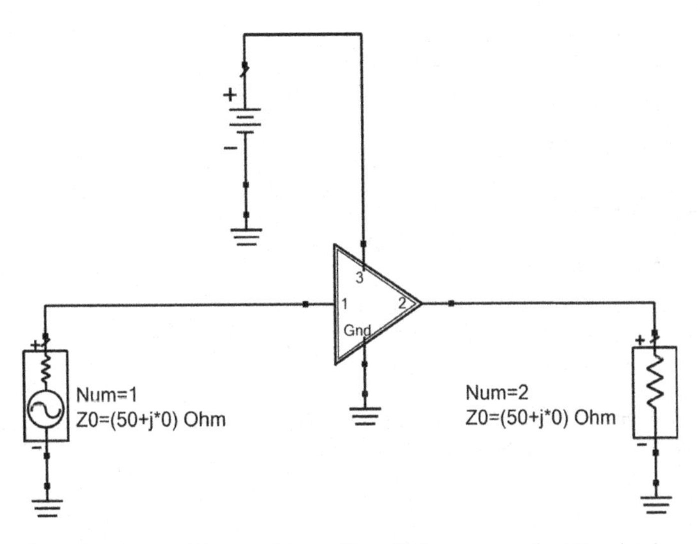

Figure 3.6 An amplifier working with a 50 Ω source and a 50 Ω load.

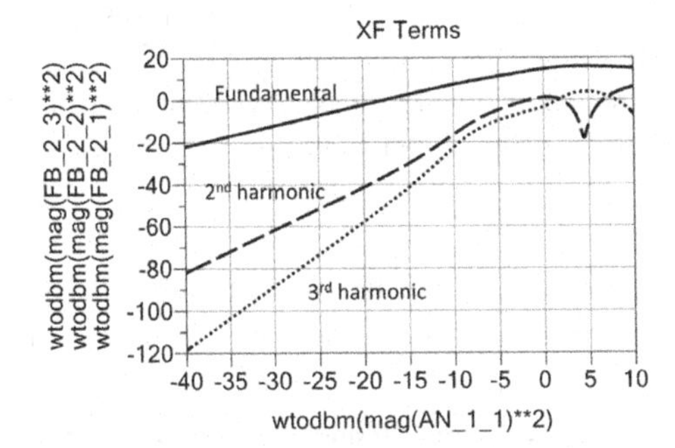

Figure 3.7 Large-signal reference response versus level of the large-signal stimulus, as captured by the X-parameters.

As the level of the large-signal stimulus decreases, the levels of the harmonics decrease faster than the fundamental. Below a certain level, the harmonics become negligible with respect to the fundamental (say below −30 dBm), and the behavior of the system should be well predicted by the linear network theory (S-parameters).

Also, as the level of the large-signal stimulus decreases, the response at each harmonic should reach a point from where it should vary with a constant slope proportional to the order of the harmonic. Figure 3.8 shows the slope with respect to the level of the large-signal stimulus versus the level of the large-signal stimulus for the first three harmonics.

As expected, the slopes become constant at small signals (below approximately −30 dBm) and equal to 1 dB/dB, 2 dB/dB, and 3 dB/dB for the first, second, and third harmonic, respectively. This is the behavior expected in the linear region of operation.

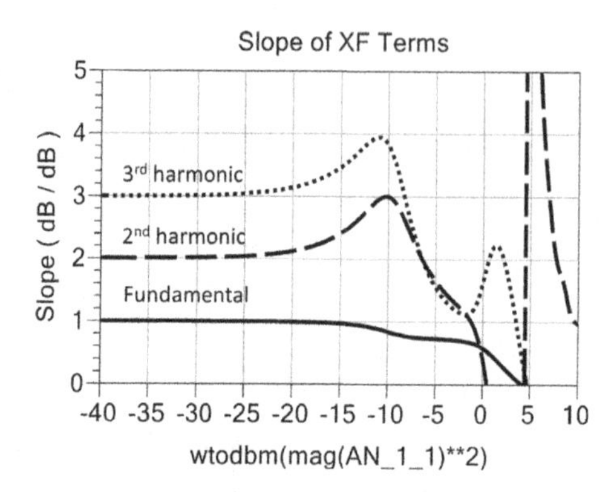

Figure 3.8 Slope with respect to the level of the large-signal stimulus versus large signal stimulus of the $X^{(F)}$ terms for the first three harmonics.

Figure 3.9 shows the typical variation of the magnitude of $X^{(S)}$ and $X^{(T)}$ parameters versus the level of the large-signal stimulus. Two representative cases are shown:

- parameters representing transfer functions at the same frequency (small-signal response at the same frequency as the small-signal stimulus), and
- parameters representing cross-frequency transfer functions (small-signal response at a frequency different from that of the small-signal stimulus).

The two cases shown in Figure 3.9 are for:

- the $X^{(S)}_{2,1;2,1}$ and $X^{(T)}_{2,1;2,1}$, which represent transfer functions from the wave at the fundamental frequency incident at the output port to the wave at the fundamental frequency scattered from the output port, and
- the $X^{(S)}_{2,2;2,1}$ and $X^{(T)}_{2,2;2,1}$, which represent cross-frequency transfer functions from the wave at the fundamental frequency incident at the output port to the wave at the second-harmonic frequency scattered from the output port.

At low levels, the magnitude of $X^{(S)}_{2,1;2,1}$ becomes constant, while the magnitude of $X^{(T)}_{2,1;2,1}$ keeps decreasing, thus becoming negligible with respect to the $X^{(S)}_{2,1;2,1}$.

Similarly, the cross-frequency terms, $X^{(S)}_{2,2;2,1}$ and $X^{(T)}_{2,2;2,1}$ decrease continuously in magnitude, thus becoming also negligible with respect to $X^{(S)}_{2,1;2,1}$.

It is now important to observe what happens to the phase of the same-frequency $X^{(S)}$ terms. The example of $X^{(S)}_{2,1;2,1}$ is shown in Figure 3.10. As expected, its phase also becomes constant, independent of the level of the large-signal stimulus.

All of the above proves that at low levels of the large-signal stimulus the behavior becomes linear, and this is accurately captured by the X-parameter model.

It is also expected that the values of the $X^{(S)}$ terms at the same frequency become identical with the corresponding classical S-parameters. In our example, the output reflection coefficient and reverse isolation, both with respect to small-signal perturbations around the LSOP, are the $X^{(S)}_{2,1;2,1}$ and $X^{(S)}_{1,1;2,1}$ parameters, which become equal to the S_{22} and S_{12} when the large-signal drive level becomes small enough.

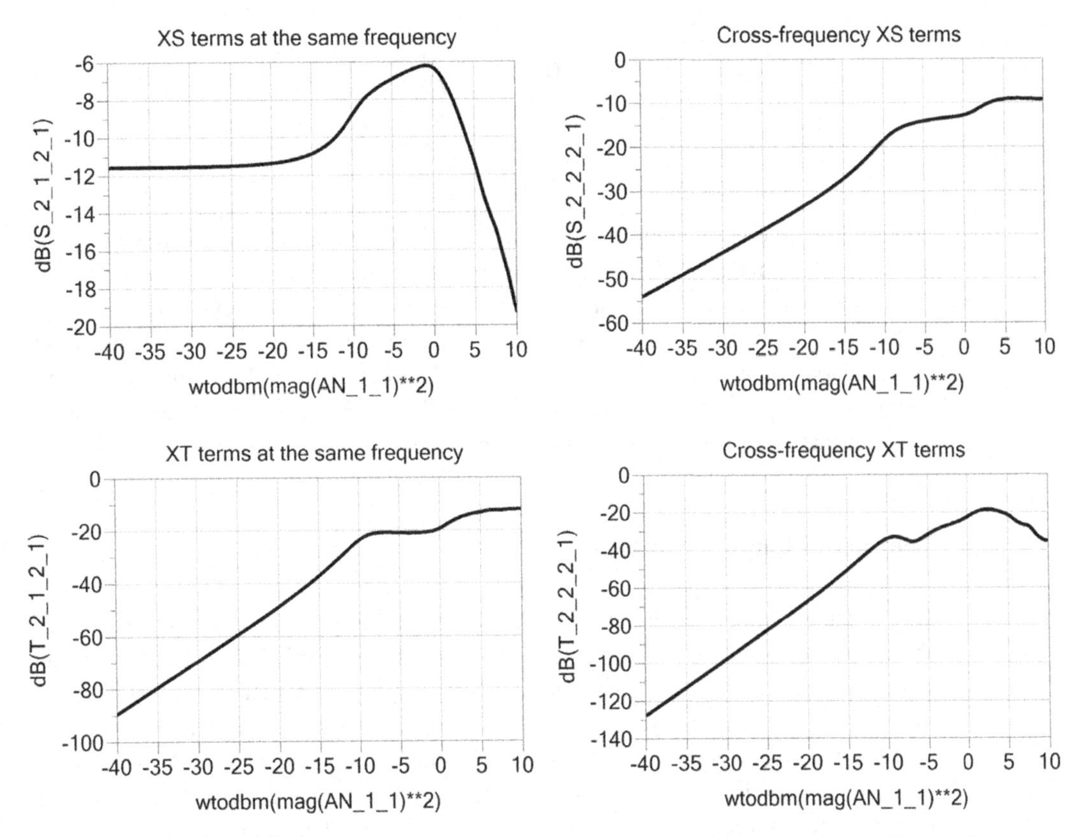

Figure 3.9 $X^{(S)}$ terms at the same frequency become constant in magnitude, and all cross-frequency terms, both $X^{(S)}$ and $X^{(T)}$, become negligible at low levels of the large-signal stimulus.

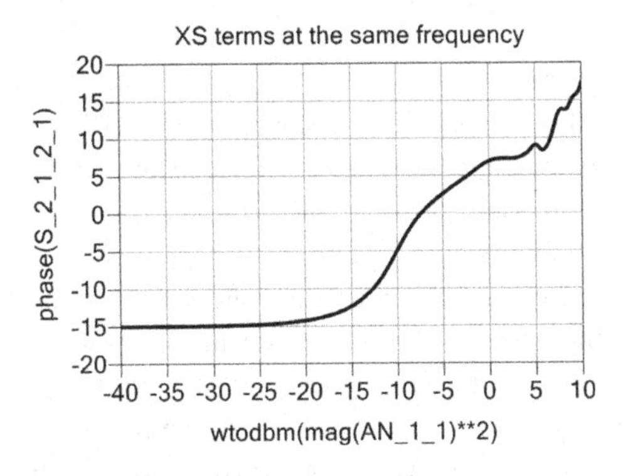

Figure 3.10 The phase of the $X^{(S)}$ terms at the same frequency becomes constant at low levels of the large-signal stimulus.

In addition, the large-signal transfer functions also become equal to the corresponding S-parameter terms when the large-signal drive level becomes small enough. For our particular example, the large-signal gain, $G_{21}^{(LS)}$, and the large-signal input reflection coefficient, $\Gamma_{in}^{(LS)}$, are defined as shown in (3.43):

$$G_{21}^{(LS)} = \frac{X_{2,1}^{(F)}}{AN_{1,1}},$$
$$\Gamma_{in}^{(LS)} = \frac{X_{1,1}^{(F)}}{AN_{1,1}}. \tag{3.43}$$

Figure 3.11 shows a comparison between the X-parameters at low levels of the large-signal stimulus and an independent measurement of the S-parameters. The X-parameters not shown in Figure 3.11 become zero at low drive levels.

It is thus clear that X-parameters represent a complete description of the behavior of the network, in both the nonlinear and linear regions of operation. X-parameters are thus a superset of S-parameters. At low drive levels all X-parameters become zero except for the ones that correspond to the S-parameters.

3.6 Two-stage amplifier design

One of the important design tasks is to predict the behavior of cascaded systems when the models for the individual systems are known. As an example, the result of cascading two amplifiers, as depicted in Figure 3.12, is very well predicted when the system operates in the linear region. This is achieved using the S-parameter formalism.

When the system is pushed into nonlinear operation, S-parameters are insufficient for an accurate description of the cascade, but X-parameters have the full capability of covering the compression region, in addition to the linear (small-signal) region.

Figure 3.12 shows a power sweep that extends up to approximately 3 dB gain compression of the overall system.

The cascade of the X-parameter models of the two amplifiers, shown in Figure 3.13, predicts very well all harmonics, in both magnitude and phase, for all power levels where the models have been measured.

As an example, Figure 3.14 shows the first three harmonics scattered from the output of the system into the load, in both magnitude and phase. The response predicted by the cascaded X-parameter models is superimposed over the actual response of the system, showing the excellent accuracy of the models.

It is important to understand and control the inter-stage interaction during the design and manufacturing process of the system. The behavioral models must predict with accuracy both the forward- and reverse-propagating waves at the interface between the two amplifiers. Figure 3.15 shows the first three harmonics of the forward-propagating waves, and Figure 3.16 shows the reverse-propagating waves. Both magnitudes and phases are accurately predicted by the cascade of the X-parameter models.

This level of accuracy is also achieved under simultaneous source and load mismatch. Such a performance has not been previously achieved by any other behavioral model preceding X-parameters.

One of the questions often asked is how important are the $X^{(T)}$ parameters for the accuracy of the model. One of the relatively simple tests that can emphasize their

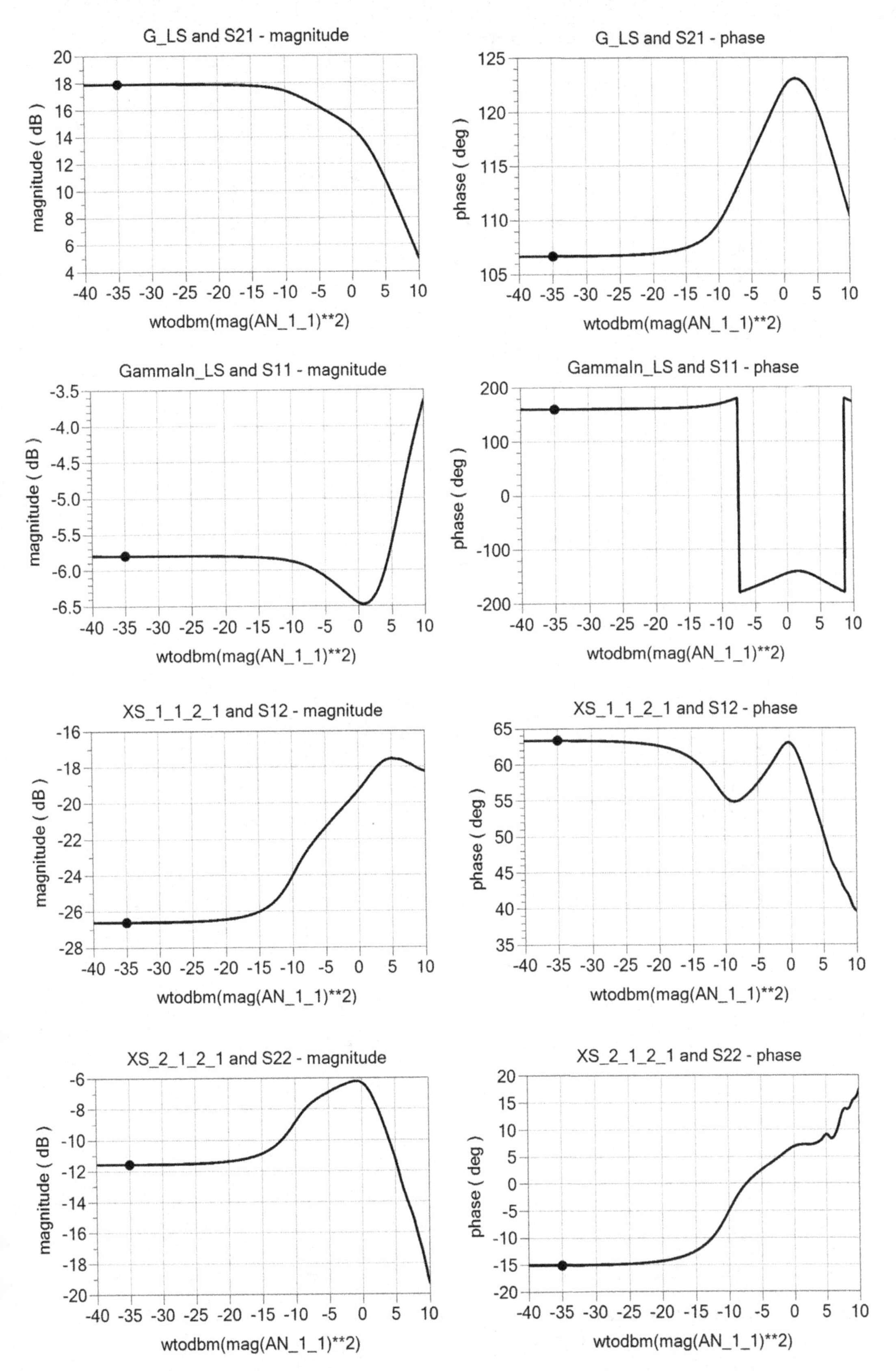

Figure 3.11 X-parameters (lines) become S-parameters (dots) for low levels of the large-signal stimulus.

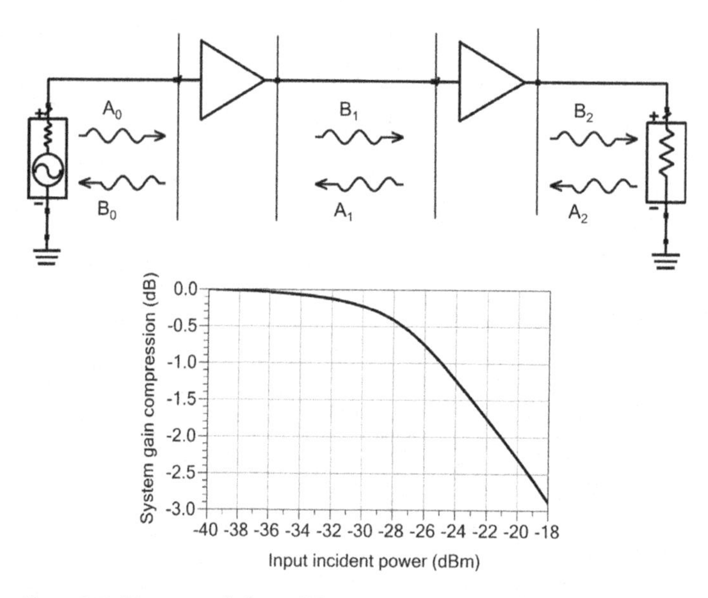

Figure 3.12 Two cascaded amplifiers – system gain in compression.

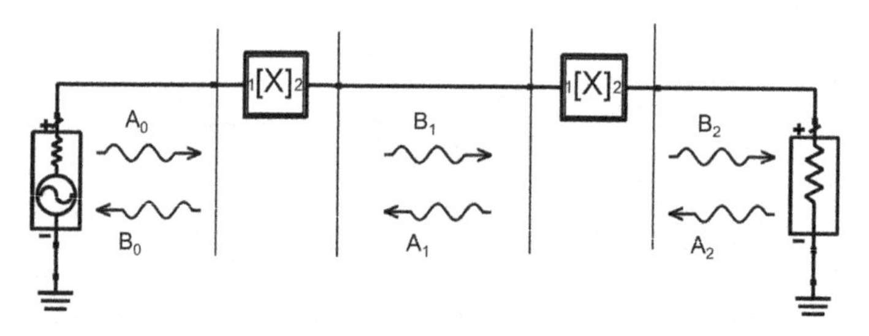

Figure 3.13 Two cascaded amplifiers – represented by their corresponding X-parameter models.

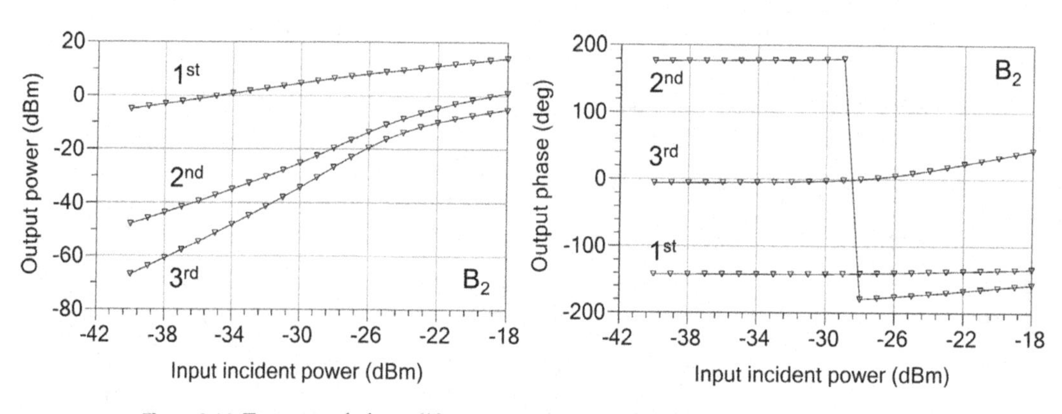

Figure 3.14 Two cascaded amplifiers – output harmonic waves predicted by X-parameters (line) versus system response (symbols).

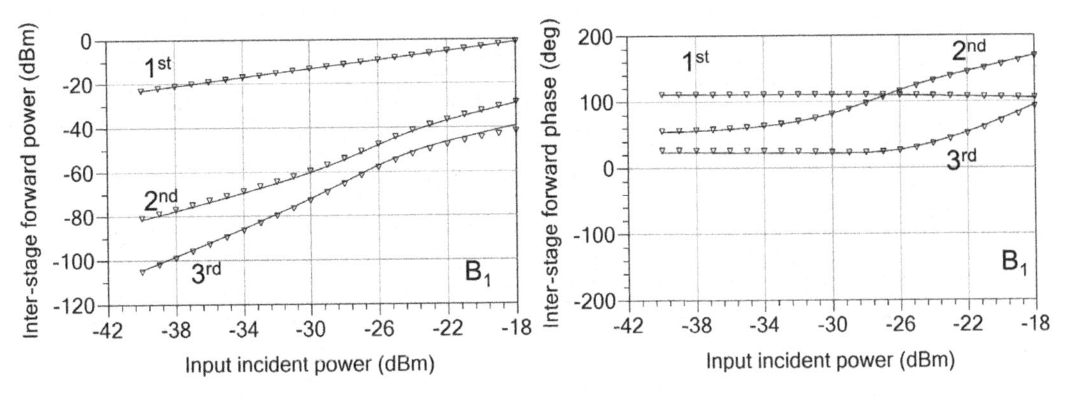

Figure 3.15 Two cascaded amplifiers – inter-stage forward waves predicted by X-parameters (line) versus system response (symbols).

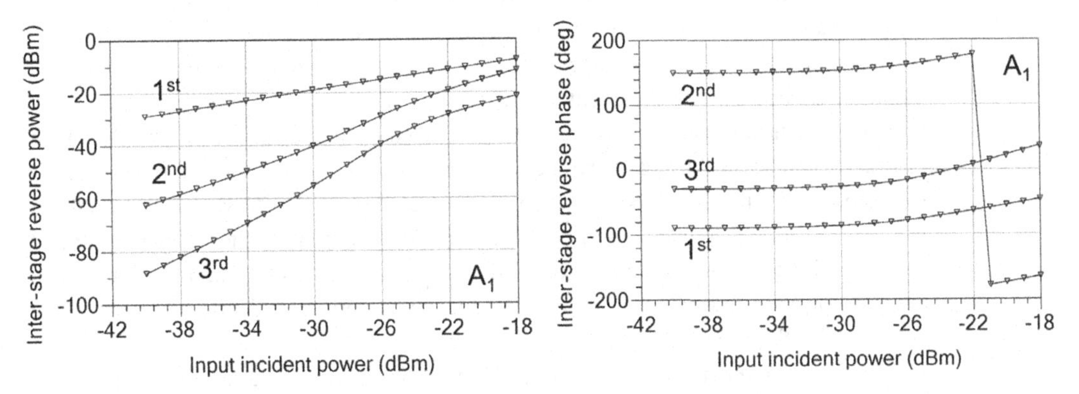

Figure 3.16 Two cascaded amplifiers – inter-stage reverse waves predicted by X-parameters (line) versus system response (symbols).

importance is to use a model that does not have them. In the above example, the X-parameter models of the two amplifiers have been simplified by removing all the $X^{(T)}$ parameters from the original models. A simulation has been run using the simplified models with a source impedance of $Z_S = (80 - 15j)\ \Omega$ and a load impedance of $Z_L = (30 + 25j)\ \Omega$. The results for the inter-stage forward propagating waves are shown in Figure 3.17.

There is a significant inaccuracy in predicting the behavior of the system when the $X^{(T)}$ parameters are not considered. This is true for all nodes in the cascade and for both forward- and reverse-propagating waves. Two main reasons contribute to these inaccuracies.

The first reason is the way the model handles the load interactions. The impact of the load mismatch across all harmonics is improperly evaluated by the model without the $X^{(T)}$ terms. The first amplifier works on the input impedance of the second amplifier, which may show significant mismatch, and the second amplifier works on the system

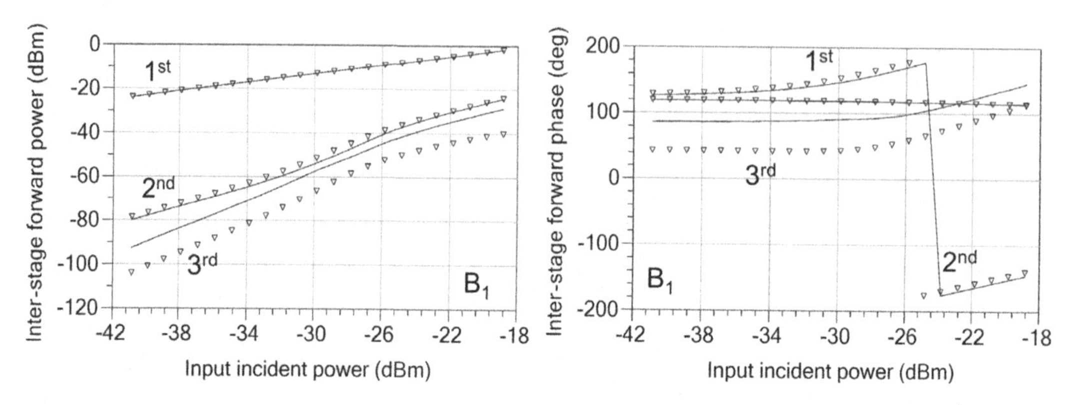

Figure 3.17 Two cascaded amplifiers – inter-stage forward waves predicted without $X^{(T)}$ parameters (line) versus system response (symbols).

load, which may also show significant mismatch at all harmonics. Thus, both models cannot ensure good accuracy with respect to the load without the $X^{(T)}$ terms.

The second source of inaccuracies is the impact of the source harmonics. The first amplifier creates harmonic content in the input scattered wave (the B_0 in Figure 3.13), due to its nonlinear behavior, which is then reflected back into the system by the source mismatch, thus contributing to the incident wave A_0 (in Figure 3.13). In a similar way, the second amplifier contributes to the harmonic content of the B_1 wave, which reflects back on the output impedance of the first amplifier and adds to the harmonic content generated by the nonlinear behavior of the first amplifier itself. The model needs to process correctly the entire harmonic content of the source, and this cannot be done accurately without the $X^{(T)}$ parameters.

The above comments explain, and are themselves re-enforced by, the inaccuracies observed in Figure 3.17.

Such attempts at formulating a simple extension of S-parameters into the nonlinear region have been used in the past with limited success. Two such examples are "large-signal S-parameters" and "hot S-parameters," which are distinct concepts. Each of them may describe the system with reasonable accuracy, but only over very restricted conditions of applicability, such as very good terminations and inter-stage match, which are rarely seen in practice. Two of the sources of their inaccuracy under broad range conditions come from the missing $X^{(T)}$ terms and the missing cross-frequency interaction terms.

The full formulation of the X-parameters (i.e. including both $X^{(S)}$ and $X^{(T)}$ terms) is thus essential to ensure the accuracy required by a successful design process.

3.7 Amplifier matching under large-signal stimulus

One of the strengths of the X-parameter model is the capability of accurately handling the potential mismatch that occurs at various nodes in the system for both the input and output ports of an amplifier.

3.7.1 Output matching and hot-S_{22}

Performance of an amplifier under small-signal conditions is completely determined by its S-parameters. The source and load conditions (source and load match) are uniquely identified for optimum operation, and a well-defined design process based on S-parameters has been successfully used for many years.

A similar approach has been sought for the nonlinear behavior, but it has not been successful until recently due to the inaccuracy of the modeling capabilities offered by technologies prior to X-parameters.

One important design objective is to understand how to achieve the output match and how best to measure and model it.

The output-match measurement has to be carried out under the so-called "hot" conditions. This means that the output reflection response to an output incident signal must be measured when the amplifier is driven into the nonlinear behavior by a signal at its input port. This requirement poses the practical difficulty of separating the two components of the output signal:

- the large signal that appears due to the input signal, and
- the output signal that appears due to the reflection of the output incident wave on the output impedance (the output reflection coefficient) of the amplifier.

This is difficult because the two components occur at the same frequency, and older network analyzers do not have the capability of separating these waves. This feature is now supported by the nonlinear network analyzer, which is the basic instrument supporting X-parameter measurement.

Figure 3.18 depicts the output waves of an amplifier under large-signal drive conditions from a single tone applied at its input port.

A small-signal incident tone is applied as an incident wave on the output port at the same frequency with the large signal scattered from the output due to the large-signal stimulus at port 1. This is identified as small-signal wave $a_{2,1}$. The phase of $a_{2,1}$ is swept for a constant magnitude, and thus its complex envelope vector describes a circle.

The scattered wave response at port 2 varies in both magnitude and phase, and its complex amplitude vector describes an ellipse. The X-parameter model accurately predicts this response, as shown in Figure 3.18.

The small-signal response at port 2, identified as $b_{2,1}$, is the additional response occurring at wave $B_{2,1}$ on top of the LSOP (established by the large-signal drive) due to the incident wave $a_{2,1}$. As the phase of $a_{2,1}$ varies, $b_{2,1}$ varies in both magnitude and phase; its complex amplitude vector describes an ellipse centered on the tip of the complex amplitude vector representing the LSOP response.

Knowing the incident and reflected waves, $a_{2,1}$ and $b_{2,1}$, respectively, the output reflection coefficient of the amplifier under large-signal drive conditions can be determined. The result is displayed in Figure 3.18.

It is very interesting to observe that the output reflection coefficient is not constant: it depends on the phase of the incident wave $a_{2,1}$, which is a condition outside the

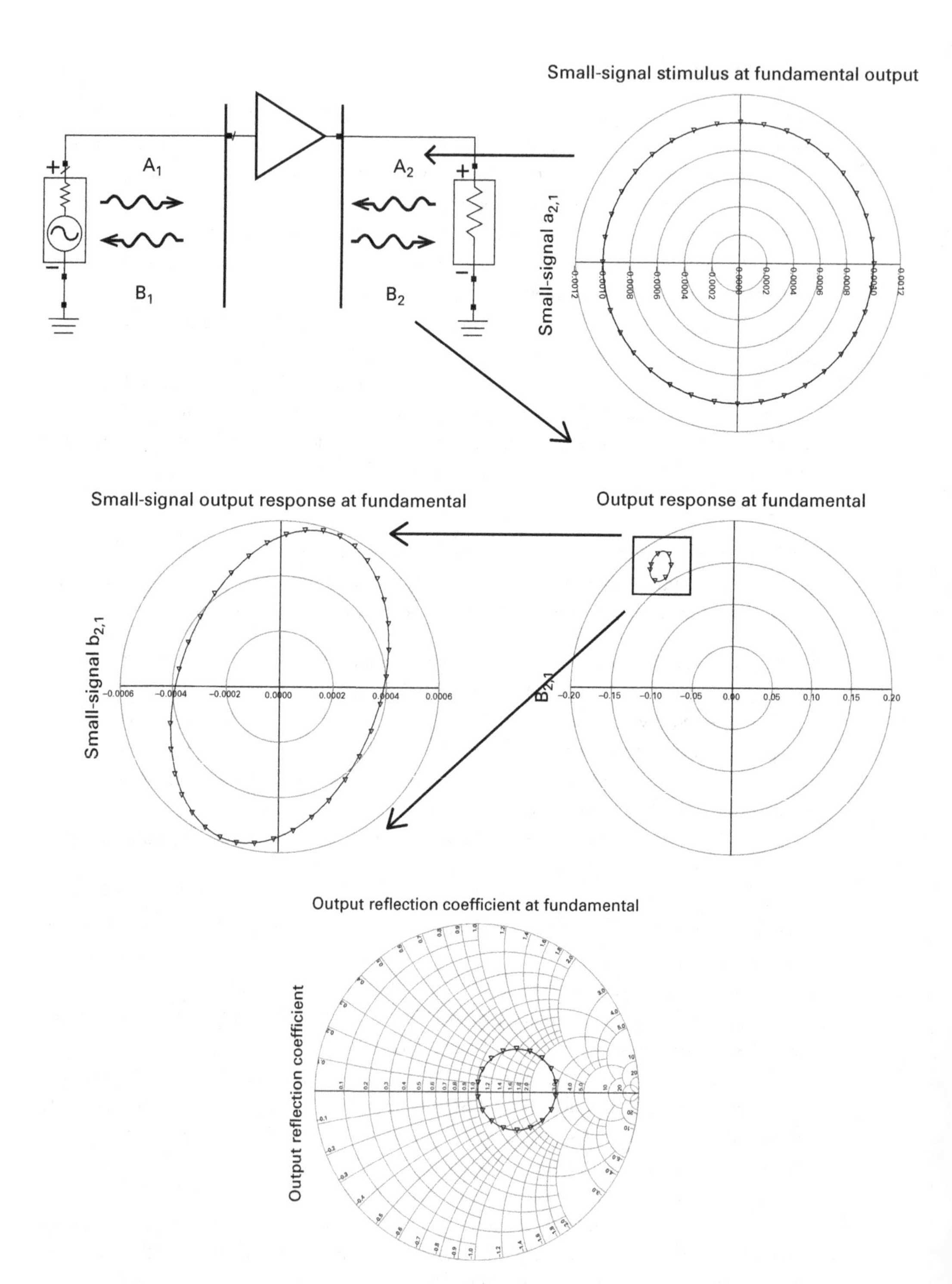

Figure 3.18 Output reflection coefficient at fundamental frequency of an amplifier under single-tone large-signal drive, predicted by X-parameters (line) versus system response (symbols).

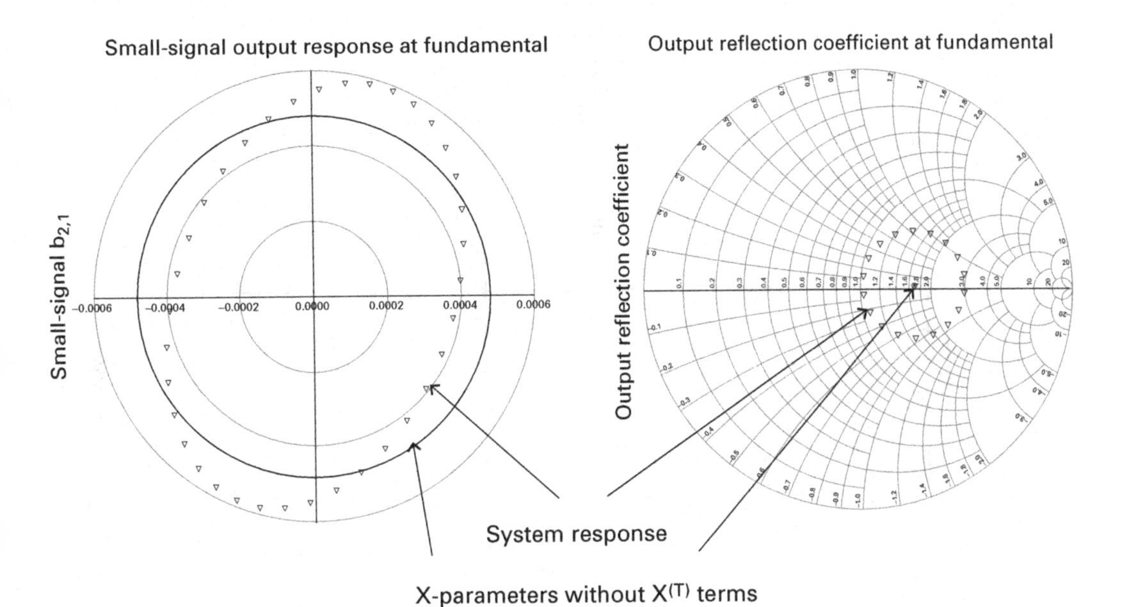

Figure 3.19 Left: response to small-signal perturbations around the LSOP predicted without $X^{(T)}$ parameters (line) versus system response (symbols). Right: output reflection coefficient predicted without $X^{(T)}$ parameters (single point) versus system response (symbols) – limitations of hot S_{22}.

amplifier itself. The output reflection coefficient describes a circle in the Smith chart as a function of the phase of $a_{2,1}$.

It is thus clear that the output reflection coefficient is not an intrinsic property of the nonlinear amplifier under large-signal drive because it depends on variables outside the amplifier itself (it depends on the load reflection coefficient, Γ_L). This raises important questions regarding what design process to follow when implementing the output-matching networks for nonlinear systems.

It is also interesting to observe how far away the model would be if the $X^{(T)}$ parameters were not considered. This scenario is shown in Figure 3.19.

This is a condition similar to the hot-S_{22} measurement that was intensively used before the X-parameter model was introduced. As shown, the small-signal response, $b_{2,1}$, describes a circle as a function of the phase of $a_{2,1}$, and the output reflection coefficient of the amplifier under large-signal drive becomes a single point, thus incorrectly suggesting it would be an intrinsic property of the amplifier.

Based on this discussion, it is clearly concluded that the output reflection coefficient is not an intrinsic property of nonlinear systems and that load mismatch is one of the external conditions that might influence its value.

The consideration of both $X^{(S)}$ and $X^{(T)}$ parameters together makes it possible for the X-parameter formalism to characterize accurately and completely the output reflection properties with respect to small-signal perturbations around the LSOP (like the load mismatches).

Once again, the full formulation of the X-parameters proves to be essential in ensuring the necessary accuracy in modeling the output reflection coefficient of nonlinear systems.

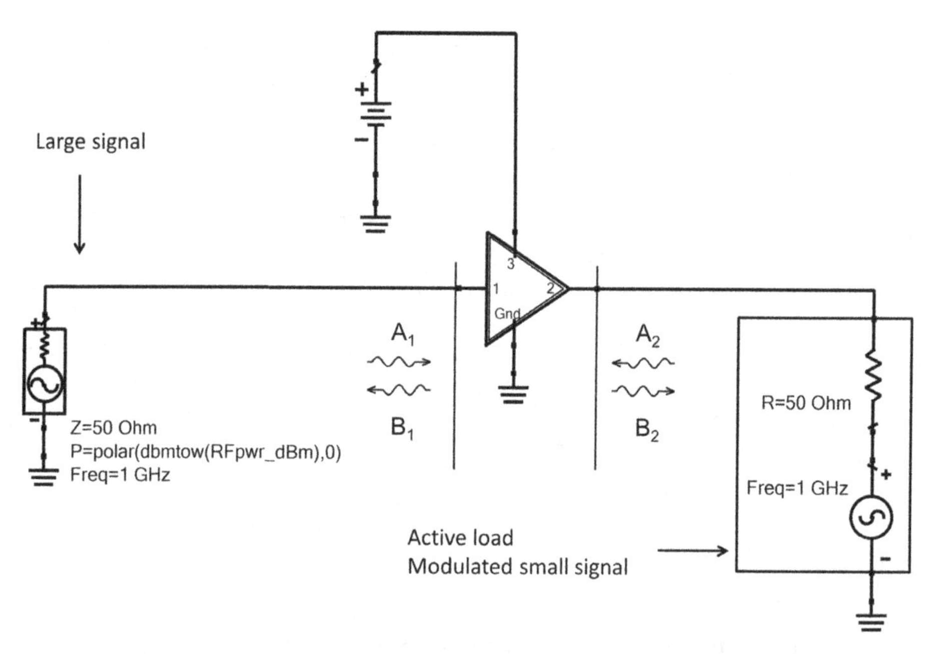

Figure 3.20 Modulated signal applied on the output of an amplifier driven by a large input tone.

The small-signal output reflection coefficient of an amplifier also depends on the LSOP set by the large-signal stimulus. As an example, let us consider the reflection of a modulated signal, with an arbitrary complex modulation, applied to the output of an amplifier when the amplifier is driven by a large single tone applied at its input. The amplifier is thus working on an active load with internal impedance of 50 Ω and with small-signal modulated available power, as shown in Figure 3.20.

The time-domain waveform and the trajectory described by the complex envelope of the modulated power wave incident on the output port of the amplifier are shown in Figure 3.21. In order to illustrate the complete generality of these concepts, the modulation is fictitious and it has been synthesized such that the trajectory (the polar plot) of its complex envelope resembles a "smiley face."

The time-domain modulation waveform is a periodic signal, and the time-domain representation in Figure 3.21 captures one period.

It is interesting to observe how the modulation of the reflected signal (the small-signal scattered wave at port 2, part of the $B_{2,1}$ wave) is distorted as the large signal drives the amplifier into nonlinear behavior (into compression). This is shown in Figure 3.22 (a summary of which is depicted on the front cover).

The plots shown in Figure 3.22 represent the behavior of the actual amplifier. It is important to demonstrate that the X-parameter model shown in (3.25) accurately reproduces this behavior. The response in Figure 3.22 is the $B_{2,1}$ scattered wave.

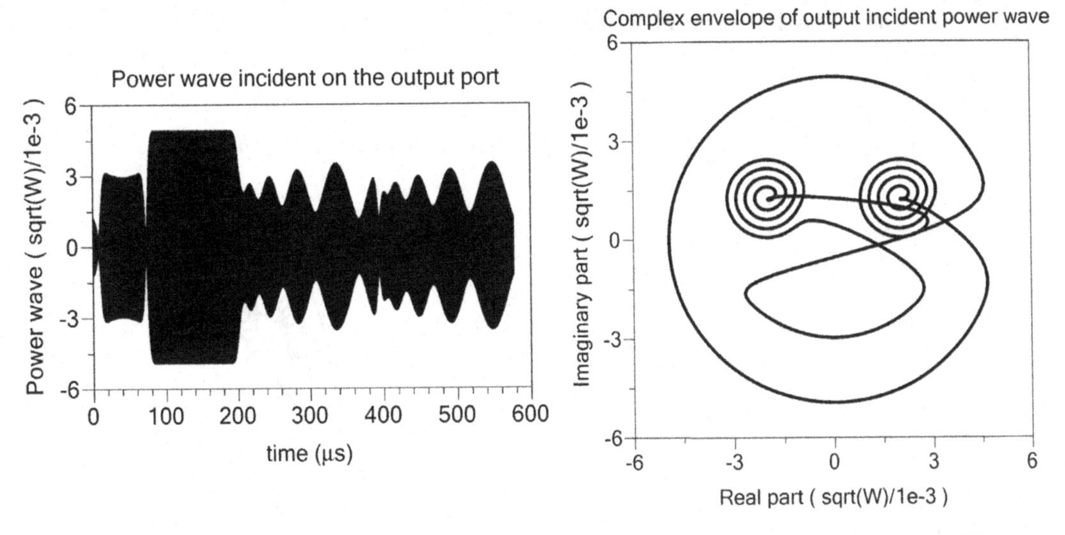

Figure 3.21 Small-signal modulated power wave incident on the output port of the amplifier.

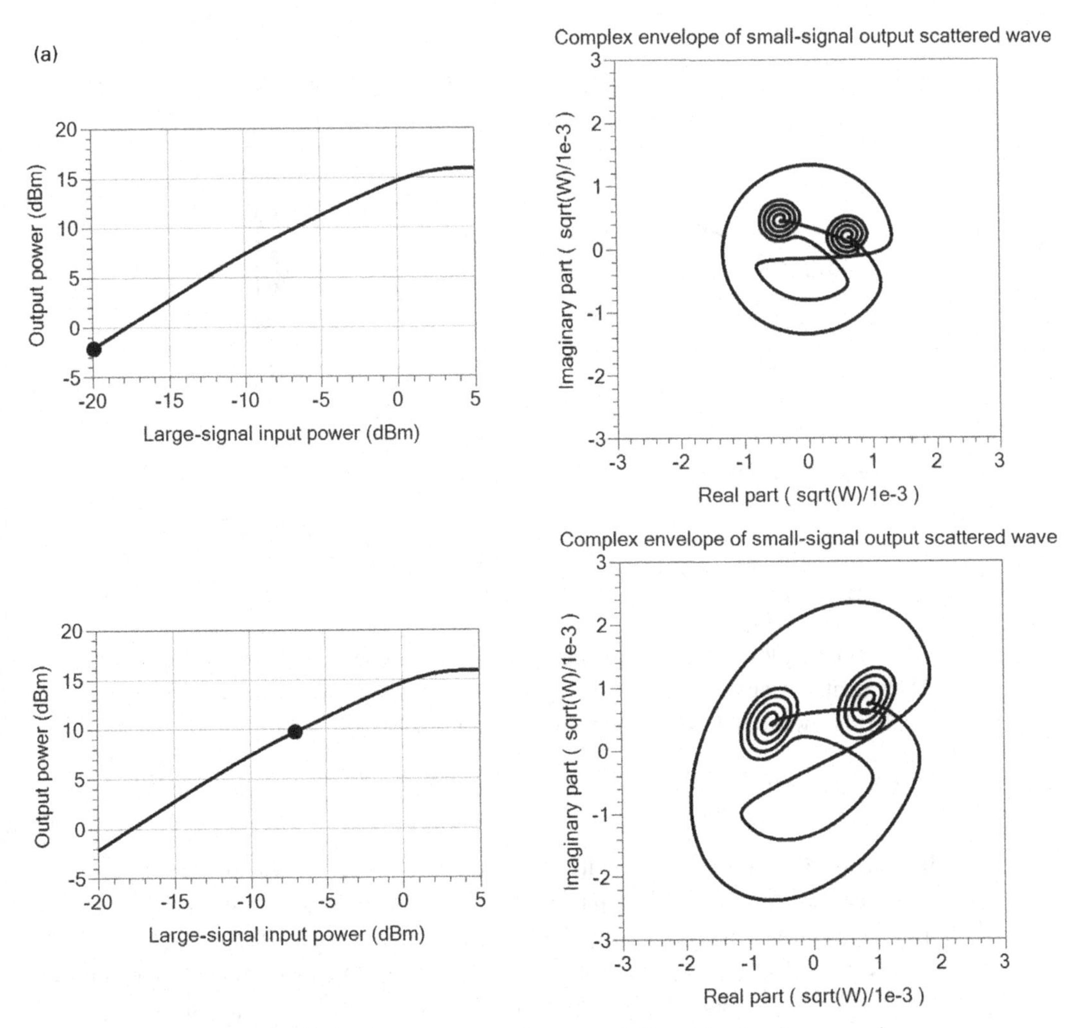

Figure 3.22 Small-signal output scattered wave at different levels of compression.

(b)

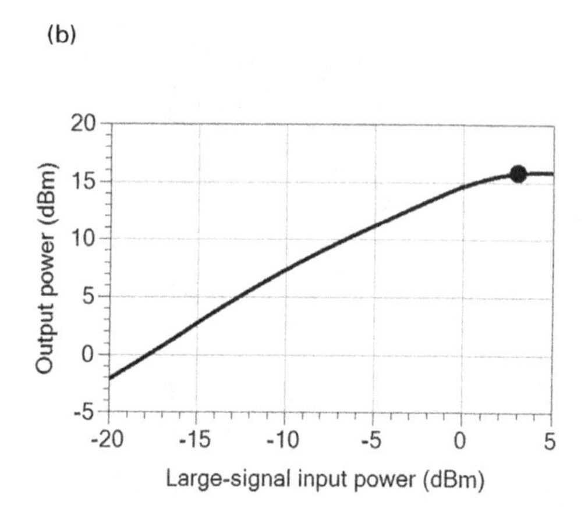

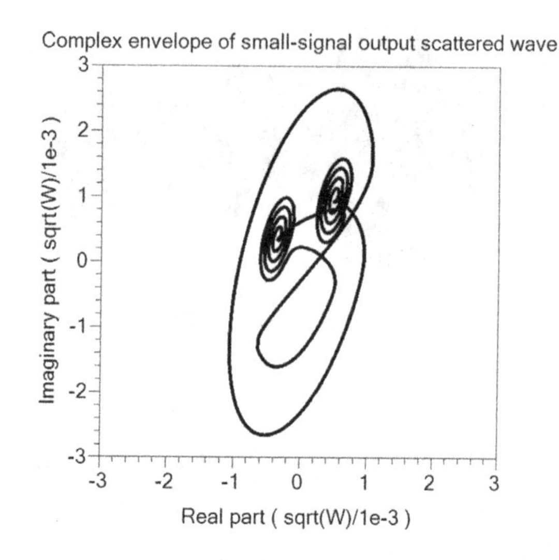

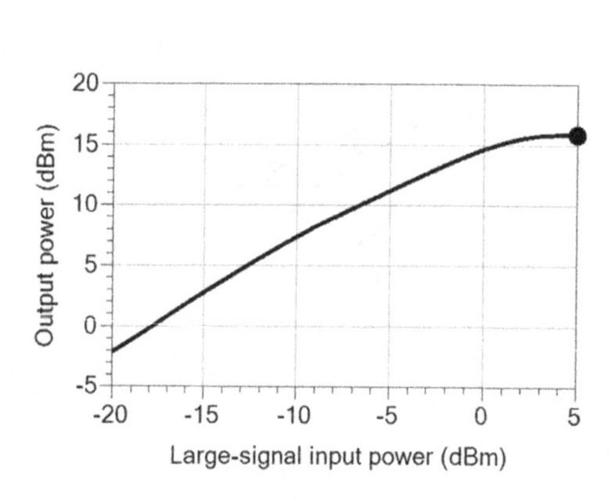

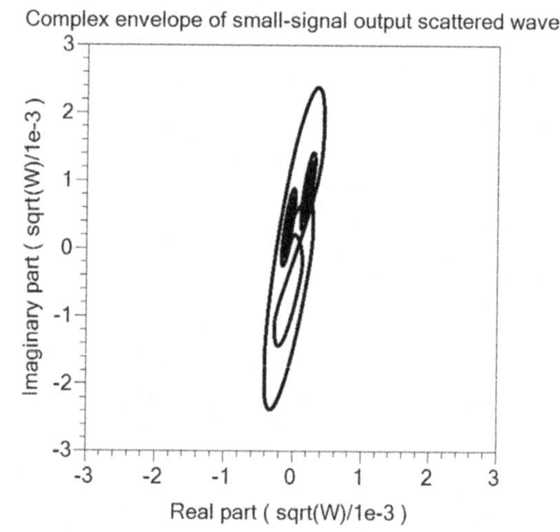

Figure 3.22 *(cont.)*

Considering that in a hot-S_{22} experiment a small-signal stimulus is applied only at the fundamental frequency, all $A_{q,1}$ waves are zero, except for $A_{2,1}$. The X-parameter model for the $B_{2,1}$ wave reduces to the form shown in (3.44):

$$B_{2,1} \cong X^{(FB)}_{2,1}(refLSOPS)P^1 + X^{(S)}_{2,1;2,1}(refLSOPS)A_{2,1}P^0 + X^{(T)}_{2,1;2,1}(refLSOPS)A^*_{2,1}P^2. \tag{3.44}$$

In order to demonstrate this capability, the X-parameter model of the amplifier in Figure 3.20 is extracted as a function of the input incident power level. The variation of the $X^{(S)}_{2,1;2,1}$ and $X^{(T)}_{2,1;2,1}$ parameters versus the power level of the input drive signal is shown in Figure 3.23.

As expected, at low drive levels, the $X^{(S)}_{2,1;2,1}$ has the dominant value when compared with the $X^{(T)}_{2,1;2,1}$ parameter. As the drive level increases, the $X^{(T)}_{2,1;2,1}$ parameter increases at a faster rate than the $X^{(S)}_{2,1;2,1}$ and becomes comparable with it. The continuous

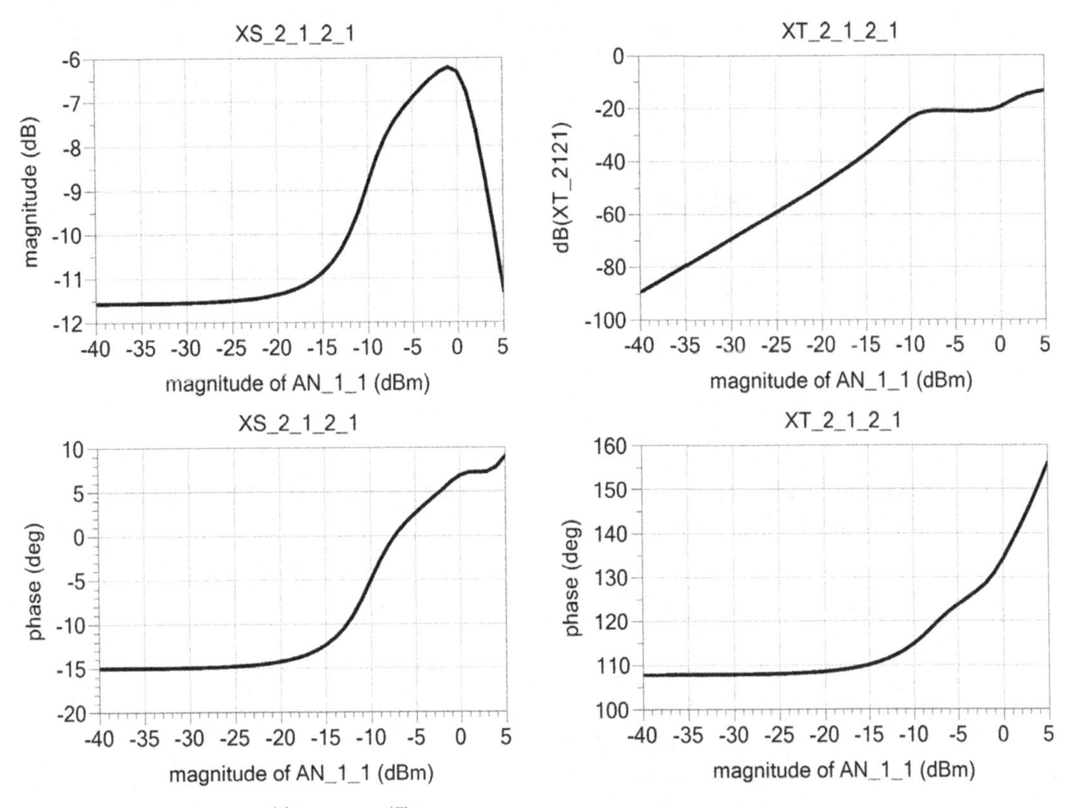

Figure 3.23 $X^{(S)}_{2,1;2,1}$ and $X^{(T)}_{2,1;2,1}$ versus input drive level.

variation of the relative values between these two parameters, combined with the formulation of the model itself, is the reason why the X-parameter model is capable of reproducing the nonlinear behavior of the amplifier with accuracy, as described in the following figures.

The wave $A_{2,1}$ incident on port 2 at the fundamental frequency is only the small signal modulated with the fictitious modulation of the "smiley face" shown in Figure 3.21.

The small-signal response of the amplifier to the small-signal stimulus $A_{2,1}$ is predicted by the model, as shown in (3.45):

$$B_{2,1} - X^{(FB)}_{2,1}(refLSOPS)P^1 \cong X^{(S)}_{2,1;2,1}(refLSOPS)A_{2,1}P^0 + X^{(T)}_{2,1;2,1}(refLSOPS)A^*_{2,1}P^2. \tag{3.45}$$

At the power level of -20 dBm, the $X^{(S)}$ and $X^{(T)}$ parameters have the values shown in (3.46):

$$\begin{aligned} AN_{1,1} &= -20 \text{ dBm}, \\ X^{(S)}_{2,1;2,1} &= 0.270/-14 \text{ deg}, \\ X^{(T)}_{2,1;2,1} &= 0.004/108 \text{ deg}. \end{aligned} \tag{3.46}$$

The small-signal response predicted by the model in (3.45) at -20 dBm input drive level is shown in Figure 3.24. The amplifier has a predominantly linear behavior. The

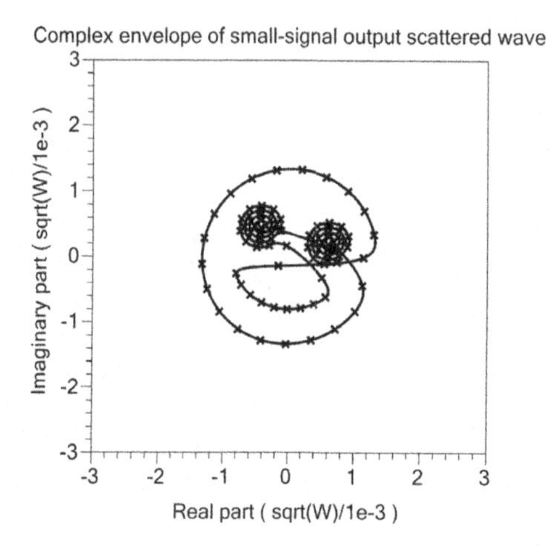

Figure 3.24 Small-signal response at -20 dBm input drive level predicted by X-parameter model (line) versus amplifier response (symbols)

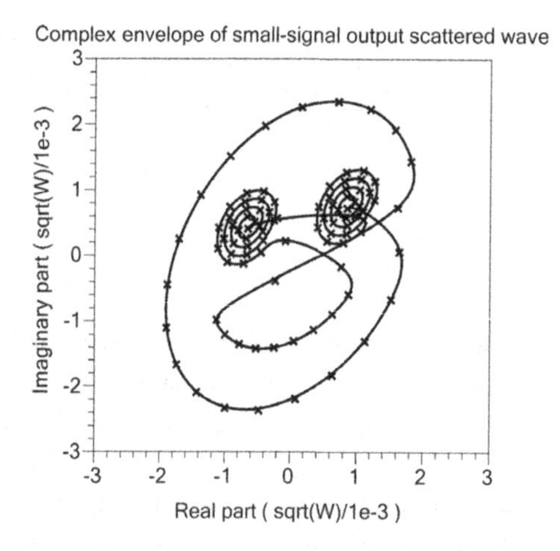

Figure 3.25 Small-signal response at -7 dBm input drive level predicted by X-parameter model (line) versus amplifier response (symbols).

$X^{(T)}_{2,1;2,1}$ has negligible value as compared to $X^{(S)}_{2,1;2,1}$, so the response is a linear transformation of the stimulus, showing a magnification and phase rotation.

At the power level of -7 dBm, the $X^{(S)}$ and $X^{(T)}$ parameters have the values shown in (3.47):

$$\begin{aligned} AN_{1,1} &= -7 \text{ dBm}, \\ X^{(S)}_{2,1;2,1} &= 0.425/0 \text{ deg}, \\ X^{(T)}_{2,1;2,1} &= 0.090/120 \text{ deg}. \end{aligned} \tag{3.47}$$

The small-signal response predicted by the model in (3.45) at -7 dBm input drive level is shown in Figure 3.25. The amplifier has a slightly nonlinear behavior. The $X^{(T)}_{2,1;2,1}$ is

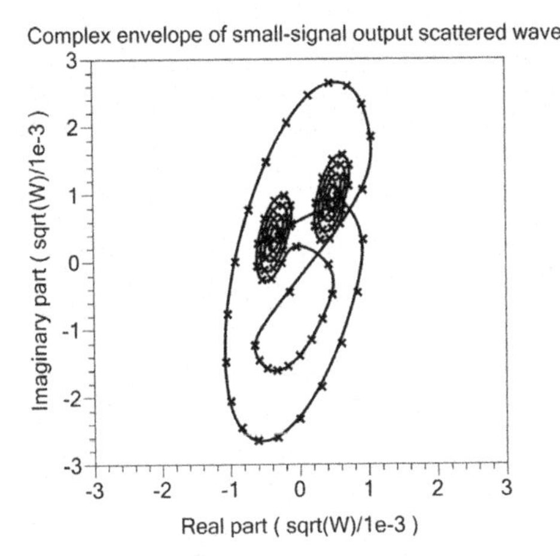

Figure 3.26 Small-signal response at 3 dBm input drive level predicted by X-parameter model (line) versus amplifier response (symbols).

no longer negligible compared to $X^{(S)}_{2,1;2,1}$, so the response is a more complex transformation of the stimulus, showing a magnification and a phase rotation that both depend on the instantaneous phase of the stimulus. The "smiley face" is now taller than it is wide.

At the power level of 3 dBm, the $X^{(S)}$ and $X^{(T)}$ parameters have the values shown in (3.48):

$$\begin{aligned} AN_{1,1} &= 3 \text{ dBm}, \\ X^{(S)}_{2,1;2,1} &= 0.367/7 \text{ deg}, \\ X^{(T)}_{2,1;2,1} &= 0.183/146 \text{ deg}. \end{aligned} \tag{3.48}$$

The small-signal response predicted by the model in (3.45) at 3 dBm input drive level is shown in Figure 3.26. At this drive level, the amplifier has a significantly nonlinear behavior. The $X^{(S)}_{2,1;2,1}$ and $X^{(T)}_{2,1;2,1}$ parameters have comparable values, which result in a significant distortion of the waveform.

At the power level of 5 dBm, the $X^{(S)}$ and $X^{(T)}$ parameters have the values shown in (3.49); they are almost equal in amplitude and have almost opposite phases (there is a 146° phase difference between them):

$$\begin{aligned} AN_{1,1} &= 5 \text{ dBm}, \\ X^{(S)}_{2,1;2,1} &= 0.272/9 \text{ deg}, \\ X^{(T)}_{2,1;2,1} &= 0.214/155 \text{ deg}. \end{aligned} \tag{3.49}$$

The small-signal response predicted by the model in (3.45) at 5 dBm input drive level is shown in Figure 3.27. At this drive level, the amplifier has a strongly nonlinear behavior. The response looks flattened due to the almost equal and almost out-of-phase values of the $X^{(S)}_{2,1;2,1}$ and $X^{(T)}_{2,1;2,1}$ parameters.

In all cases, the X-parameter model accurately predicts the behavior of the amplifier, which proves once again the strength of this technology.

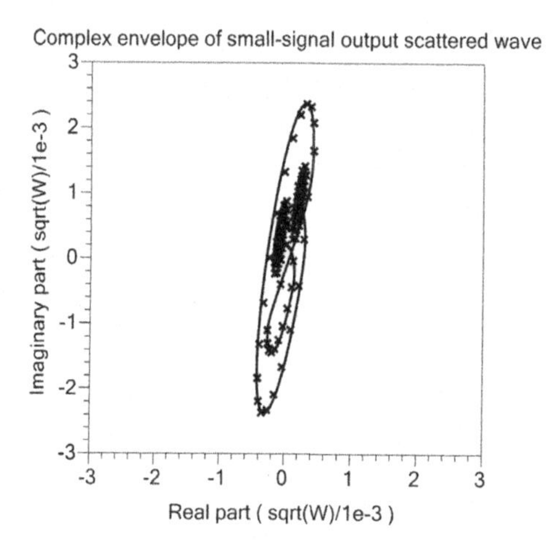

Figure 3.27 Small-signal response at 5 dBm input drive level predicted by X-parameter model (line) versus amplifier response (symbols).

3.7.2 Input matching

In a very similar manner, it can be proven that the input match of a nonlinear system is also dependent on the phase of the incident waves. While this was generally accepted for the frequency at which the large signal was applied, it can now be proven for all the harmonic frequencies.

Figure 3.28 shows the input waves on the second harmonic at the input of an amplifier under a single-tone large-signal drive.

A small-signal incident wave is applied, a_{12}, and the response on the second harmonic, B_{12}, is monitored.

The response describes an ellipse when the phase of the small-signal stimulus $a_{1,2}$ varies. The X-parameter model accurately predicts this response, as shown in Figure 3.28.

Separating the small-signal component of the response, $b_{2,1}$, allows for an accurate calculation for the input reflection coefficient on the second harmonic frequency.

As shown in Figure 3.28, the reflection coefficient is not constant with respect to the phase of the incident wave, which leads to the conclusion that the input reflection coefficient is not an intrinsic property of the nonlinear system.

It would again be interesting to observe how different the result would be if the $X^{(T)}$ parameters were not considered in the model. The results are shown in Figure 3.29.

This is a condition similar to the extension of S-parameters measured under hot conditions that was attempted with limited success before the X-parameter model was introduced. As shown, the small-signal response, $b_{1,2}$, describes a circle as a function of the phase of $a_{1,2}$, and the input reflection coefficient becomes a single point, thus incorrectly suggesting it would be an intrinsic property of the amplifier.

Based on this discussion, it is clearly concluded that the input reflection coefficient is not an intrinsic property of nonlinear systems. Source mismatch is one of the external conditions that might influence its value.

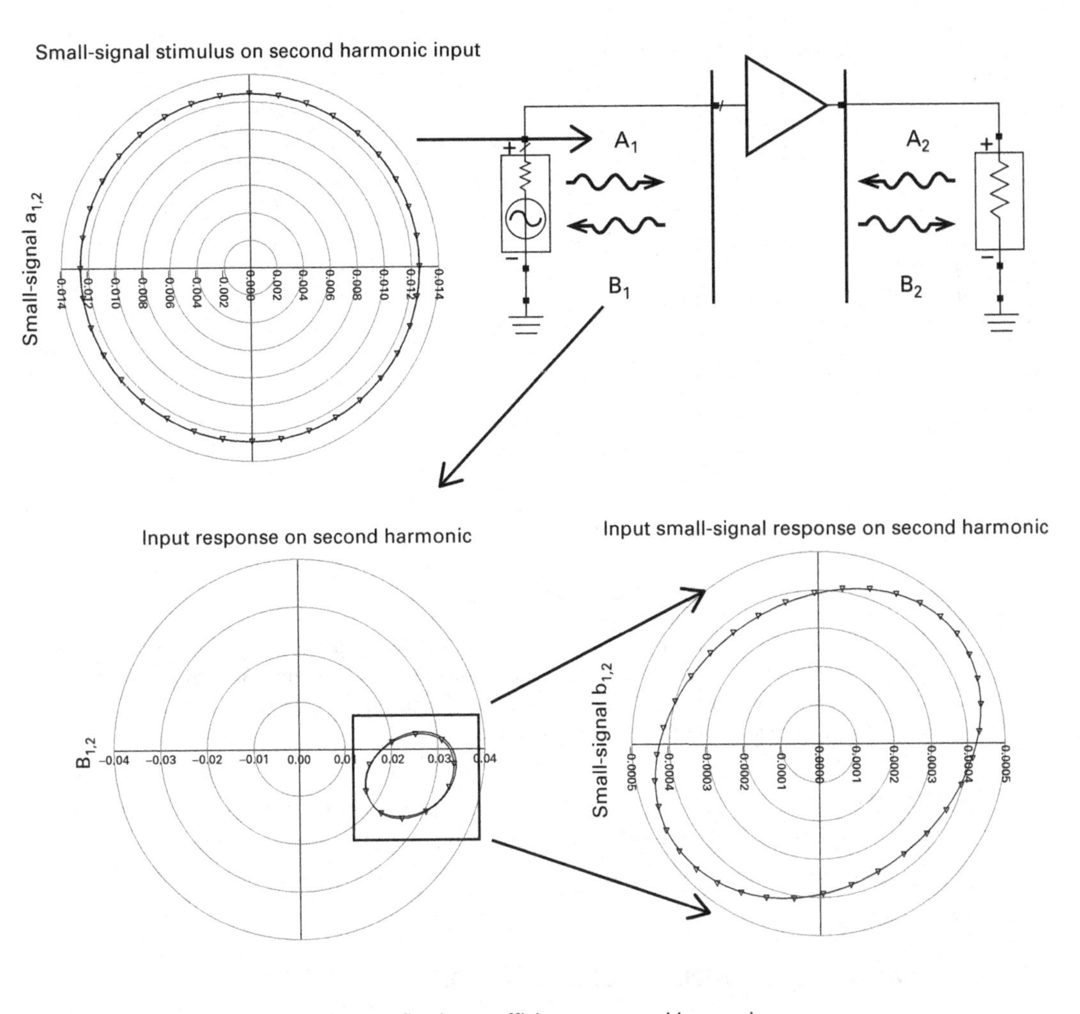

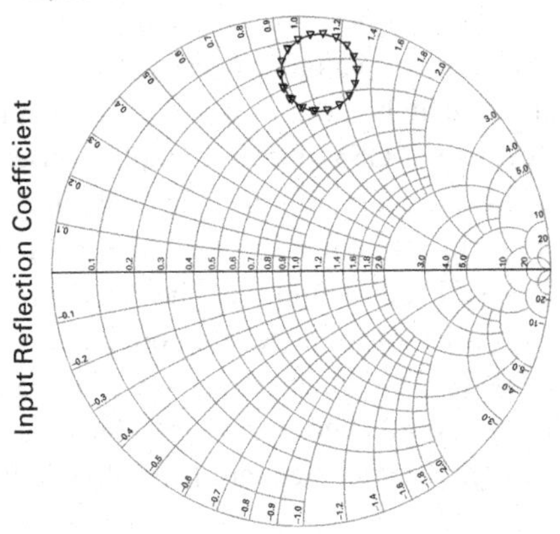

Figure 3.28 Input reflection coefficient on second harmonic frequency of an amplifier under large-signal drive, predicted by X-parameters (line) versus system response (symbols).

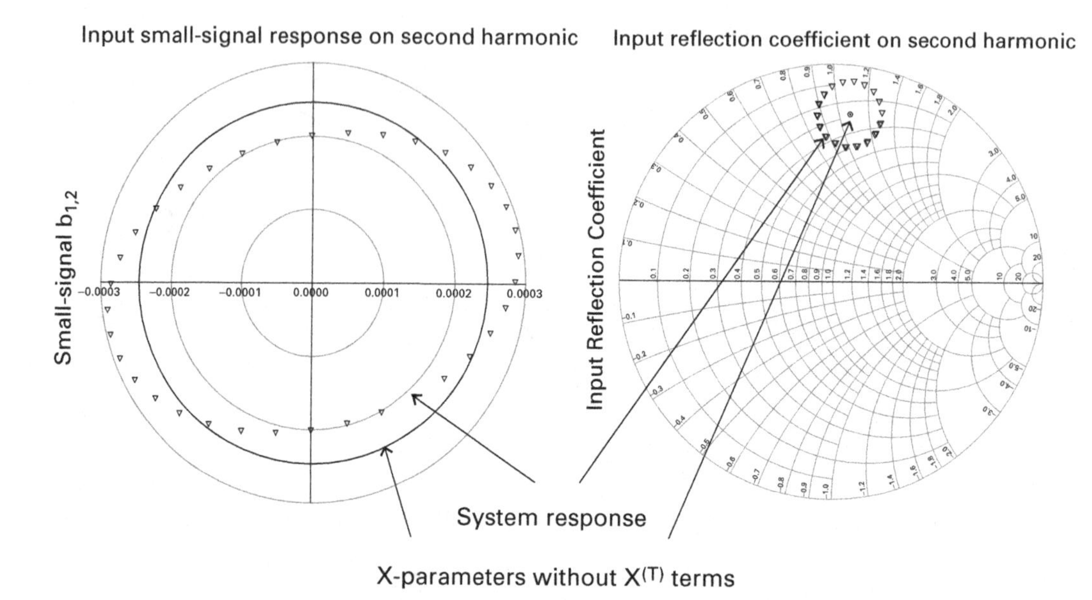

Figure 3.29 Left: response on second harmonic to small-signal perturbations around the LSOP predicted without $X^{(T)}$ parameters (line) versus system response (symbols). Right: input reflection coefficient at second harmonic predicted without $X^{(T)}$ parameters (single point) versus system response (symbols) – limitations of hot-S-parameters.

All the examples presented so far show that the X-parameter model does represent the intrinsic properties of the nonlinear system and that it accurately reproduces its behavior under various external conditions.

3.8 Practical application – a GSM amplifier

The ability of X-parameters to describe correctly the output match of devices under nonlinear operating conditions is valuable because there are many practical examples of devices that are not adequately described by other methods such as hot-S_{22}. One example of such a device is a GSM handset amplifier. Because GSM uses a constant envelope, amplifiers are typically run in saturated mode in order to achieve high efficiency. This results in the amplifier always operating in a nonlinear state, regardless of output power level. As a handset amplifier, the output is connected to an antenna, and the load seen by the amplifier varies depending on the environment in which the phone is operated. Because of this, the characterization of the amplifier's behavior under a wide range of match conditions is very important. Due to the highly nonlinear operating condition of the amplifier, X-parameters are needed in order to describe this behavior accurately.

This section describes the use of X-parameters to characterize the Skyworks SKY77329 Power Amplifier Module, a commercially available module used in mass-produced phones such as the Sony Ericsson W810. It is a multi-chip module which

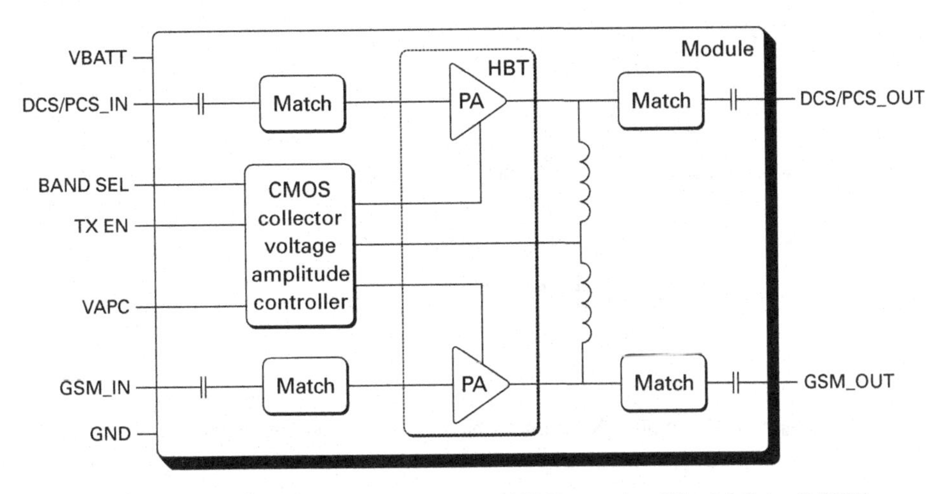

Figure 3.30 Functional block diagram of the SKY77329 Power Amplifier Module. © 2005 Skyworks Solutions Inc. Reproduced, with permission, from [2].

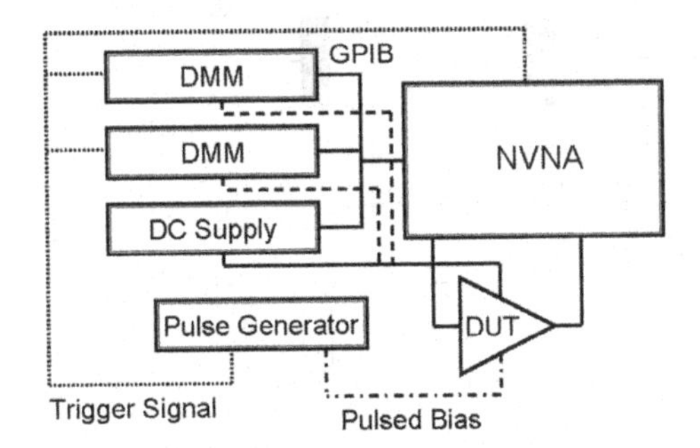

Figure 3.31 Measurement system used to characterize the SKY77329.

contains two RF amplifiers as well as analog and digital control circuitry. Additional details about the module and its characterization can be found in [1]. Figure 3.30 shows the block diagram as well as the available control lines.

Since GSM handsets operate under pulsed conditions, the device should be characterized under pulsed conditions as well. A nonlinear vector network analyzer (NVNA) system was configured with pulse generators and triggering to provide a 1/8 duty cycle pulse with a period of 4.615 μs to provide GSM-like operating conditions. Figure 3.31 shows the measurement-system configuration. This system was used to measure the X-parameters of the module, as well as the behavior of the module under different load conditions.

Measurements of DC voltage and current, taken during the pulse along with the RF measurements, were included in both the X-parameters and the validation measurements. This is required in order to characterize and validate power-added efficiency, an important metric for handset power amplifiers. Separate measurements were taken for the high-band and low-band amplifiers, including sweeping the battery voltage on pin

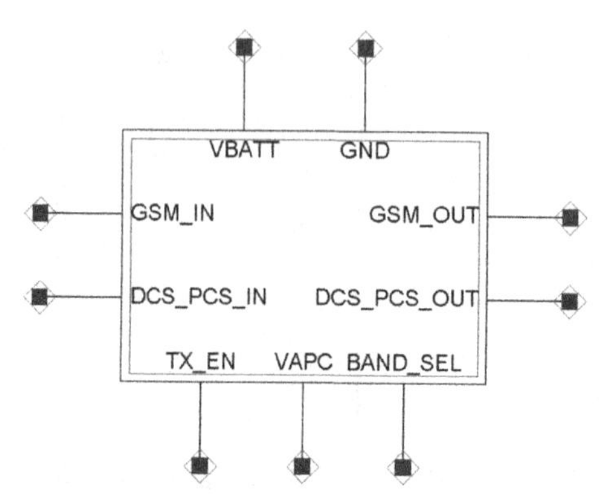

Figure 3.32 Complete X-parameter-based representation of the module in the circuit simulator.

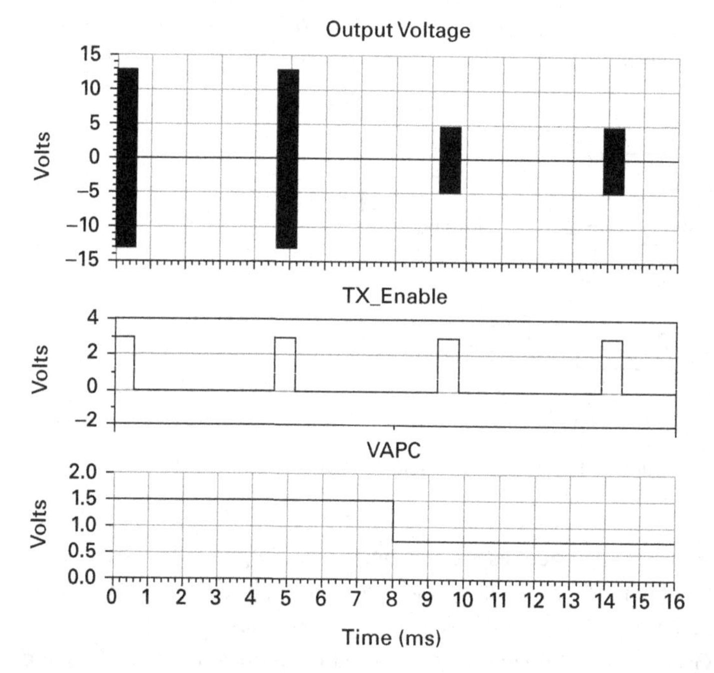

Figure 3.33 Accurate reproduction of the module output behavior in response to digital and analog control and modulation.

VBATT and the amplitude control voltage on pin VAPC to characterize the device across its entire range of power levels. The two models were then combined in the circuit simulator to create a single component representing the module, as shown in Figure 3.32.

By adding simple logic on the digital control pins TX_EN and BAND_SEL, the component is able to reproduce accurately the behavior of the module, even under relatively complex stimulus conditions such as those shown in Figure 3.33.

Because the X-parameters include measurements of the device behavior under matched conditions, simulation results correlate extremely well with validation

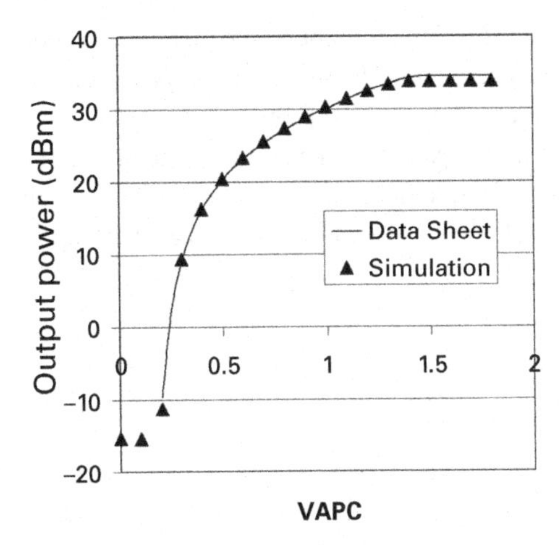

Figure 3.34 Comparison of simulated and specified output power versus VAPC.

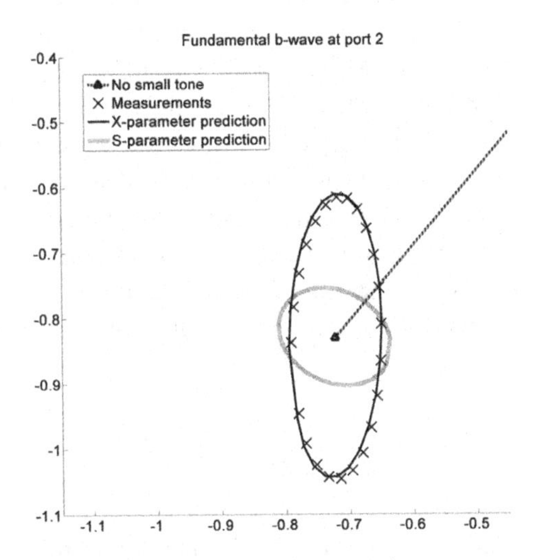

Figure 3.35 Comparison of measured, X-parameter-predicted, and hot-S_{22}-predicted response to a phase-swept stimulus tone at the output port.

measurements. In order to verify the measurements, simulation results were also compared to the datasheet. Figure 3.34 shows a comparison of simulated and specified output power versus voltage on the VAPC amplitude control pin.

In order to validate that the X-parameters accurately capture the device dependence on load match, independent validation measurements were made with an additional tone applied to the output. The phase of this tone was swept while the magnitude was kept constant, resulting in a set of stimulus conditions similar to that described in Figure 3.18. The response of the device was measured and compared to the response predicted by the measured X-parameters as well as the response predicted by measured hot-S_{22}. Figure 3.35 shows the results of this comparison.

It is immediately apparent from the elliptical shape of the measured response that hot-S_{22} is not capable of predicting this behavior. The measured X-parameters, however, are in excellent agreement with the measurements and provide an accurate characterization of the device behavior. The resulting X-parameter-based model is therefore able to predict accurately the behavior of the module under a wide range of environmental and stimulus conditions, making it a valuable tool for tasks such as evaluating the module or designing it into a system.

3.9 Summary

This chapter presented a useful and systematic simplifying approximation to the multi-variate time-invariant spectral maps defined on the harmonic grid that were introduced in Chapter 2. The complicated multi-variate nonlinear maps were approximated by much simpler nonlinear maps defined on a small number of selected large tones only, and simple linear maps accounting for the contributions of the many tones with small amplitudes. This approach results in a dramatic reduction of complexity while providing an excellent description of many important practical cases. One-tone X-parameters are derived by linearizing the nonlinear spectral map around the periodic large-signal operating point established by the DUT response to a large single-tone excitation, and applying the principle of time invariance to the nonlinear mapping. Examples of cascading two nonlinear amplifiers demonstrate the accuracy of the approximations and the advantages over methods based on ad hoc generalization of S-parameters. The new $X^{(T)}$ "conjugate terms" that arise in linearized X-parameter theory are shown to contribute significantly to output matching under strong input drive, even for small mismatch conditions.

Exercise

3.1 The ellipse described by the sum of $X^{(S)}$ and $X^{(T)}$ parameters. Consider a non-linear multi-port under the stimulus of a single large-signal tone of frequency ω_0, applied at port 1, $A_{1,1}$. Prove that the tip of the vector $B_{p,k}$ representing the response at port p on the frequency $\omega_k = k\omega_0$, describes an ellipse when a small-signal perturbation is applied at port q on frequency ω_l with constant amplitude and variable phase.

The solution is presented in Appendix E, but the main steps (1)–(5) are summarized here.

(1) The response at port p on the harmonic k is described by the X-parameter model, as shown in (3.50):

$$B_{p,k} \cong X^{(FB)}_{p.k}(refLSOPS)P^k + X^{(S)}_{p,k;q,l}(refLSOPS)A_{q,l}P^{k-l} + X^{(T)}_{p,k;q,l}(refLSOPS)A^*_{q,l}P^{k+l}. \quad (3.50)$$

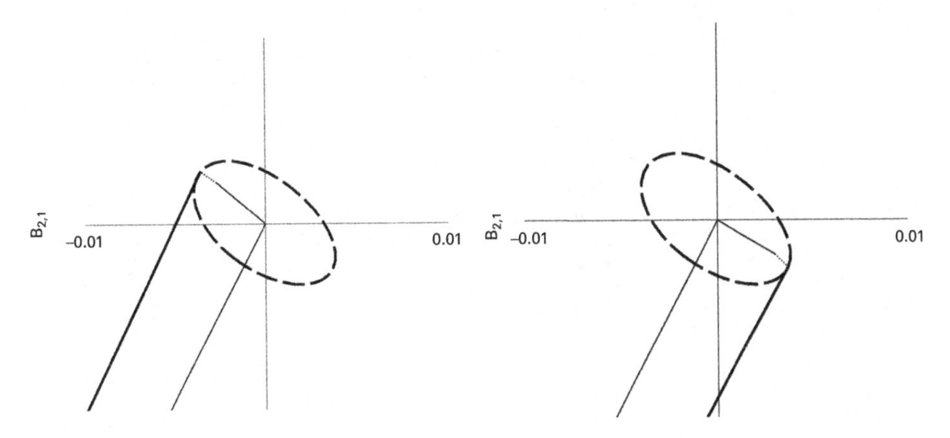

Figure 3.36 Small-signal terms almost aligned on the major axis (center is on the LSOP response).

There are three terms in the response $B_{p,k}$: one large-signal and two small-signal terms.

(2) There is a specific phase of $A_{q,l}$ for which the two small-signal terms in (3.50) are aligned, resulting in a maximum displacement from the large-signal term. Prove that the direction on which they are aligned has the phase shown in (3.51):

$$phaseMajorAxis = \frac{phase\left(X_{p,k;q,l}^{(S)}\right) + phase\left(X_{p,k;q,l}^{(T)}\right)}{2} + k \cdot phase(A_{1,1}). \tag{3.51}$$

The maximum distance from the large-signal term is $\left(\left|X_{p,k;q,l}^{(S)}\right| + \left|X_{p,k;q,l}^{(T)}\right|\right) \cdot \left|A_{q,l}\right|$, the sum of the magnitudes of the two small-signal terms. This makes for a maximum distance on the trajectory as shown in (3.52), which is the major axis of the ellipse, as shown in Figure 3.36:

$$MajorAxis = 2\left(\left|X_{p,k;q,l}^{(S)}\right| + \left|X_{p,k;q,l}^{(T)}\right|\right) \cdot |A_{q,l}|. \tag{3.52}$$

(3) There is a specific phase of $A_{q,l}$ for which the two small-signal terms in (3.50) are counter-aligned, resulting in a minimum displacement from the large-signal term. Prove that the direction on which they are aligned has the phase shown in (3.53):

$$phaseMinorAxis = \frac{phase\left(X_{p,k;q,l}^{(S)}\right) + phase\left(X_{p,k;q,l}^{(T)}\right)}{2} + k \cdot phase(A_{1,1}) + \frac{\pi}{2}. \tag{3.53}$$

The phases in (3.51) and (3.53) identify two perpendicular axes.

The minimum distance from the large-signal term is $\left(\left|X_{p,k;q,l}^{(S)}\right| - \left|X_{p,k;q,l}^{(T)}\right|\right) \cdot \left|A_{q,l}\right|$, the difference of the magnitudes of the two small-signal terms.

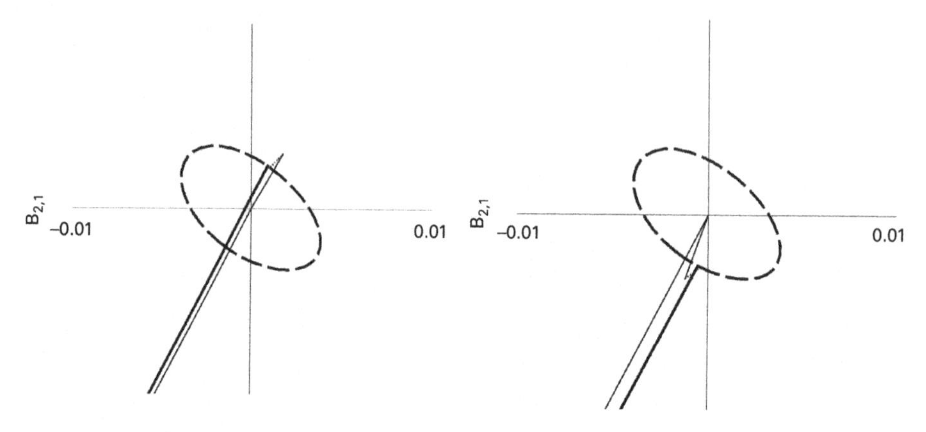

Figure 3.37 Small-signal terms almost counter-aligned on the minor axis (center is on the LSOP response).

This makes for a minimum distance on the trajectory as shown in (3.54), which is the minor axis of the ellipse, as shown in Figure 3.37:

$$MinorAxis = 2\left(\left|X_{p,k;q,l}^{(S)}\right| - \left|X_{p,k;q,l}^{(T)}\right|\right) \cdot |A_{q,l}|. \tag{3.54}$$

(4) Find the two points located on the major axis for which the sum of the distances from one of the extremes of the minor axis to them is equal to the sum of the distances from one extreme of the major axis to them and equal to $2\left(\left|X_{p,k;q,l}^{(S)}\right| + \left|X_{p,k;q,l}^{(T)}\right|\right) \cdot \left|A_{q,l}\right|$. Prove that these two points are symmetrically placed with respect to the intersection of the two axes at distances of $2\left|A_{q,l}\right|\sqrt{\left|X_{p,k;q,l}^{(S)}\right| \cdot \left|X_{p,k;q,l}^{(T)}\right|}$. These are the foci of the ellipse.

(5) Prove that the distances from the tip of the total-response vector, $B_{p,k}$, to the two foci are determined by the expressions in (3.55):

$$\begin{aligned} d_1 =& \left|A_{q,l}\right|\left(\left|X_{p,k;q,l}^{(S)}\right| + \left|X_{p,k;q,l}^{(T)}\right|\right) \\ &-2\left|A_{q,l}\right|\sqrt{\left|X_{p,k;q,l}^{(S)}\right| \cdot \left|X_{p,k;q,l}^{(T)}\right|}\cos\left(phase(A_{q,l}) - phase\left(X_{p,k;q,l}^{(T)}\right)\right), \\ d_2 =& \left|A_{q,l}\right|\left(\left|X_{p,k;q,l}^{(S)}\right| + \left|X_{p,k;q,l}^{(T)}\right|\right) \\ &+2\left|A_{q,l}\right|\sqrt{\left|X_{p,k;q,l}^{(S)}\right| \cdot \left|X_{p,k;q,l}^{(T)}\right|}\cos\left(phase(A_{q,l}) - phase\left(X_{p,k;q,l}^{(T)}\right)\right). \end{aligned} \tag{3.55}$$

The sum of these two distances is thus constant with respect to the phase of the small-signal perturbation, $\phi^{(a)}$, and has the value

$$d = d_1 + d_2 = 2\left|A_{q,l}\right|\left(\left|X_{p,k;q,l}^{(S)}\right| + \left|X_{p,k;q,l}^{(T)}\right|\right). \tag{3.56}$$

This proves that the tip of $B_{p,k}$ describes an ellipse when the phase of $A_{q,l}$ varies. The characteristics of this ellipse have been determined during the exercise.

References

[1] J. Horn, J. Verspecht, D. Gunyan, L. Betts, D. E. Root, and J. Eriksson, "X-parameter measurement and simulation of a GSM handset amplifier," in *EuMIC 2008*, Amsterdam, Oct. 2008.

[2] Skyworks, "SKY77329 PA module for quad-band GSM / EDGE," SKY77329 Datasheet, Oct. 2005.

Additional reading

D. E. Root, J. Horn, T. Nielsen, *et al.*, "X-parameters: the emerging paradigm for interoperable characterization, modeling, and design of nonlinear microwave and RF components and systems," in *IEEE Wamicon2011 Tutorial*, Clearwater, FL, Apr. 2011.

D. E. Root, J. Verspecht, D. Sharrit, J. Wood, and A. Cognata, "Broad-band, poly-harmonic distortion (PHD) behavioral models from fast automated simulations and large-signal vectorial network measurements," in *IEEE Trans. Microw. Theory Tech.*, vol. **53**, no. 11, pp. 3656–3664, Nov. 2005.

J. Verspecht, "Describing functions can better model hard nonlinearities in the frequency domain than the Volterra theory," Ph.D. thesis annex, Vrije Univ. Brussel, Belgium, Nov. 1995; available at http://www.janverspecht.com/skynet/Work/annex.pdf.

J. Verspecht and D. E. Root, "Poly-harmonic distortion modeling," *IEEE Microwave*, vol. **7**, no. 3, pp. 44–57, June 2006.

J. Verspecht, M. V. Bossche, and F. Verbeyst, "Characterizing components under large signal excitation: defining sensible 'large signal S-Parameters'?!," in *49th ARFTG Conf. Dig.*, Denver, CO, 1997, pp. 109–117.

4 X-parameter measurement

One of the key features that led to the wide adoption of S-parameters was the availability of hardware and calibration techniques capable of making quick, accurate, and repeatable S-parameter measurements. S-parameters can also be easily extracted in simulation from device or circuit models. In either case, the resulting S-parameters can immediately be used in simulation or design tools. In order to achieve similar success in the nonlinear domain, X-parameters must be easily measured and also easily extracted from simulation.

4.1 Measurement hardware

X-parameters, like S-parameters, represent the steady-state behavior of a device in the frequency domain. The linearity assumption of S-parameters, however, greatly simplifies the measurement requirements. The additional capabilities of X-parameters come at the cost of additional complexity in the measurement system as well as the modeling paradigm.

4.1.1 Hardware requirements

X-parameters include cross-frequency terms that capture the distortion products generated by device nonlinearities. In order to measure these terms, the measurement hardware must be capable of measuring coherent cross-frequency phase. Because S-parameters include no cross-frequency interaction (a consequence of the linearity assumption), each frequency may be measured independently with no need for a consistent time base or cross-frequency phase. Since all S-parameters are ratios of waves, there is also no need for accurate measurement of absolute power – only the relative power is needed. As a result, the hardware and calibration techniques developed for S-parameter measurement generally are not sufficient for X-parameter measurement and must be extended to include cross-frequency phase and calibrated absolute power.

4.1.2 Mixer-based systems

The most common instrument architecture used to measure S-parameters is the mixer-based vector network analyzer (VNA). The mixer-based VNA (Figure 4.1) owes its popularity to its extremely high achievable dynamic range and modern calibration

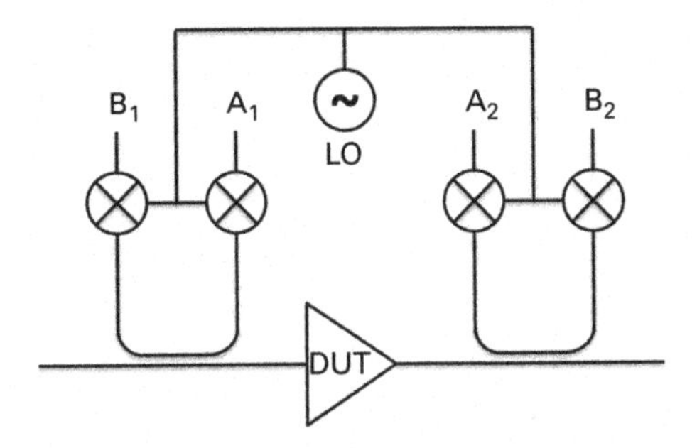

Figure 4.1 Mixer-based VNA diagram.

techniques that enable highly accurate and repeatable measurements. The basic idea behind the architecture is the use of mixers to convert high-frequency signals that are difficult to measure to a lower intermediate frequency (IF) which is much easier to handle.

The measurement frequency is selected by adjusting the frequency of the local oscillator (LO) driving the mixers so the IF remains constant. Since only one frequency is measured at a time, narrow-band filters can be used to eliminate noise and spurs at other frequencies and enable very high dynamic-range measurements. Calibration is critical for accurate measurements, since the effect of the mixers must be removed in addition to the cables and couplers.

Although this approach is extremely well suited to S-parameter measurements, it is missing the phase coherence required to measure the cross-frequency terms present in X-parameters. Because each frequency is measured individually, there is no consistent time base or phase relationship between consecutive measurements. Even if each measurement were accurately time stamped, the phase shift of the LO is generally unknown as it shifts to a new frequency. This causes measurement phase coherence to be lost.

Figure 4.2 shows measurements taken of the fundamental, second, and third harmonics at arbitrary times and the resulting measured phases of each signal. If these phase measurements are used to reconstruct the waveform, it looks nothing like the actual signal. If a phase-reference signal with known harmonic phase relationships (in this case, a pulse train with known phase of 0 at all harmonics) is measured at the same time as each measurement, however, the phase offset of each measurement can be determined and removed. The reconstructed waveform using corrected phases matches the original waveform.

Equation (4.1) shows how this correction is accomplished:

$$\Phi_n^{corr} = \Phi_n^{meas} - \left(\Phi_n^{ref\_meas} - \Phi_n^{ref\_char}\right). \tag{4.1}$$

The phase correction factor, defined as the difference between the measured phase of the reference at harmonic n, $\Phi_n^{ref\_meas}$, and the known or pre-characterized phase of the reference at harmonic n, $\Phi_n^{ref\_char}$, is subtracted from the measured phase of the DUT at harmonic n, Φ_n^{meas}, to obtain the corrected phase of the DUT at harmonic n, Φ_n^{corr}.

In order to enable measurement of coherent cross-frequency phase in a mixer-based system, an additional receiver can be added to measure a "phase-reference" signal that

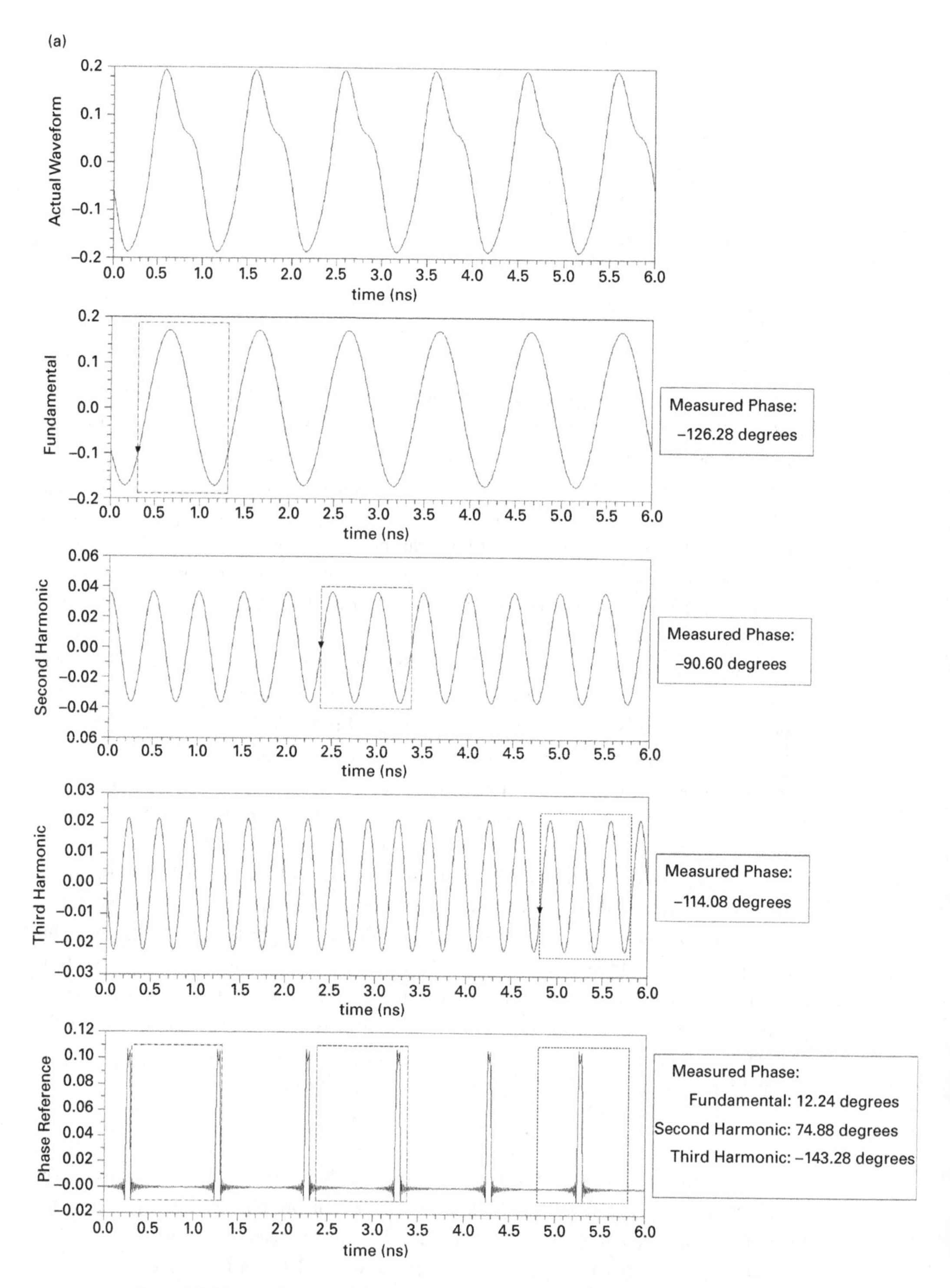

Figure 4.2 Phase coherence. (a) Measurements of fundamental, second, and third harmonics; also shown are the original waveform and the phase-reference signal. (b) Waveform reconstructed from raw phase measurements. (c) Waveform reconstructed from phase-corrected measurements; this matches the actual waveform in (a). Corrected phase = measured phase − measured phase-reference phase.

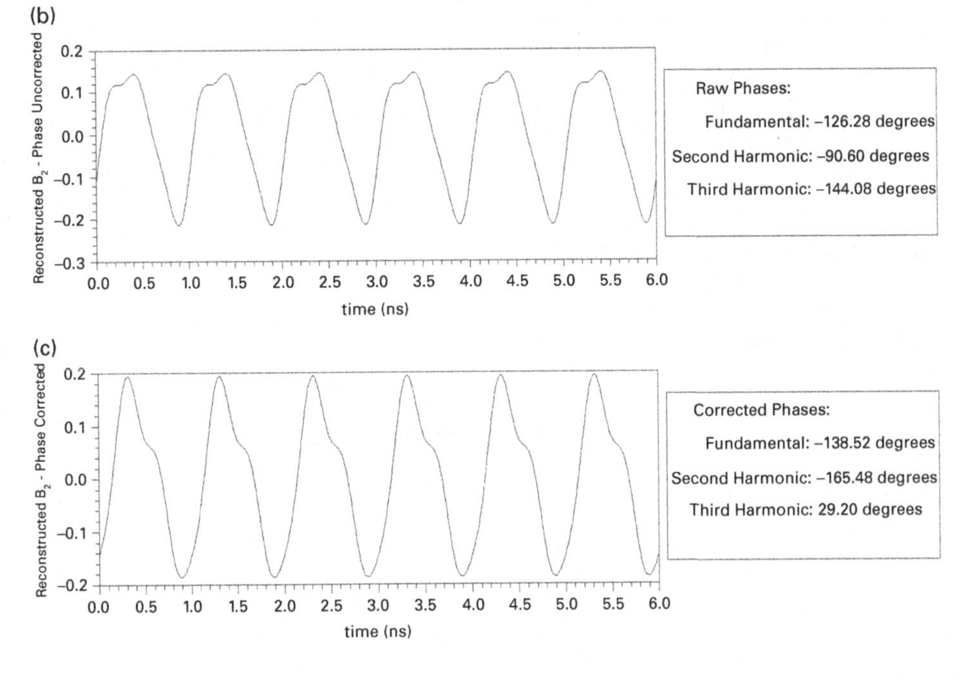

Figure 4.2 *(cont.)*

has spectral content at all frequencies of interest with a consistent phase relationship. By measuring the phase of the incident and scattered waves at each frequency relative to this phase reference, phase coherence is restored. Although it is necessary to know the cross-frequency phase response of the phase reference during calibration in order to make accurate cross-frequency phase measurements, only consistency is required to restore phase coherence in the measurement system. Calibration and phase-reference selection will be discussed in detail in Sections 4.2 and 4.3.

This type of system, shown in Figure 4.3, is referred to as a nonlinear vector network analyzer (NVNA) because it is capable of making the cross-frequency phase measurements necessary to characterize nonlinear devices. A VNA with enough receivers can be converted to an NVNA by adding a phase reference and the necessary software to handle measurement and calibration. The NVNA architecture has many of the characteristics of the VNA architecture from which it is derived, including the high-dynamic-range capability and repeatability.

4.1.3 Sampler-based systems

Mixer-based systems measure signals by accurately capturing the magnitude and phase of all spectral components of a signal. Once the complete spectrum is known, the inverse Fourier transform can be used to compute the waveforms in the time domain. Conversely, if the time-domain waveforms are known, the Fourier transform can be used to compute the complete spectrum – including coherent cross-frequency phase.

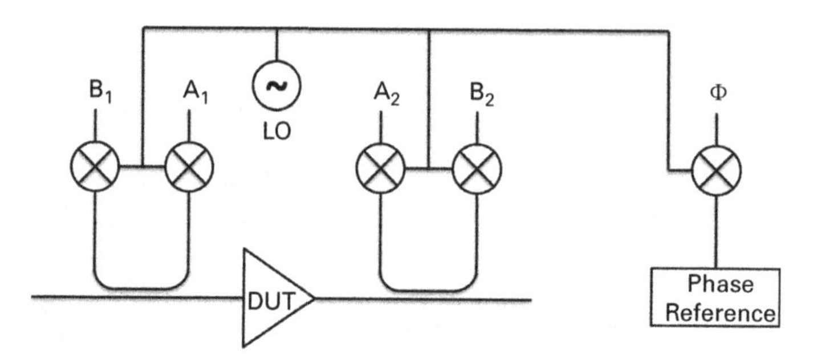

Figure 4.3 Mixer-based NVNA diagram.

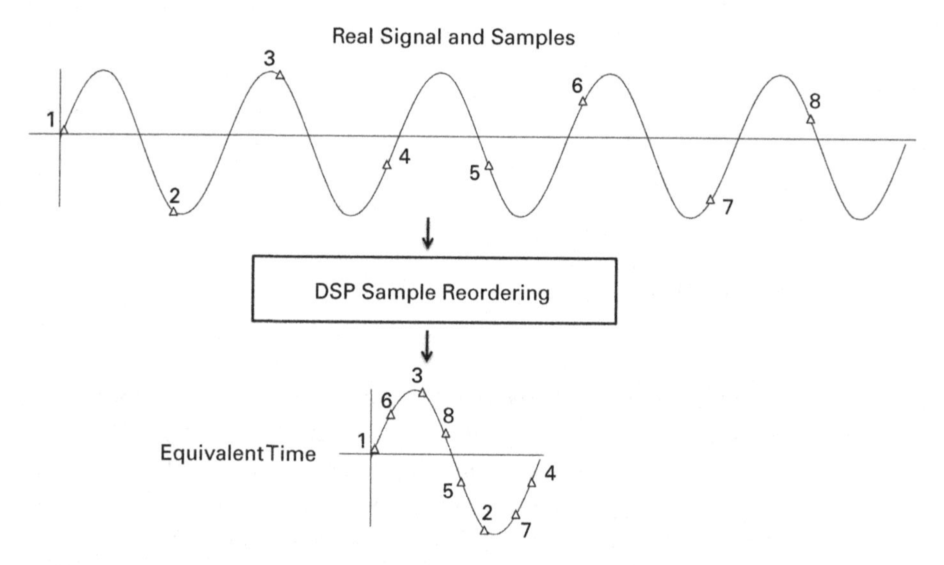

Figure 4.4 Equivalent-time samplers.

Therefore any measurement system that is capable of capturing time-domain waveforms can be used for X-parameter measurements.

Sampler-based systems measure signals by sampling the time-domain waveforms, either in real time or in equivalent time. Real-time samplers measure sequential samples of a signal and, assuming the sampling rate is sufficiently high, can reconstruct the signal using some form of interpolation. Real-time samplers are capable of measuring a repetitive signal within a single period of the signal, and can also be used to measure non-repetitive signals or transient events. Equivalent-time samplers, which can only measure repetitive signals, measure one or more samples per period over the span of many periods of the signal, then reconstruct a single period of the signal by mapping the actual time of each sample to the equivalent time within the single period (see Figure 4.4).

In either case, the sampler-based system captures complete time-domain waveforms. These waveforms can be transformed into the complete spectrum (magnitude and phase

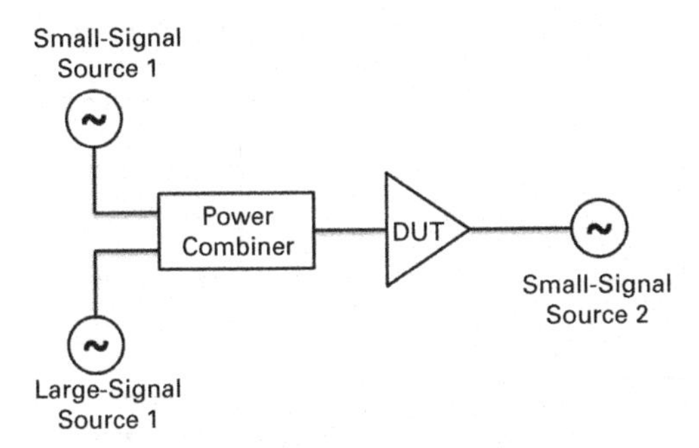

Figure 4.5 Example X-parameter stimulus configuration.

of all components) that is required for X-parameter measurements. Calibration is typically still required at high frequencies in order to account for the effects of the test set and non-ideal frequency response of the sampler. Although a phase reference is not required in a sampler-based system in order to make coherent cross-frequency phase measurements, a well-characterized calibration phase reference may be required for accuracy at high frequencies.

4.1.4 Stimulus requirements

Independent of the hardware architecture chosen to measure the stimulus and response spectrum and/or waveforms, the measurement system must also contain hardware capable of providing the appropriate stimulus for X-parameter measurements. This includes both the large-signal stimulus appropriate for the device under test (DUT) and an additional stimulus needed to inject the small-signal stimulus that is required for extraction of the small-signal ($X^{(S)}$, $X^{(T)}$, $X^{(Y)}$, and $X^{(Z)}$) terms.

The large-signal stimulus required depends on the DUT and the application for which the X-parameters are being measured. It may be as simple as DC only for an oscillator, or may include several DC and RF sources as well as source and load tuners for more complex circuits. The small-signal stimulus may be provided by a single-tone continuous wave source with a frequency range covering all frequencies of interest for the DUT. It must be able to be applied at any port simultaneously with the large-signal stimulus. This can be achieved by switching a single source between ports and combining with the large-signal stimulus, or by using multiple sources to cover the different ports, as shown in Figure 4.5.

4.2 Calibration

Calibration, or the mathematical correction for systematic measurement errors, is required for accurate measurements at high frequencies. Calibration procedures typically require a system-error model, a method for identifying error terms in that model, and a method of removing the modeled system error from measurements.

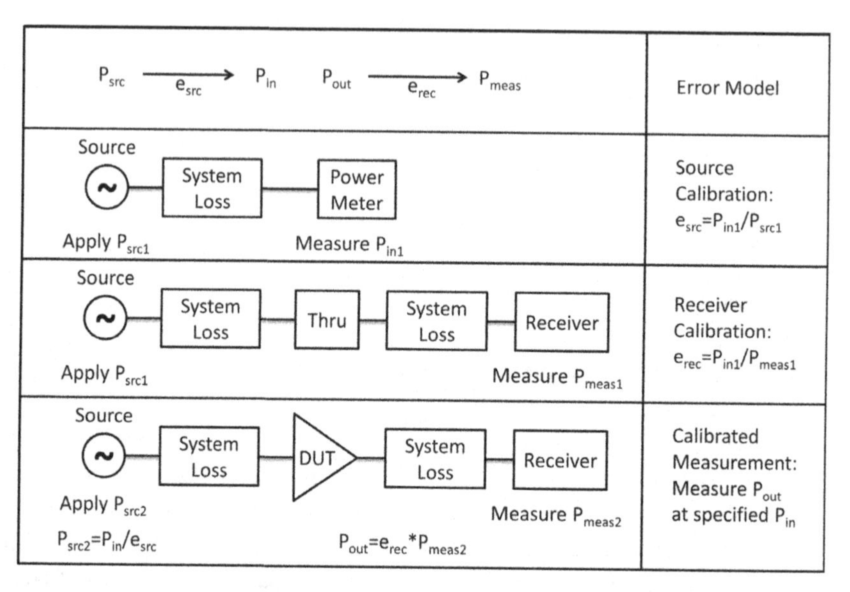

Figure 4.6 Scalar-loss correction.

4.2.1 Scalar-loss correction

A simple example of a calibration method commonly used for large-signal measurements is scalar-loss correction, in which the path between the DUT and the measurement hardware is modeled as a simple scalar loss. Figure 4.6 shows an example scalar-correction error model as well as a method used to identify the error terms and apply the correction to a measurement.

This correction only affects the magnitude of the measurement, and neglects errors due to effects other than loss (such as mismatch). More complex error models can be used to improve calibration quality, but more complex measurements are typically required to identify the error terms in these models.

4.2.2 S-parameter calibration

Two key reasons for the success of S-parameters are that they represent the intrinsic DUT characteristics (independent of the measurement system used) and they can be measured relatively easily with a high degree of accuracy. This is largely due to the calibration techniques that have been developed to ensure accurate removal of system error, as well as the high level of automation of the calibration procedure. A calibration wizard that guides users through the measurements needed to identify error terms and automatic measurement calibration is a standard feature in all modern network analyzers.

Figure 4.7 shows a network analyzer error model with 12 error terms. Because only ratioed waves are needed for S-parameter measurements, the absolute power and phase

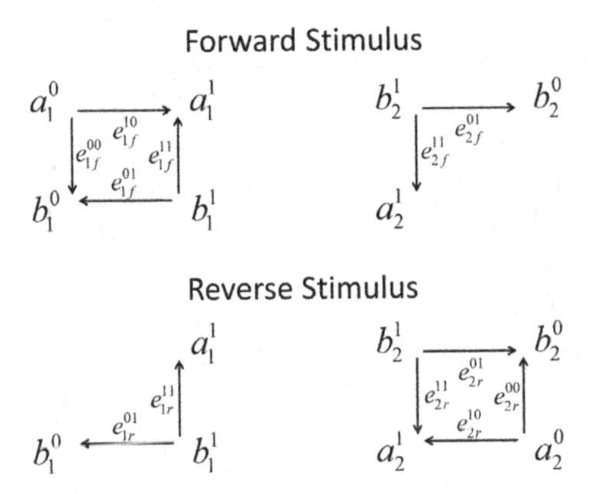

Figure 4.7 S-parameter error-correction model without normalization.

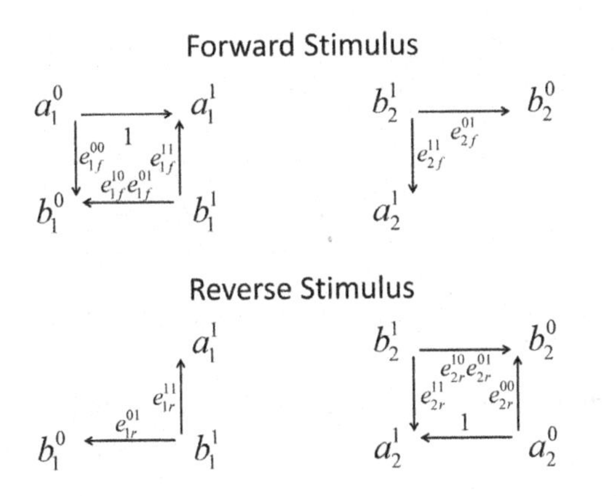

Figure 4.8 Ten-term S-parameter error-correction model.

of the waves do not need to be measured. One term can therefore be normalized to unity and some terms do not need to be isolated, reducing the number of unknown terms to ten and thus simplifying the identification process. This ten-term error model, shown in Figure 4.8, is commonly used for two-port S-parameter calibration in modern network analyzers.

Several methods exist for identifying the terms of the error models by measuring known standards such as open circuits, shorts, loads, and transmission lines. This error model is capable of accounting for test set losses, phase shifts, imperfect directivity, and mismatches as well as many systematic errors caused by the measurement hardware itself (mixers, amplifiers, filters, etc.). It is typically applied in such a way that the calibration and S-parameter extraction from the measurements are combined into one step. In this case, the imperfect load match seen by the DUT is also accounted for. The application of the error model to a measurement is shown in (4.2):

$$S_{11}^{1}=\frac{\left(\frac{S_{11}^{0}-e_{1f}^{00}}{e_{1f}^{10}e_{1f}^{01}}\right)\left(1+\frac{S_{22}^{0}-e_{2r}^{00}}{e_{2r}^{10}e_{2r}^{01}}e_{2r}^{11}\right)-e_{2f}^{11}\left(\frac{S_{21}^{0}S_{12}^{0}}{e_{2f}^{01}e_{1r}^{01}}\right)}{\left(1+\frac{S_{11}^{0}-e_{1f}^{00}}{e_{1f}^{10}e_{1f}^{01}}e_{1f}^{11}\right)\left(1+\frac{S_{22}^{0}-e_{2r}^{00}}{e_{2r}^{10}e_{2r}^{01}}e_{2r}^{11}\right)-e_{1r}^{11}e_{2f}^{11}\left(\frac{S_{21}^{0}S_{12}^{0}}{e_{2f}^{01}e_{1r}^{01}}\right)},$$

$$S_{21}^{1}=\frac{\left(\frac{S_{21}^{0}}{e_{2f}^{01}}\right)\left(1+\frac{S_{22}^{0}-e_{2r}^{00}}{e_{2r}^{10}e_{2r}^{01}}\left(e_{2r}^{11}-e_{2f}^{11}\right)\right)}{\left(1+\frac{S_{11}^{0}-e_{1f}^{00}}{e_{1f}^{10}e_{1f}^{01}}e_{1f}^{11}\right)\left(1+\frac{S_{22}^{0}-e_{2r}^{00}}{e_{2r}^{10}e_{2r}^{01}}e_{2r}^{11}\right)-e_{1r}^{11}e_{2f}^{11}\left(\frac{S_{21}^{0}S_{12}^{0}}{e_{2f}^{01}e_{1r}^{01}}\right)},$$

$$S_{12}^{1}=\frac{\left(\frac{S_{12}^{0}}{e_{1r}^{01}}\right)\left(1+\frac{S_{11}^{0}-e_{1f}^{00}}{e_{1f}^{10}e_{1f}^{01}}\left(e_{1f}^{11}-e_{1r}^{11}\right)\right)}{\left(1+\frac{S_{11}^{0}-e_{1f}^{00}}{e_{1f}^{10}e_{1f}^{01}}e_{1f}^{11}\right)\left(1+\frac{S_{22}^{0}-e_{2r}^{00}}{e_{2r}^{10}e_{2r}^{01}}e_{2r}^{11}\right)-e_{1r}^{11}e_{2f}^{11}\left(\frac{S_{21}^{0}S_{12}^{0}}{e_{2f}^{01}e_{1r}^{01}}\right)}, \tag{4.2}$$

$$S_{22}^{1}=\frac{\left(\frac{S_{22}^{0}-e_{2r}^{00}}{e_{2r}^{10}e_{2r}^{01}}\right)\left(1+\frac{S_{11}^{0}-e_{1f}^{00}}{e_{1f}^{10}e_{1f}^{01}}e_{1f}^{11}\right)-e_{1r}^{11}\left(\frac{S_{21}^{0}S_{12}^{0}}{e_{2f}^{01}e_{1r}^{01}}\right)}{\left(1+\frac{S_{11}^{0}-e_{1f}^{00}}{e_{1f}^{10}e_{1f}^{01}}e_{1f}^{11}\right)\left(1+\frac{S_{22}^{0}-e_{2r}^{00}}{e_{2r}^{10}e_{2r}^{01}}e_{2r}^{11}\right)-e_{1r}^{11}e_{2f}^{11}\left(\frac{S_{21}^{0}S_{12}^{0}}{e_{2f}^{01}e_{1r}^{01}}\right)}.$$

Note that this combined correction/extraction relies on the assumption that the DUT behaves linearly, and is not valid for measurement of devices behaving nonlinearly.

4.2.3 NVNA calibration

A nonlinear vector network analyzer obviously does not assume that the DUT is behaving linearly, so some modification of the calibration procedures used for S-parameters is clearly required. In addition to the linearity issue, an NVNA should be able to measure absolute power and cross-frequency phase, not just ratioed waves, so all terms must be individually identified, and normalization to unity is not allowed for any term. Furthermore, many nonlinear measurements (including X-parameters) include injection of signals at both ports simultaneously, so separate models for forward and reverse sweeps that assume constant load match are no longer sufficient.

To address these issues, a single eight-term error model that can be applied to forward, reverse, and combined stimulus configurations can be used (see Figure 4.9). The extraction of terms must also be modified to include absolute magnitude and the appropriate cross-frequency phase corrections, as well as isolation of all terms. This can be achieved by measuring an absolute power reference (such as a power meter) and a well-characterized phase reference in addition to the standards required for S-parameter calibration.

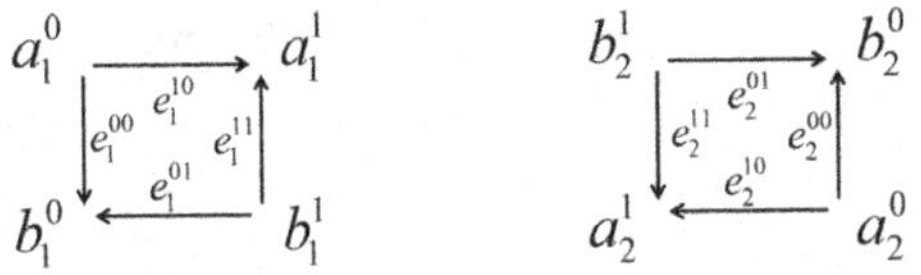

Figure 4.9 Eight-term error-correction model.

The resulting error model can be used to correct wave measurements, as shown in (4.3):

$$\begin{bmatrix} a_n^1 \\ b_n^1 \end{bmatrix} = \begin{bmatrix} \dfrac{e_n^{10}e_n^{01} - e_n^{00}e_n^{11}}{e_n^{01}} & \dfrac{e_n^{11}}{e_n^{01}} \\ \dfrac{-e_n^{00}}{e_n^{01}} & \dfrac{1}{e_n^{01}} \end{bmatrix} \begin{bmatrix} a_n^0 \\ b_n^0 \end{bmatrix}. \tag{4.3}$$

Both the stimulus and the response at both ports are measured. Corrections for any imperfections in the stimulus, such as source harmonics or imperfect load match, are not made at this time. Such corrections, which essentially predict what the device will do under a condition other than that under which it was measured, involve modeling assumptions about the device and are left to the model extraction. For X-parameters, this will be discussed in Section 4.4.

4.3 Phase references

The procedure for NVNA calibration includes measuring the same characterized passive standards used for S-parameter calibration, measuring an absolute power standard such as a power sensor, and measuring a phase standard. Unlike S-parameter calibration and power calibration, phase calibration is not commonly used in small- or large-signal measurements outside of NVNA, and may be a new topic even for those with a great deal of experience with high-frequency measurements. The key component of phase calibration is a well-characterized "phase reference" with an output spectrum that includes tones at all frequencies of interest with consistent, known phase relationships.

4.3.1 Phase-reference signals

From a theoretical standpoint, any well-defined signal with a spectrum that contains energy at multiple frequencies can be used as a phase reference. Figure 4.10 shows the amplitude spectrum of several well-known types of waveforms. The waveforms shown in Figure 4.10 are "realistic" in that they have finite rise times of 10 ps and are therefore fairly representative of what can be generated by currently available technologies.

For example, the spectrum of an ideal square wave has harmonic content at all odd harmonics in addition to the fundamental frequency. The spectral components at all frequencies are well defined, both in magnitude and phase. A square wave could therefore be used as a phase-reference signal as long as it is very nearly ideal or any

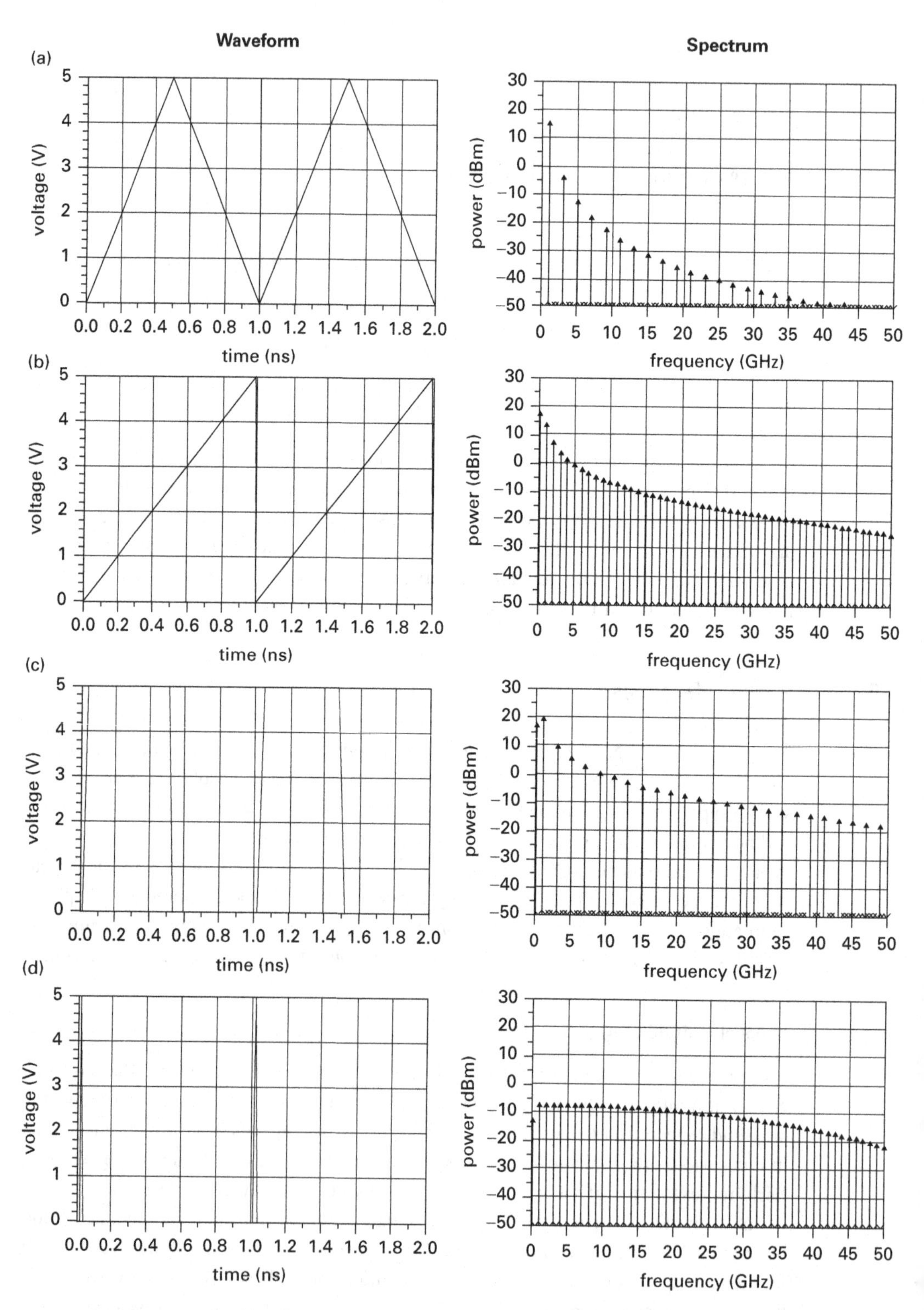

Figure 4.10 Spectra of various signals. (a) Triangle wave; (b) saw wave; (c) square wave; (d) pulse train.

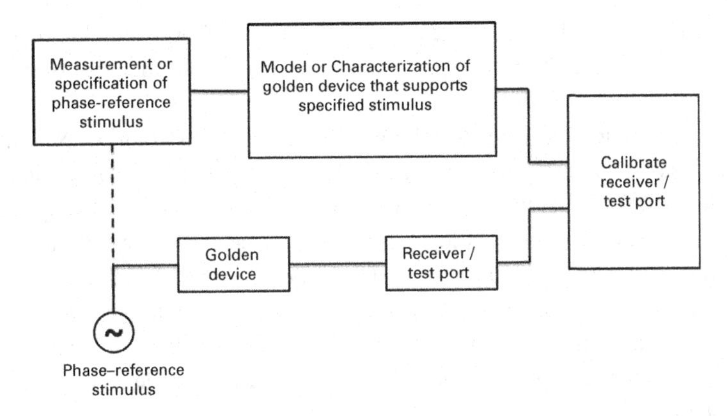

Figure 4.11 "Golden" device as phase reference.

non-idealities (such as finite rise/fall times) have been well characterized. Since the spectrum of the square wave contains only odd harmonics, however, it is difficult or even impossible to choose a square wave that has energy at all frequencies of interest for many types of measurements. Triangle waves share this limitation. Saw waves and repetitive pulse trains contain energy at all harmonics, making them better candidates for practical application. Pulse trains have the advantage that an ideal pulse train has harmonic content that does not roll off with frequency. Although real pulse trains do roll off due to finite rise times and non-ideal pulse shapes, they typically perform better at high harmonic orders than available alternatives.

Although the signals discussed so far are defined in terms of their time-domain waveforms and have spectra that can be analytically calculated, other types of signals can be used as well. For example, a well-characterized "golden" nonlinear component, such as a saturated amplifier, can be used to generate a reference signal (see Figure 4.11). As long as the output of the component is known, including the phase of harmonics and/or intermodulation products, the resulting signal can be used as a phase reference.

4.3.2 Measurement considerations

Although many signals can theoretically be used as phase references, there are practical considerations that must be taken into account for real measurements. In order to make a good phase reference, the signal must be able to be consistently and repeatably generated by a real source or component, consistently and repeatably measured by a real measurement system, and well characterized at the calibration connector.

Real signal generators do not generate ideal signals, so non-idealities of the source (such as distortion and phase noise or jitter) and their impact on the quality of the reference signal must also be considered. Distortion that is repeatable and known or can be characterized has no impact on the quality of the phase reference, but any variation in distortion will translate to uncertainty in the phase reference. Phase noise and jitter also introduce uncertainty, and there may be a large multiplicative factor for high-frequency measurements with a low-frequency reference signal. For example, 0.36 degrees of phase noise at 10 MHz or 100 ps of jitter translate to 360 degrees of phase noise at 10 GHz.

The phase-reference signal must be known in order to be used for NVNA calibration, so characterization of the real signal generator is important regardless of the signal chosen. The complexity of the characterization required, however, may depend on both the signal chosen and the technology used to produce the signal. A saw wave, for example, may have different distortion that must be characterized at different repetition frequencies. The signal generator may also depend on other factors such as ambient temperature or, if applicable, input power. The more factors the signal generator is sensitive to, the more complex characterization is needed to enable its use as an accurate phase reference.

Real measurement systems have limited dynamic range and a noise floor. Signals that have too little energy at frequencies of interest will not be measurable on a real measurement system, and therefore are not appropriate phase-reference signals. The spectral power of saw, square, and triangle waves decreases as harmonic index increases, so these signals may be difficult to use when many harmonics are required. Pulse trains have constant spectral power versus frequency, but there may be tradeoffs between pulse width and minimum/maximum frequency of operation. The pulse height is limited by the maximum input of the measurement system, so pulse width determines maximum power per pulse. Wider pulses give more power at low pulse repetition frequencies, but limit the maximum frequency since the signal must settle between pulses.

4.3.3 Practical phase references

Repetitive-pulse generators are the most commonly used phase references in nonlinear measurement systems for a number of reasons. Several technologies exist that are capable of generating high-quality, repeatable pulse trains that can be used at a variety of pulse-repetition frequencies and contain sufficient energy at high frequency to enable accurate measurements. A pulse train is also well characterized if the shape of a single pulse is known, regardless of the pulse-repetition frequency. For technologies that are insensitive to other factors such as temperature and input power, a relatively simple characterization is therefore sufficient for a wide range of uses (see Figure 4.12).

Well-characterized "golden" devices are not commonly used in commercial nonlinear measurement systems because accurate characterization of such devices over wide ranges of stimulus conditions is difficult. There are several advantages to this approach, however, that justify its use in some cases. The "golden" device can be stimulated at the same frequencies as the DUT, guaranteeing that spectral energy is present in the reference signal at all frequencies of interest. When using a repetitive-pulse generator, the stimulus frequencies are restricted to be on a harmonic grid with some minimum tone spacing. These restrictions are removed when using a "golden" device.

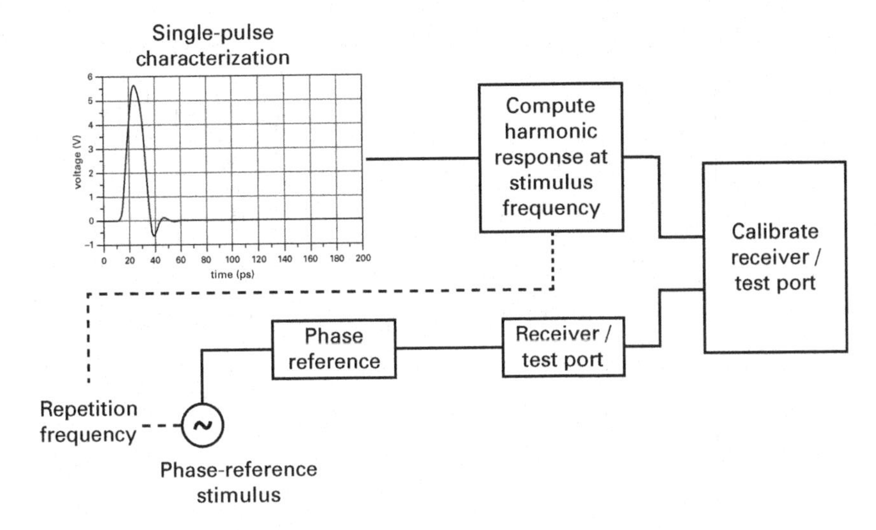

Figure 4.12 Phase-reference characterization.

4.4 Measurement techniques

In order to measure X-parameters, a test plan that defines the stimulus conditions and measurements from which the X-parameters will be extracted as well as an algorithm to extract X-parameters from those measurements are required. A method for measuring the large-signal terms ($X^{(F)}$, $X^{(FI)}$, and $X^{(FV)}$) is presented in this section, along with two alternative methods for measuring the small-signal terms ($X^{(S)}$, $X^{(T)}$, $X^{(Y)}$, and $X^{(Z)}$). The methods assume the measurements to be phase normalized to *refLSOP*, as discussed in Chapter 3. All methods are initially described as they would operate on an ideal measurement system, assuming that the desired stimulus is applied precisely with no unwanted spectral content; practical measurement considerations that apply to all methods are discussed at the end of this section.

4.4.1 Large-signal-response measurements

The large-signal response contained in the $X^{(F)}$ terms is conceptually the easiest to measure and identify. The measurement consists of simply applying the desired large-signal stimulus with no small-signal stimulus present and measuring the response at all frequencies of interest. The $X^{(F)}$ term at any specified port and frequency corresponding to the applied large-signal stimulus is equal to the measured response at that frequency.

4.4.2 Small-signal-response measurements

From a theoretical perspective, any set of small-signal stimuli that is sufficiently rich to stimulate all small-signal behavior of interest and sufficiently varied that all terms can be independently identified can be used to extract X-parameters. Practical

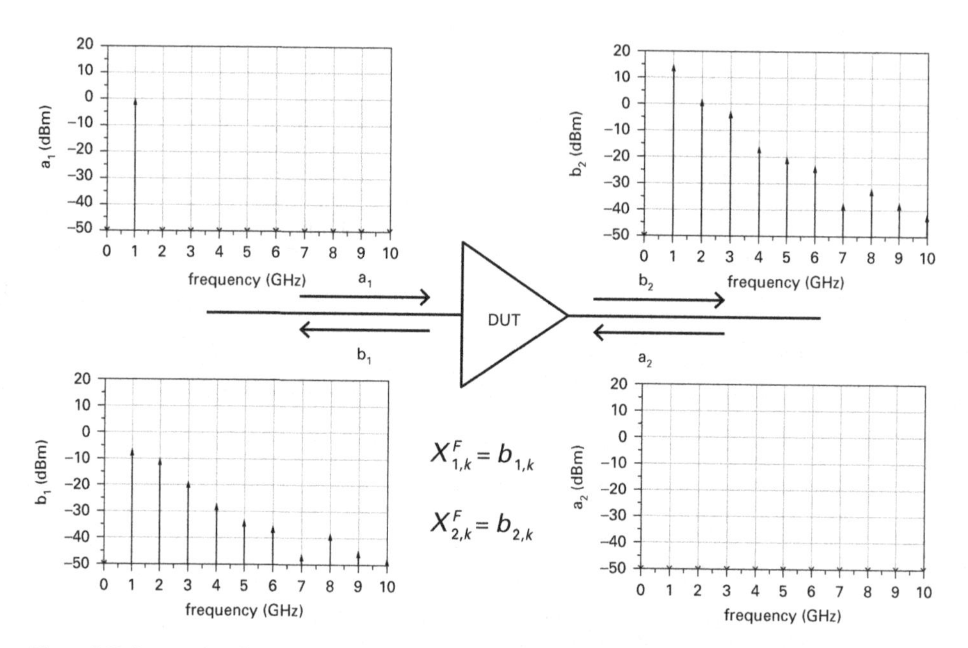

Figure 4.13 Large-signal-response measurements and identification.

measurement considerations, however, have led to the adoption of the offset-frequency and offset-phase measurement techniques for general use. Both of these techniques are optimal in the sense that X-parameters can be extracted from the theoretical minimum number of required measurements, and both have been demonstrated to be capable of making accurate X-parameter measurements on real measurement systems.

4.4.2.1 Offset-frequency measurements

The small-signal stimulus used in the offset-frequency method consists of, as the name suggests, tones slightly offset in frequency from the large-signal stimulus and its harmonics and/or mixing products. Since the small-signal stimulus is applied at the same time as the large-signal stimulus, the DUT is operating in a nonlinear state and mixing occurs. A single tone slightly offset from one large-signal response frequency will generate output at upper and lower sidebands of all large-signal response frequencies, as shown in Figure 4.14.

Because of this slight offset, the DUT response to the small-signal stimulus also falls at frequencies offset from the large-signal stimulus, and can easily be separated from the large-signal response by filtered tuning of the measurement frequency. Furthermore, at frequencies other than DC, each measured upper sideband response corresponds directly to a single $X^{(S)}$ term, and each lower sideband response corresponds directly to a single $X^{(T)}$ term, so all terms can be independently identified as simple ratios of (phase-normalized) scattered waves to incident waves, as shown in Figure 4.15 and (4.4):

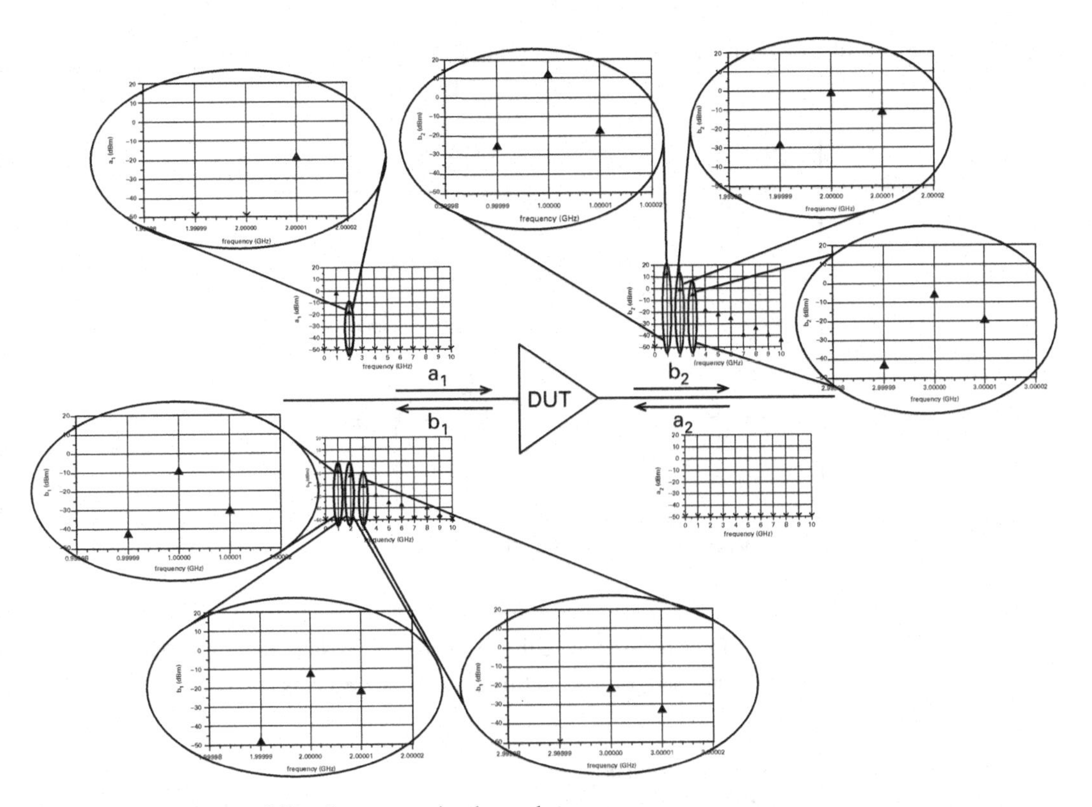

Figure 4.14 Offset-frequency stimulus and response.

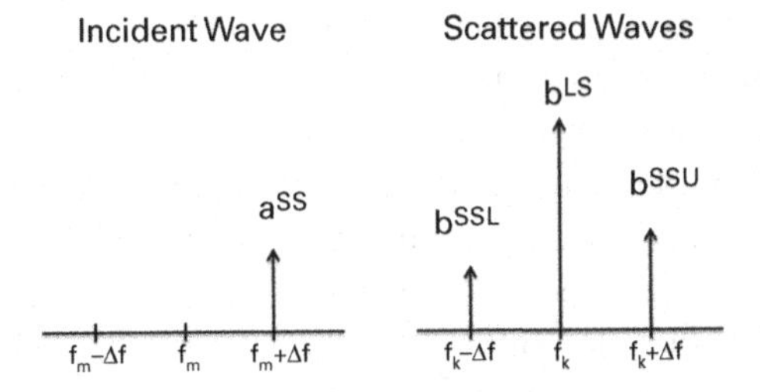

Figure 4.15 Offset-frequency measurements and identification.

$$
\begin{aligned}
X^{(S)}_{p,k,q,m} &= \frac{b^{SSU}_{p,k}}{a^{SS}_{q,m}}, \\
X^{(T)}_{p,k,q,m} &= \frac{b^{SSL}_{p,k}}{\left(a^{SS}_{q,m}\right)^{*}}.
\end{aligned}
\tag{4.4}
$$

The offset-frequency measurement technique relies on the fact that the on-frequency small-signal terms are equal to the limit of the offset-frequency terms as Δf goes to 0. As long as the offset chosen is small enough, good agreement is guaranteed. What qualifies as "small enough," however, is device dependent, so, without a-priori knowledge about the device being tested, there is always some risk that too large an offset might be used.

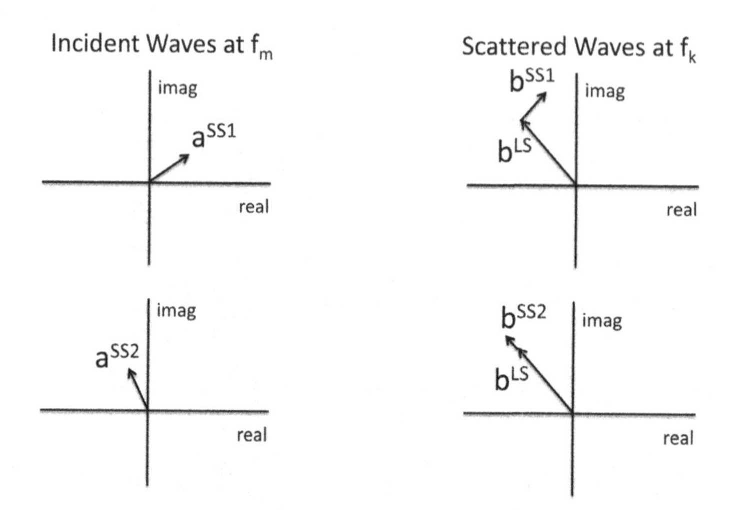

Figure 4.16 Offset-phase measurements and identification.

Very small offsets can also complicate the measurement system requirements and result in longer measurements due to filter settling time requirements.

4.4.2.2 Offset-phase measurements

Unlike the offset-frequency method, the small-signal stimulus of the offset-phase method is always applied at precisely the same frequencies at which the large-signal stimulus and its harmonics and/or mixing products fall. Since the response to the small-signal stimulus falls at the same set of frequencies as the response to the large-signal stimulus, separation of the large- and small-signal response must be handled by the extraction instead of filtering in the measurement system. Furthermore, the on-frequency small-signal response may contain contributions from both the $X^{(S)}$ and $X^{(T)}$, and these contributions must be identified separately in the extraction. This is achieved in the offset-phase method by subtracting the large-signal stimulus and response from the combined response to identify the small-signal part, and by making measurements under two small-signal conditions where a phase offset is introduced into the second small-signal stimulus, as shown in Figure 4.16.

The $X^{(S)}$ and $X^{(T)}$ terms can then be identified by solving (4.5):

$$\begin{aligned} b_{p,k}^{SS1} &= X_{p,k,q,m}^{(S)} a_{q,m}^{SS1} + X_{p,k,q,m}^{(T)} \left(a_{q,m}^{SS1}\right)^*, \\ b_{p,k}^{SS2} &= X_{p,k,q,m}^{(S)} a_{q,m}^{SS2} + X_{p,k,q,m}^{(T)} \left(a_{q,m}^{SS2}\right)^*. \end{aligned} \tag{4.5}$$

Although this measurement technique has twice as many small-signal stimulus conditions as the offset-frequency technique, it still has the same optimal number of measurements since half as many measurements (on-frequency versus upper- and lower-sideband measurements) are taken at each stimulus condition. This method is described in detail in [1].

4.4.3 Practical measurement considerations

Both methods discussed in the preceding sections would work flawlessly on an ideal measurement system capable of producing exactly the requested stimulus and making precise, noise-free, spur-free measurements. Unfortunately, real measurement systems are not ideal, so some consideration must be given to making the measurement technique robust to the errors present in real measurement systems.

Non-idealities in the measurement system can come from two main sources: measurement errors (such as noise and spurs) and stimulus errors (such as source harmonics and mismatch). Good measurement technique, including both experiment design and extraction, can minimize the sensitivity of the extracted parameters to measurement errors and correct for stimulus errors to the greatest extent possible. Minimizing sensitivity to measurement errors, while important, will not be discussed in great detail here. It is a problem common to many measurements, and the techniques used to address it for X-parameter measurement (taking multiple measurements and averaging/fitting, designing experiments for optimal conditioning, etc.) are common. Correcting for imperfect stimulus is of particular interest, however, since it can be handled in a unique way in X-parameter measurements.

Linear measurements are routinely corrected for imperfect stimulus through load-match correction built into most standard S-parameter measurement routines. This correction mathematically removes the effect of the non-ideal load match seen by the DUT during the S-parameter measurement *assuming* that the DUT is behaving linearly. As a result, true 50 Ω (or other specified impedance) S-parameters can be obtained on a system without an ideal 50 Ω port match.

Most nonlinear measurements are not capable of carrying out stimulus correction because not enough is known about the DUT. For example, in order to correct an amplifier third-harmonic measurement specified into a 50 Ω impedance that is measured under actual impedances that vary between 45 and 55 Ω at the fundamental, second, and third harmonics, the measurement system must know how the third harmonic responds to incident power at the fundamental, second, and third harmonics. There is simply not enough information to determine this from a simple forward measurement into a passive load.

X-parameters, on the other hand, contain exactly the needed information in the $X^{(S)}_{2,3;2,1}$, $X^{(T)}_{2,3;2,1}$, $X^{(S)}_{2,3;2,2}$, $X^{(T)}_{2,3;2,2}$, $X^{(S)}_{2,3;2,3}$, and $X^{(T)}_{2,3;2,3}$ terms. Since these terms are extracted from small-signal stimulus to which the DUT is responding in a spectrally linear way, linear correction techniques (similar to S-parameter load match) can be used to extract them in a way that fully corrects for imperfect stimulus. Once extracted, these terms can be applied to the large-signal-stimulus measurements to extract $X^{(F)}$ terms that are fully corrected for imperfect stimulus and represent the DUT behavior under the specified conditions.

X-parameters enable, for the first time, fully corrected measurements of nonlinear figures of merit under specified conditions. This can either translate to better measurements with the same quality of measurement system currently used, or similar measurement quality with less expensive measurement equipment. It also results in significantly

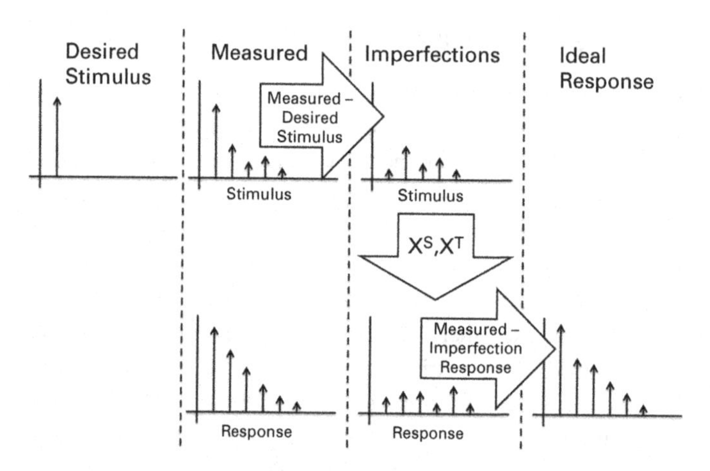

Figure 4.17 X-parameter correction for imperfect stimulus.

less system-to-system variation of measurements, since the system-dependent stimulus imperfections have been corrected out to identify the intrinsic behavior of the DUT.

4.4.4 Simulation-based extraction

X-parameters can be extracted in simulation by using any of the measurement techniques discussed in the preceding sections. Practical measurement considerations can often be ignored in high-dynamic-range simulators, so the simple extraction techniques described generally work very well on their own. In some simulators, however, there are more efficient ways to extract X-parameters.

When simulating the DUT under the large-signal stimulus, harmonic-balance simulators have access to the harmonic Jacobian in addition to the waveforms at the solution. Since the small-signal terms of the X-parameters are the partial derivatives of scattered waves at various ports and frequencies with respect to incident waves at various ports and frequencies, there is a direct relationship between these terms and the harmonic Jacobian which is described in [2]. By applying the correct boundary conditions, the small-signal terms can be extracted directly from the solution of the large-signal-stimulus simulation without the need to perform additional nonlinear circuit simulations with small-signal stimulus present. This can result in significant performance improvements.

4.5 X-parameter files

In order to ensure seamless interoperability between measurement hardware, simulation-based extraction tools, and circuit simulators capable of using X-parameters, a standard file format is required. The open .xnp file format serves this purpose, providing the flexibility needed to contain complex, multi-dimensional X-parameter as well as the structure needed to enable automatic interpretation of the data by circuit simulators or other processing tools.

Table 4.1 Independent variable naming conventions used in the .xnp file format

Independent-variable name	Meaning
fund_k	*k*th fundamental frequency
VDC_p	DC voltage applied to port *p* as part of LSOPS
IDC_p	DC current applied to port *p* as part of LSOPS
AN_p_f	magnitude of a large-*reference-signal* incident wave applied to port *p* at frequency *f* as part of LSOPS, where *f* is represented by its harmonic or mixing indices (AN_p_n for one fundamental frequency, AN_p_n1_n2_. . . for multiple fundamental frequencies)
AM_p_f AP_p_f	magnitude (AM) and phase in degrees (AP) of a large-signal incident wave (which is not a *reference signal*) applied to port *p* at frequency *f* as part of LSOPS
GM_p_f GP_p_f	magnitude (GM) and phase in degrees (GP) of the reflection coefficient of the load at port *p* at frequency *f* as part of LSOPS
GX_p_f GY_p_f	real (GX) and imaginary (GY) parts of the reflection coefficient of the load presented at port *p* at frequency *f* as part of LSOPS
ZM_p_f ZP_p_f	magnitude (ZM) and phase in degrees (ZP) of the load impedance presented at port *p* at frequency *f* as part of LSOPS
ZX_p_f ZY_p_f	real (ZX) and imaginary (ZY) parts of the load impedance presented at port *p* at frequency *f* as part of LSOPS

4.5.1 Structure

The .xnp file format is based on and strictly adheres to the generic MDIF file format, documented in [3], which is capable of storing multi-dimensional data with arbitrary numbers of independent and dependent variables. Three types of blocks are used in .xnp files: the XParamAttributes block, the XParamPortData block, and the XParamData block. An .xnp file contains one XParamAttributes block that includes the file format version of the file, the number of ports of the device described by the file, and the number of fundamental frequencies included in the LSOP. One XParamPortData block is also included in the file, and this contains the reference impedance and name of each port. These two blocks, which are informational only, are typically included at the beginning of the file. Comments describing the measurement or simulation setup may also be included. Finally, the X-parameters are contained in one or more XParamData blocks in the file.

4.5.2 Naming conventions

Because of the flexibility of the X-parameter framework, different files may contain different types of data. For example, a measurement on an amplifier may include X-parameters at multiple DC-bias voltages with current output, while a measurement on a mixer may have no bias at all but multiple fundamental frequencies. Naming conventions for independent and dependent variables are used to identify what information is present in the file. These conventions are described in Table 4.1 and Table 4.2.

Table 4.2 Dependent variable naming conventions used in the .xnp file format

Dependent-variable Name	Meaning
FB_p_f	$X^{(FB)}$ – scattered wave at port p, frequency f due to LSOP
FI_p	$X^{(FI)}$ – DC current at port p due to LSOP
FV_p	$X^{(FV)}$ – DC voltage at port p due to LSOP
S_pOut_fOut_pIn_fIn T_pOut_fOut_pIn_fIn	$X^{(S)}$ and $X^{(T)}$ terms relating input wave at port *pIn*, frequency *fIn* to scattered wave at port *pOut*, frequency *fOut*
XY_pOut_pIn_fIn	$X^{(Y)}$ term relating input wave at port *pIn*, frequency *fIn* to DC current at port *pOut*
XZ_pOut_pIn_fIn	$X^{(Z)}$ term relating input wave at port *pIn*, frequency *fIn* to DC voltage at port *pOut*

```
! fund_1 = 1 GHz NumPts = 1
! VDC_3 = 10   NumPts=1
! GM_2_1 = [100e-03->800e-03]   NumPts=8
! GP_2_1 = [100->170]   NumPts=7
! AN_1_1 = [10e-06(-20.000000dBm)->100e-06(-10.000000dBm)]   NumPts=6
```

Figure 4.18 Comment lines.

4.5.3 Example file

The current version of the X-parameter file format is version 2.0, as of this publication. The older version 1.0 may still be found in the industry, from older measurements.

The major difference between version 2.0 and its predecessor version 1.0 is the type of waves that are used for the X-parameter description: version 1.0 used voltage waves, whereas version 2.0 uses generalized power waves.

In addition, only the X-parameter data blocks (identified as XParamPortData in version 2.0) existed in version 1.0. The XParamAttributes block and the XParamPort-Data block did not exist in version 1.0.

Explanatory comment lines were also not very common in the earlier version 1.0, although they were supported. Beginning with version 2.0, comment lines are commonly used to clarify the measurement conditions under which the X-parameters have been extracted.

An example of a version 2.0 X-parameter file of an amplifier with one DC power supply is presented here. The content of a typical X-parameter file is described in the order in which the information is commonly listed in an actual file.

Comment lines, identified by an exclamation mark, may be used for conveying information in a human readable format, and they are ignored by the simulator. An example is shown in Figure 4.18.

```
BEGIN XParamAttributes
% Index(int)    Version(real)    NumPorts(int)    NumFundFreqs(int)
 0              2.0              3                1
END
```

Figure 4.19 XParamAttributes block.

```
BEGIN XParamPortData
% PortNumber (int)  RefZ0(complex)          PortName(string)
 1                  50            0          "RFin"
 2                  50            0          "RFout"
 3                  50            0          "DCbias"
END
```

Figure 4.20 XParamPortData block.

Comment lines may be included anywhere in the file (at the beginning, in the body, or at the end of the file).

Based on the data in Figure 4.18, a user can determine the following:

- the model has one fundamental frequency (it was extracted with a single large tone present in the stimulus) at 1 GHz;
- a 10 V bias was applied at port 3;
- a load-dependent model was extracted versus the load presented to port 2 of the DUT;
- the load at port 2 was specified as a reflection coefficient;
- the magnitude of the reflection coefficient was swept from 0.1 to 0.8 with eight points;
- the phase of the reflection coefficient was swept from 100° to 170° with seven points;
- a tone was applied at port 1 at the fundamental frequency and it was used as the phase reference;
- the power of the tone applied at port 1 was swept from −20 dBm to −10 dBm with six points.

The XParamAttributes block specifies the file version, the number of ports of the DUT, and the number of fundamental frequencies used in the X-parameter model. An example is shown in Figure 4.19.

The XParamPortData block specifies the reference impedance and the port name for each of the ports of the DUT. Complex reference impedances are supported, and they are specified in units of measure of ohms, in real–imaginary format. An example is shown in Figure 4.20.

```
VAR fund_1(real) = 1000000000
VAR VDC_3(real) = 10
VAR GM_2_1(real) = 0.2
VAR GP_2_1(real) = 30
BEGIN XParamDate
% AN_1_1(real) FI_3(real) FB_1_1(complex) FB_1_2(complex) FB_2_1(complex)
% FB_2_2(complex) S_1_1_1_1(complex) T_1_1_1_1(complex) S_1_2_1_1(complex) T_1_2_1_1(complex)
% S_2_1_1_1(complex) T_2_1_1_1(complex) S_2_2_1_1(complex) T_2_2_1_1(complex) XY_3_1_1(complex)
% S_1_1_1_2(complex) T_1_1_1_2(complex) S_1_2_1_2(complex) T_1_2_1_2(complex) S_2_1_1_2(complex)
% T_2_1_1_2(complex) S_2_2_1_2(complex) T_2_2_1_2(complex) XY_3_1_2(complex) S_1_1_2_1(complex)
% T_1_1_2_1(complex) S_1_2_2_1(complex) T_1_2_2_1(complex) S_2_1_2_1(complex) T_2_1_2_1(complex)
% S_2_2_2_1(complex) T_2_2_2_1(complex) XY_3_2_1(complex) S_1_1_2_2(complex) T_1_1_2_2(complex)
% S_1_2_2_2(complex) T_1_2_2_2(complex) S_2_1_2_2(complex) T_2_1_2_2(complex) S_2_2_2_2(complex)
% T_2_2_2_2(complex) XY_3_2_2(complex)
...
Numerical values
...
END
```

Figure 4.21 XParamData block.

The X-parameter data blocks follow next. The values of all independent variables are listed for each block, as shown in the example in Figure 4.21.

4.6 Summary

Given the proper tools, X-parameters can be extracted from models in simulation or from measurements taken on the device itself. A measurement system must be able to provide the appropriate large-signal stimulus and small-signal stimulus, and must be able to measure cross-frequency phase (or, equivalently, time-domain waveforms) in order to support X-parameter measurements. At the time of publication, several such systems are commercially available. X-parameter measurement can be highly automated and requires little user expertise beyond that required for S-parameter measurement. This enables the X-parameters to provide for nonlinear devices a powerful behavioral modeling paradigm with ease of use approaching that of the standard paradigm for linear devices, S-parameters.

References

[1] D. B. Gunyan, D. E. Root, L. C. Betts, and J. M. Horn, "Large signal scattering functions from orthogonal phase measurements," U.S. Patent 7 671 605, Mar. 2, 2010.

[2] D. E. Root, D. D. Sharrit, and J. Wood, "Behavioral model generation," U.S. Patent 8 170 854, May 1, 2012.

[3] Agilent Technologies, Inc., "Advanced Design System" documentation (ADS2009U1 or later) – Working with data files – X-parameter GMDIF format. Available at http://edocs.soco.agilent.com/display/ads2009U1/Working+with+Data+Files#WorkingwithDataFiles-XparameterGMDIFFormat.

Additional reading

"Applying error correction to network analyzer measurements," Agilent AN 1287–3, Agilent Technologies, Inc., Santa Clara, CA, 2002.

T. Van den Broeck and J. Verspecht, "Calibrated vectorial nonlinear-network analyzers," in *1994 MTT-S IMS Dig.*, May 1994, vol. **2**, pp. 1069–1072.

J. Verspecht, "Method and a test setup for measuring large-signal S-parameters that include the coefficients relating to the conjugate of the incident waves," U.S. Patent 7 038 468, May 2, 2006.

5 Multi-tone and multi-port cases

5.1 Introduction

In this chapter, the general X-parameter formalism is applied in two different cases of great practical interest beyond the simplest case of nearly matched components under large drive considered in Chapter 3. Specifically, X-parameter expressions are developed by spectral linearization around an LSOP defined by the DUT in response to two large incident signals, as opposed to the case of just one large signal considered in Chapter 3.

The first case is an example of two signals whose frequencies, f_1 and f_2, are the same, but are incident with distinct phases at different ports. This is an important subcase of a commensurate frequency relationship, namely $f_1/f_2 = n/m$ for n and m integers. The second case considers incommensurate frequencies, where the ratio of the two signal frequencies is irrational. Whether or not the signals are commensurate affects the dimension of the LSOP as well as the form of the time-invariant X-parameter equations, as will be evidenced below.

The first extension, to arbitrary load-dependent X-parameters, describes the case where the DUT behaves fully nonlinearly with respect to both $A_{1,1}$ and $A_{2,1}$, the phasors corresponding to waves at the same fundamental frequency, incident at two distinct ports, port 1 and port 2, respectively. This is usually the condition of operation for bare transistors and highly mismatched high-power amplifiers under strong input drive. The spectrum generated by a DUT subject to such excitations still lies on a harmonic grid, but now the LSOP, and therefore the X-parameter nonlinear functions, are more complicated than the case considered in Chapter 3. This extension includes most aspects of continuous-wave (CW) load pull, historically used to perform scalar characterization of the behavior of a nonlinear DUT as a function of the complex load impedance. It also includes waveform measurements at DC, fundamental and harmonic frequencies, sometimes referred to as time-domain load pull. These "load-dependent X-parameters" can be measured using an NVNA with passive tuners, or active injection, or some combination of tuners and active injection (hybrid load pull). X-parameters provide a powerful measurement-based nonlinear design flow. They can be used immediately to design nonlinear multi-stage amplifiers, including high-efficiency power amplifiers such as Doherty and other amplifier classes that may even require harmonic tuning.

The second extension compared to Chapter 3 deals with two large signals at incommensurate fundamental frequencies incident at the same port or at distinct ports of the

DUT. Here too the LSOP must be defined in terms of two tones. The frequencies at which the X-parameter functions must be defined do not lie on a harmonic grid in this case, but rather on a more complicated, discrete but dense, frequency spectrum corresponding to all intermodulation products of the fundamental tones. An extension of the notation labeling the spectral components and the corresponding X-parameter functions is introduced. The example chosen to illustrate this capability is that of three terminal mixers, where the local oscillator (LO) and a radio-frequency (RF) signal are treated as incommensurate large input signals at distinct ports and frequencies, producing intermodulation terms at the IF and the other ports. Powerful measurement-based nonlinear three-port models of mixers, including intermodulation products with both magnitude and phase, as nonlinear functions of bias, and power of each distinct tone, can be handled with this approach. Mismatch, for example at the IF port of a mixer, can be accounted for by spectral linearization around the multi-tone LSOP.

The two extensions of X-parameters presented in this chapter, when used together, enable end-to-end measurement-based design approaches to nonlinear RF circuits and systems.

5.2 Commensurate signals – large $A_{1,1}$ and large $A_{2,1}$: load-dependent X-parameters

There are several cases of great practical importance where it becomes necessary to relax the simplifying approximation made in Chapter 3 that the scattered waves, $B_{p,k}$, depend nonlinearly only on the applied bias conditions and $A_{1,1}$, the single large RF component incident at port 1 at the fundamental frequency. An example is a bare transistor presented with a very large output mismatch at the fundamental frequency. In this case, the large $B_{2,1}$ wave generated by the device in response to a large incident $A_{1,1}$ will reflect from the mismatch and create a large incident wave at the output port, also at the fundamental frequency, given by (5.1):

$$A_{2,1} = \Gamma_{2,1} B_{2,1}. \tag{5.1}$$

In (5.1), $\Gamma_{2,1}$ is the complex reflection coefficient at port 2 at the fundamental frequency. The nearly matched case, treated in Chapter 3, is sufficient for $|\Gamma_{2,1}| \ll 1$, since, for small enough reflection coefficient, the magnitude of $A_{2,1}$, from (5.1), will be small enough such that the spectral linearization of the DUT's response with respect to $A_{2,1}$ around $A_{2,1} = 0$ will be valid. Thus, under well-matched conditions, $A_{2,1}$ can be small enough, even for devices with large power gain, that its influence on the LSOP of the DUT can be considered negligible, and its contribution to the scattered waves, $B_{p,k}$, can be modeled as approximately linear in $A_{2,1}$ and $A^*_{2,1}$, with coefficient functions $X^{(S)}_{p,k;2,1}$ and $X^{(T)}_{p,k;2,1}$, respectively.

On the other hand, large mismatch means that the condition $|\Gamma_{2,1}| \ll 1$ is no longer satisfied and values of $|\Gamma_{2,1}|$ even greater than unity can occur with active loads. For large $A_{1,1}$ (strong input drive), a large output mismatch means $A_{2,1}$ can be sufficiently large, from (5.1), that it changes the DUT LSOP, and hence the spectral linearization used previously is not valid.

It is important to emphasize, however, that large mismatch (large $|\Gamma_{2,1}|$) by itself does not necessarily mean that the spectral linearization with respect to $A_{2,1}$ around $A_{2,1} = 0$ of Chapter 3 is not valid. For example, if the DUT is biased in an off state (e.g. a transistor biased at pinchoff), $B_{2,1}$ can be very small, so, even for a large $\Gamma_{2,1}$, $A_{2,1}$ can still be small enough from (5.1) that linearization around $A_{2,1} = 0$ can still be valid. If the DUT's scattered waves depend very weakly on $A_{2,1}$, then even a large $|A_{2,1}|$ may not significantly affect the DUT LSOP; the formalism of Chapter 3 can still apply. It is necessary to go beyond the scope of Chapter 3 only when *both* the mismatch is large *and* the sensitivity of the DUT to mismatch is large simultaneously [1].

It is impossible to know a priori how large an $|A_{2,1}|$ is required to invalidate the considerations of Chapter 3 for a given device. Of course, this is quite analogous to the question of how large a source power level can be used to make an accurate set of conventional S-parameter measurements of an active device before the assumption that the device is operating linearly breaks down. Certainly, for unmatched power transistors and high efficiency power amplifiers at moderate signal levels and above, the nonlinear dependence on $A_{2,1}$ of the DUT response must be taken into account.

One approach for large $|A_{2,1}|$ is to take terms of higher order than unity in $A_{2,1}$ and $A_{2,1}^*$ to approximate their nonlinear contributions to $B_{p,k}$ [2]. Instead, in what follows, the full nonlinear dependence on $A_{2,1}$ will be treated in a manner consistent with that of $A_{1,1}$. That is, $A_{2,1}$ becomes an argument (independent variable) of the X-parameter functions. No polynomial basis is required nor is an expansion coefficient extraction method needed. The accuracy of the resulting model depends only on measurement accuracy and sampling density in the space of the independent variables.

5.2.1 Time invariance, phase normalization, and commensurate two-tone LSOP

The condition created by the large $A_{1,1}$ and large $A_{2,1}$ is shown in Figure 5.1. Two large signals, $A_{1,1}$ and $A_{2,1}$, and the DC bias are considered as part of the LSOPS, as shown in (5.2):

$$LSOPS = \left(DCS_p, A_{1,1}, A_{2,1}\right). \tag{5.2}$$

The response to additional incident waves at harmonics of the fundamental frequency, considered small, is treated by spectral linearization. The treatment of nonlinear spectral maps of a time-invariant DUT on a frequency grid, presented in Chapter 2, is fully applicable here. Since both the $A_{1,1}$ and $A_{2,1}$ signals are at the same frequency, they always have a fixed relative phase, $\phi_{2,1}$, independent of time shift, given by (5.3), with $P = e^{j\phi(A_{1,1})}$:

$$e^{j\phi_{2,1}} \equiv e^{j\left[\phi(A_{2,1}) - \phi(A_{1,1})\right]} = e^{j\phi(A_{2,1})}P^{-1}. \tag{5.3}$$

The reference LSOP stimulus (*refLSOPS*), defined in the same way as in Chapter 2, is therefore given by

$$refLSOPS = \left(DCS_p,\ |A_{1,1}|,\ |A_{2,1}|,\ \phi_{2,1}\right) = \left(DCS_p,\ |A_{1,1}|,\ A_{2,1}P^{-1}\right). \tag{5.4}$$

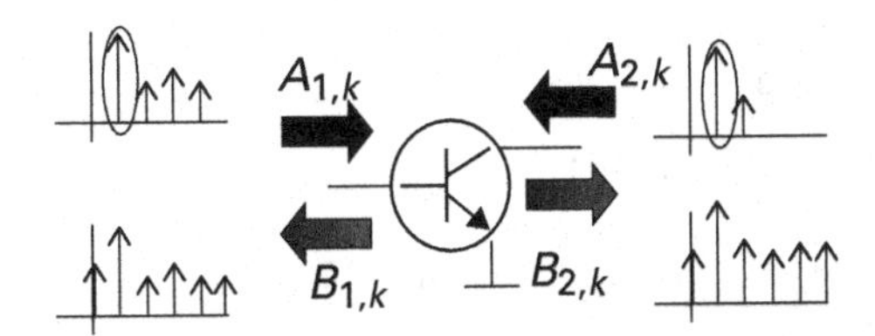

Figure 5.1 Two large tones (circled) at the same fundamental frequency and small tones at harmonic frequencies incident at ports 1 and 2 of a nonlinear DUT. Scattered waves have frequency components on the same frequency grid.

5.2.2 Spectral linearization

The general multi-tone nonlinear map from the harmonic grid into itself is now spectrally linearized around the *refLSOPS* defined by (5.4). Here both $A_{1,1}$ and $A_{2,1}$ are large, defining the LSOP, and all the other signals at each port and harmonic, $\Delta A_{p,k}$, are considered small. In addition to the large $A_{2,1}$ tone defining the operating point, the presence of an additional small $\Delta A_{2,1}$ is considered. Its significance will be apparent later. For simplicity, in this discussion $\Delta A_{1,1}$ is always taken to be zero. That is, no perturbation is considered at the input port at the fundamental frequency.

For clarification, the large-signal waves at the LSOP are $A_{1,1}^{(LSOP)}$ and $A_{2,1}^{(LSOP)}$. The total large-signal waves, including the small-signal perturbation around the LSOP, are then as shown in (5.5), considering no small-signal perturbation is applied on $A_{1,1}$ in this exercise:

$$\begin{aligned} A_{1,1} &= A_{1,1}^{(LSOP)} + \Delta A_{1,1} = A_{1,1}^{(LSOP)}, \\ A_{2,1} &= A_{2,1}^{(LSOP)} + \Delta A_{2,1}. \end{aligned} \tag{5.5}$$

The same principles as in Chapters 2 and 3 are applied. The spectral mapping must be time-invariant, which means that the nonlinear maps, $F_{p,k}$, can be evaluated at a time-shifted argument, and then time shifted back by appropriate phase multiplication of the result. The time shifts of the mappings' arguments and resulting values are applied first, and then the nonlinear functions are spectrally linearized around the *refLSOPS* given by (5.4). The result for the scattered waves is presented in (5.6), with some of the intermediate steps in the derivation included, and the results for the DC-bias responses are shown in (5.7) or (5.8):

$$\begin{aligned} B_{p,k} &= F_{p,k}\left(DCS_q, A_{1,1}, A_{2,1}^{(LSOP)} + \Delta A_{2,1}, \Delta A_{1,2}, \Delta A_{2,2}, \ldots, \Delta A_{P,K}\right) \\ &= F_{p,k}\left(DCS_q, A_{1,1}P^{-1}, \left(A_{2,1}^{(LSOP)} + \Delta A_{2,1}\right)P^{-1}, \Delta A_{1,2}P^{-2}, \Delta A_{2,2}P^{-2}, \ldots, \Delta A_{P,K}P^{-K}\right)P^k \\ &= F_{p,k}(refLSOPS)P^k + \left.\frac{\partial F_{p,k}}{\partial\left(A_{2,1}P^{-1}\right)}\right|_{refLSOPS} \Delta A_{2,1}P^{k-1} + \left.\frac{\partial F_{p,k}}{\partial\left(A_{2,1}^{*}P\right)}\right|_{refLSOPS} \Delta A_{2,1}^{*}P^{k+1} \\ &\quad + \sum_{\substack{p'=1,\ldots,P \\ k'=2\ldots,K}} \left[\left.\frac{\partial F_{p,k}}{\partial\left(A_{p',k'}P^{-k'}\right)}\right|_{refLSOPS} \Delta A_{p',k'}P^{k-k'} + \left.\frac{\partial F_{p,k}}{\partial\left(A_{p',k'}^{*}P^{k'}\right)}\right|_{refLSOPS} \Delta A_{p',k'}^{*}P^{k+k'} \right] \\ &= X_{p,k}^{(F)}(refLSOPS)P^k \\ &\quad + X_{p,k;2,1}^{(S)}(refLSOPS)\Delta A_{2,1}P^{k-1} + X_{p,k;2,1}^{(T)}(refLSOPS)\Delta A_{2,1}^{*}P^{k+1} \\ &\quad + \sum_{\substack{p'=1,\ldots,P \\ k'=2\ldots,K}} \left[X_{p,k;p',k'}^{(S)}(refLSOPS)\Delta A_{p',k'}P^{k-k'} + X_{p,k;p',k'}^{(T)}(refLSOPS)\Delta A_{p',k'}^{*}P^{k+k'}\right]; \end{aligned} \tag{5.6}$$

$$
\begin{aligned}
I_p \cong\ & X_p^{(I)}(refLSOPS) \\
& + \mathrm{Re}\left\{X_{p;2,1}^{(Y)}(refLSOPS)\Delta A_{2,1}P^{-1}\right\} \\
& + \sum_{\substack{q=1,\ldots,P \\ l=2,\ldots,K}} \mathrm{Re}\left\{X_{p;q,l}^{(Y)}(refLSOPS)\Delta A_{q,l}P^{-l}\right\};
\end{aligned} \tag{5.7}
$$

$$
\begin{aligned}
V_p \cong\ & X_p^{(V)}(refLSOPS) \\
& + \mathrm{Re}\left\{X_{p;2,1}^{(Z)}(refLSOPS)\Delta A_{2,1}P^{-1}\right\} \\
& + \sum_{\substack{q=1,\ldots,P \\ l=2,\ldots,K}} \mathrm{Re}\left\{X_{p;q,l}^{(Z)}(refLSOPS)\Delta A_{q,l}P^{-l}\right\}.
\end{aligned} \tag{5.8}
$$

The terms proportional to $\Delta A_{2,1}$ and $\Delta A_{2,1}^*$ are listed separately in (5.6)–(5.8), for emphasis, because they are related to one of the large-tone variables, $A_{2,1}$, defining the LSOP.

For clarity, we rewrite (5.6) in the following form:

$$
\begin{aligned}
& B_{p,k}(A_{1,1}^{LSOP}, A_{2,1}^{LSOP} + \Delta A_{2,1}, \Delta A_{1,2}, \ldots, \Delta A_{p,K}) = X_{p,k}^{(F)}(refLSOPS)P^k \\
& + X_{p,k;2,1}^{(S)}(refLSOPS)\left(A_{2,1} - A_{2,1}^{LSOP}\right)P^{k-1} + X_{p,k;2,1}^{(T)}(refLSOPS)\left(A_{2,1} - A_{2,1}^{LSOP}\right)^* P^{k+1} \\
& + \sum_{\substack{p'=1,\ldots,P \\ k'=2\ldots,K}} \left[X_{p,k;p',k'}^{(S)}(refLSOPS)\Delta A_{p',k'}P^{k-k'} + X_{p,k;p',k'}^{(T)}(refLSOPS)\Delta A_{p',k'}^* P^{k+k'}\right].
\end{aligned} \tag{5.9}
$$

5.3 Establishing the LSOP using a load tuner: passive load pull

The LSOP in (5.9) need not be established by two independent signal sources. In the common method of passive load pull, for example, a single signal source is used to stimulate the device at the input port and a mechanical tuner is used at the output to reflect the DUT-produced $B_{2,1}$ back into the output port as $A_{2,1}$, using (5.1). The physical experiment controls the complex reflection coefficient, $\Gamma_{2,1}$, so each of the measurements is tabulated as a function of $\Gamma_{2,1}$. The tuner is considered "ideal" for now, meaning that only at the fundamental frequency is there a non-zero reflection coefficient, $\Gamma_{2,1}$. The constraint of constant $\Gamma_{2,1}$, for all power values incident at port 1, makes the reflection coefficient the independent variable, whereas $A_{2,1}$ becomes a dependent variable.

To be specific, consider the case $p = 2$, $k = 1$ in (5.9). Non-zero values of $\Delta A_{p',k'}$, for $k' > 1$, such as the ones that may be generated by reflections of DUT-generated harmonics, for example, cause changes in $B_{2,1}$ through the terms inside the summations in (5.9). According to (5.1), this means that there must be a corresponding small change, $\Delta A_{2,1}$, in $A_{2,1}$ around its value at the LSOP. The terms $\Delta A_{2,1}$, and therefore also $\Delta A_{2,1}^*$ by virtue of the non-analytic nature of the spectral mappings, contribute additional changes to all the $B_{p,k}$ values through the X-parameter functions $X^{(S)}_{p,k;2,1}$ and $X^{(T)}_{p,k;2,1}$, respectively.

For passive load-pull-based X-parameter measurement, re-parameterizing (5.9) by $\Gamma_{2,1}$ instead of $A_{2,1}$ results in the equivalent expression (5.10) for the X-parameter model of the DUT:

$$\begin{aligned} B_{p,k} &= \tilde{X}^{(F)}_{p,k}(|A_{1,1}|,\Gamma_{2,1})P^k \\ &+ \tilde{X}^{(S)}_{p,k;2,1}(|A_{1,1}|,\Gamma_{2,1})\left(A_{2,1}-\Gamma_{2,1}\tilde{X}^{(F)}_{2,1}(|A_{1,1}|,\Gamma_{2,1})\right)P^{k-1} \\ &+ \tilde{X}^{(T)}_{p,k;2,1}(|A_{1,1}|,\Gamma_{2,1})\left(A_{2,1}-\Gamma_{2,1}\tilde{X}^{(F)}_{2,1}(|A_{1,1}|,\Gamma_{2,1})\right)^{*}P^{k+1} \\ &+ \sum_{\substack{p'=1,\ldots,P \\ k'=2\ldots,K}} \left[\tilde{X}^{(S)}_{p,k;p',k'}(|A_{1,1}|,\Gamma_{2,1})\Delta A_{p',k'}P^{k-k'} + \tilde{X}^{(T)}_{p,k;p',k'}(|A_{1,1}|,\Gamma_{2,1})\Delta A^{*}_{p',k'}P^{k+k'}\right]. \end{aligned} \tag{5.10}$$

The following constraint (5.11) is appended to the model equations (5.10):

$$A_{2,1} = \Gamma_{2,1}B_{2,1}. \tag{5.11}$$

The functions $\tilde{X}^{(F)}(.,.)$ are related to $X^{(F)}(.,.)$ by (5.12):

$$\begin{aligned} &\tilde{X}^{(F)}_{p,k}(|A_{1,1}|,\Gamma_{2,1}) = X^{(F)}_{p,k}\left(|A_{1,1}|, A^{refLSOPS}_{2,1}\right) \\ &\text{for } A^{refLSOPS}_{2,1} = \Gamma_{2,1}\tilde{X}^{(F)}_{2,1}(|A_{1,1}|,\Gamma_{2,1}). \end{aligned} \tag{5.12}$$

The sensitivity functions $\tilde{X}^{(S)}$ and $\tilde{X}^{(T)}$ are related to $X^{(S)}$ and $X^{(T)}$ by (5.13):

$$\begin{aligned} &\tilde{X}^{(S)}_{p,k;p',k'}(|A_{1,1}|,\Gamma_{2,1}) = X^{(S)}_{p,k;p',k'}\left(|A_{1,1}|, A^{refLSOPS}_{2,1}\right), \\ &\tilde{X}^{(T)}_{p,k;p',k'}(|A_{1,1}|,\Gamma_{2,1}) = X^{(T)}_{p,k;p',k'}\left(|A_{1,1}|, A^{refLSOPS}_{2,1}\right) \\ &\text{for } A^{refLSOPS}_{2,1} = \Gamma_{2,1}\tilde{X}^{(F)}_{2,1}(|A_{1,1}|,\Gamma_{2,1}). \end{aligned} \tag{5.13}$$

The "~" symbols indicate that the functional forms when $\Gamma_{2,1}$ is the independent variable are different from the functional forms when $A_{2,1}$ is the independent variable. This is common, if somewhat precise, mathematical notation for two functions related to each other in that they have the same range but their domains are related by a coordinate transformation.

In (5.10), use was made of the fact that, at the *refLSOPS*, all the $\Delta A_{p',k'} = 0$ and, through (5.11), that $A^{refLSOPS}_{2,1} = \Gamma_{2,1}X^{(F)}_{2,1}$. It is the values of the functions $\tilde{X}^{(F)}_{p,k}$, $\tilde{X}^{(S)}_{p,k;p',k'}$, and $\tilde{X}^{(T)}_{p,k;p',k'}$ that therefore appear in the measured .xnp file when a DUT's X-parameters are extracted using passive load-pull acquisition with $\Gamma_{2,1}$ the control variable (independent variable) of the experiment. It is important to stress, however, that the tuner impedance values do not need to be directly perturbed around the specified, fixed values of $\Gamma_{2,1}$. For example, the tuner can be used just to establish the LSOP, and perturbations can be applied using active source injection of small signals. The sensitivity information can be extracted from the responses to these perturbations around the LSOP. Regardless of how the measurement is done, the $X^{(S)}$ and $X^{(T)}$ parameters represent sensitivities of the response to perturbations in the A-waves around the LSOP established by the reflection coefficient of the load.

In words, (5.12) and (5.13) mean that the numerical values of the DUT's X-parameter functions in the data file measured under fixed $\Gamma_{2,1}$ conditions are the same as if the X-parameter functions were measured without a tuner (e.g. by active source injection) at a fixed $A_{2,1}^{LSOP}$ value corresponding to $\Gamma_{2,1}\tilde{X}_{2,1}^{(F)}(|A_{1,1}|, \Gamma_{2,1})$.

Even though (5.10) and (5.11) describe the DUT parameterized by $\Gamma_{2,1}$, the equations represent the DUT properly in any circuit into which it is placed, whether or not the DUT model is presented with a fixed impedance. For example, consider the DUT model characterized under controlled and swept $\Gamma_{2,1}$ conditions, presented with applied independent signals $A_{1,1}^{app}$ and $A_{2,1}^{app}$ in a particular circuit. The simulator has to evaluate (5.10) consistently with (5.11) for $A_{1,1} = A_{1,1}^{app}$ and $A_{2,1} = A_{2,1}^{app}$. The model looks up and interpolates the values of the X-parameters as functions of $\Gamma_{2,1}$ iterating on the value of $\Gamma_{2,1}$ until (5.10) is satisfied for a particular value of reflection coefficient, $\Gamma_{2,1}^{conv}$, such that also $A_{2,1}^{app} = \Gamma_{2,1}^{conv}\tilde{B}_{2,1}(\Gamma_{2,1}^{conv})$. The values of (5.10) under these conditions are precisely the correct DUT scattered waves corresponding to incident $A_{1,1} = A_{1,1}^{app}$ and $A_{2,1} = A_{2,1}^{app}$ phasors. The equations (5.10) and (5.11) are solved simultaneously with all the other equations of the circuit into which the DUT model is placed. The result is the correct solution of the circuit consistent with the correct input–output relationships of the DUT.

5.4 Additional considerations for commensurate signals

5.4.1 Extraction of X-parameter functions under controlled loads

The constraint (5.1) makes the extraction of each X-parameter function using passive load-pull measurements more complicated than the ideal experiment design outlined in Chapter 3 in terms of independent incident A-waves. These details are beyond the scope of this work.

5.4.2 Harmonic superposition

When the X-parameter model functions are measured using independent active source injection for both large signals $A_{1,1}$ and $A_{2,1}$, then, on evaluating the model for $A_{2,1} = A_{2,1}^{app} = A_{2,1}^{LSOP}$, the X-parameter functions $X_{p,k;2,1}^{(S)}$ and $X_{p,k;2,1}^{(T)}$ do not contribute to the scattered waves in (5.9) since $A_{2,1} - A_{2,1}^{LSOP} = 0$. The linear terms in (5.9) correspond to the signal components at the harmonics of the fundamental. This is the harmonic superposition principle [3]. It too can be relaxed, if necessary, by using more independent spectral frequencies in the nonlinear mapping (less spectral linearization). It will be shown in the following how effective this approximation is, even for DUTs where independently designed harmonic terminations are important to the device operation.

5.4.3 Limitations of passive load pull for load-dependent X-parameters

Passive tuners by themselves can provide reflection coefficients no larger than unity. The active or hybrid load-pull setups are therefore necessary for some applications where the operating conditions of the amplifier change from an off-state to an on-state depending on other signals in the circuit. This is the case for Doherty amplifiers, for example, where the auxiliary amplifier is off in the low-input-power range and turns on only at large input powers [4]. This will be treated in Section 5.6

Passive load pull requires the DUT to be generating significant $B_{2,k}$ from the $A_{1,1}$ drive signal, to be reflected as $A_{2,1}$ using (5.1). If the device (transistor) is off, such as when biased at pinchoff, $B_{2,k}$ is essentially zero. For applications at or near a DUT off-state, active source injection to produce a significant $A_{2,1}$ is therefore required. This is a different paradigm compared to conventional load pull, where the DUT is usually considered operating always in an on-state. However, active injection for X-parameter characterization is completely analogous to linear S-parameter measurements where a source is used to inject an $A_{2,1}$. S-parameters are routinely measured when the device is biased off as well as on. So an active-source-injection system for X-parameter measurements is really just a natural extension of the linear case, needed to characterize fully the nonlinear DUT over its entire range of operation.

5.4.4 Sampling of the three-RF-variable space defining the *refLSOPS*

There are three independent RF variables specifying the device *refLSOPS* defined by (5.4). For fixed powers of each incident RF signal, changing the relative phase angle, $\phi_{2,1}$, changes the state of the DUT. It is generally required to vary the magnitudes of each tone, $|A_{1,1}|$ and $|A_{2,1}|$, independently over their full ranges, and also to sweep $\phi_{2,1}$ over the complete range from 0 to 2π radians. This requires correspondingly more data than for the case of the one-tone *refLSOPS* considered in Chapters 2 and 3. Sufficiently dense sampling of both the magnitudes and the phase angle is required for accuracy over the complete range of DUT operation. The actual sample density required for a fixed accuracy depends on the particular nonlinearity of the DUT's responses to these excitations, which cannot be known a priori.

5.4.5 Hardware setup for load-dependent X-parameters

A typical setup for measuring load-dependent X-parameters is shown in Figure 5.2. The passive tuner at the input is used primarily to achieve a better power match to the device from the RF source. It is not necessary, in general, if sufficient power is available from the source without an impedance transformer. The passive tuner at the output port is used to vary the load at the fundamental frequency over a range of values across the Smith chart.

5.4.6 Calibrating out uncontrolled harmonic impedances

Consider the tuners in the measurement setup of Figure 5.2. Real tuners are not "ideal" in the sense of Section 5.3. Specifically, as the fundamental load is varied from one specified state to the next, the harmonic impedances presented by the tuner may vary in

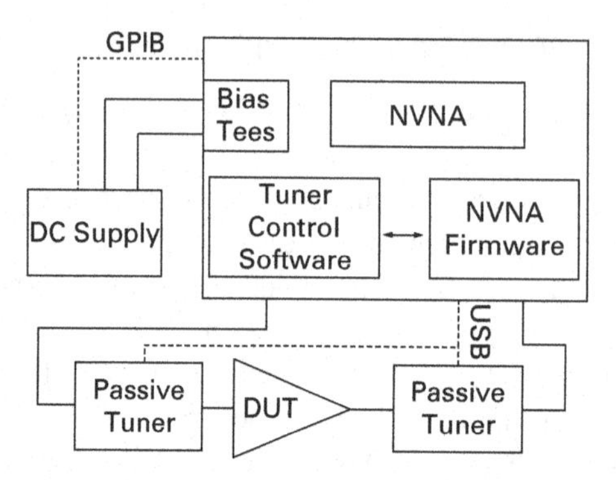

Figure 5.2 NVNA with load tuner for load-dependent X-parameter characterization.

uncontrolled ways. Provided both the DUT-generated harmonics and the harmonic reflection coefficients are not very large, these imperfections can be calibrated out by using the X-parameter characterization and identification algorithms. The technique is quite analogous to that of Chapter 3, whereby X-parameters calibrate out the source harmonic imperfections and multi-frequency mismatch of the instrument. X-parameter extraction can also correct for errors in the actual value of the fundamental load impedance of the tuner. That is, it is possible to correct the measurements to compensate for the fact that the actual measured impedance presented to the device may not be precisely the specified impedance. This enables the corrected data to be tabulated on a precise grid of specified impedances in the complex plane. This is important for the subsequent interpolation of the measured X-parameters by the simulator when using the model in a nonlinear design.

5.5 Arbitrary load-dependent X-parameters of a GaAs FET

A bare GaAs transistor, part number WJ FP2189, with typical output power of 1 W, was characterized by load-dependent X-parameters over a wide range of incident power. The loads were established by a mechanical tuner from Maury Microwave Corp. The harmonic impedances presented by the tuner at the output port were measured and recorded in a file for subsequent validation. Scattered waves were measured, including five harmonics and DC-bias values. Perturbation tones were used up to the third harmonic to obtain harmonic load sensitivities.

Delivered-power contours, in dBm, are plotted in Figure 5.3 obtained from simulations using the resulting X-parameter model and from independently measured load-pull data.

The measurements and X-parameter extraction are automated and straightforward. More complicated is the nonlinear validation. To validate the model, the actual measured harmonic loads presented by the tuner need to be presented to the model of the device in the simulator environment for an accurate comparison.

Characteristics, such as those shown in Figures 5.4–5.6, could only be simulated, using "compact" transistor models, prior to the availability of X-parameters.

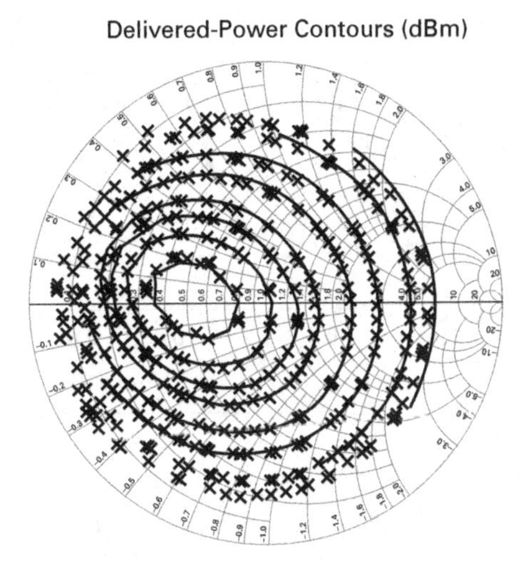

Figure 5.3 Delivered-power contours from load-dependent X-parameter model (line) and independent load-pull measurements (symbols).

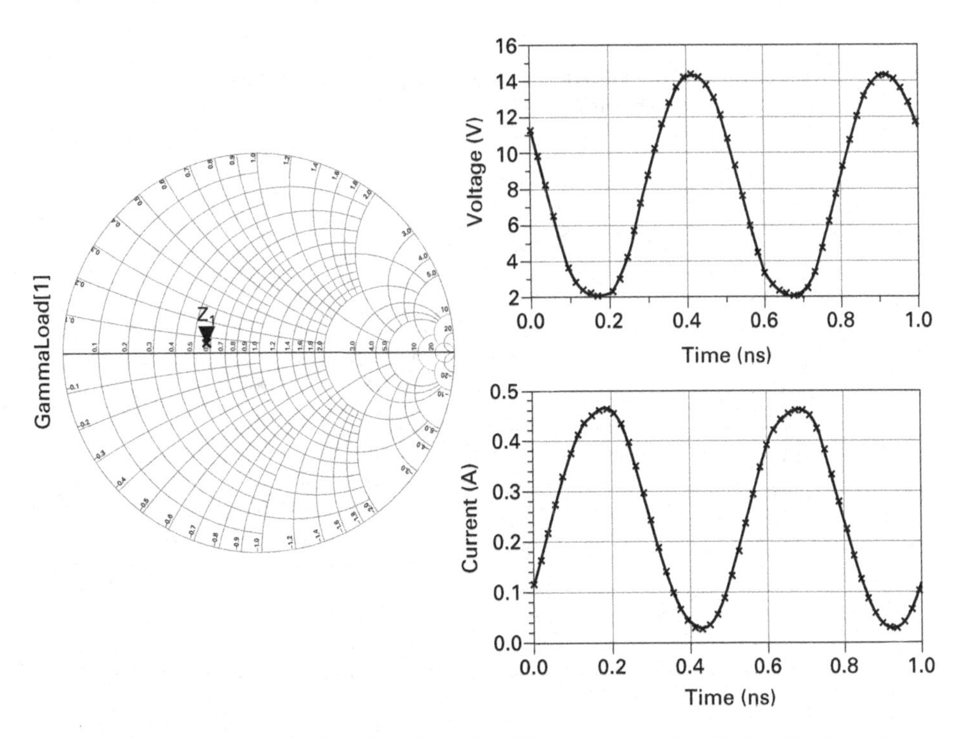

Figure 5.4 Voltage and current from load-dependent X-parameter model (line) and independent load-pull measurements (symbols) at $Z_{2,1} = Z_1$.

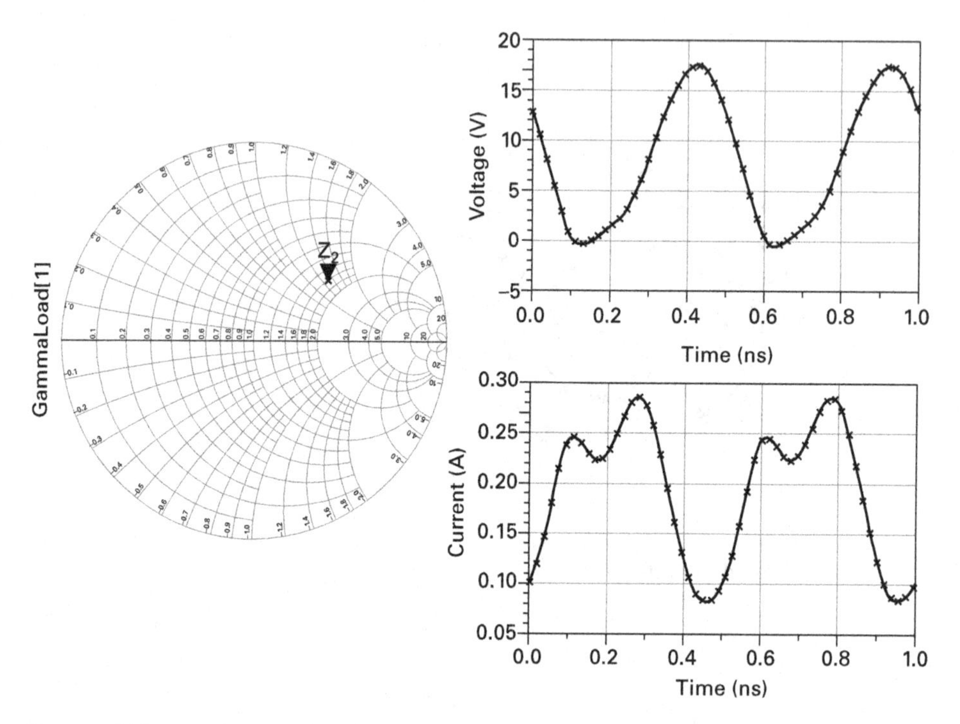

Figure 5.5 Voltage and current from load-dependent X-parameter model (line) and independent load-pull measurements (symbols) at $Z_{2,1} = Z_2$.

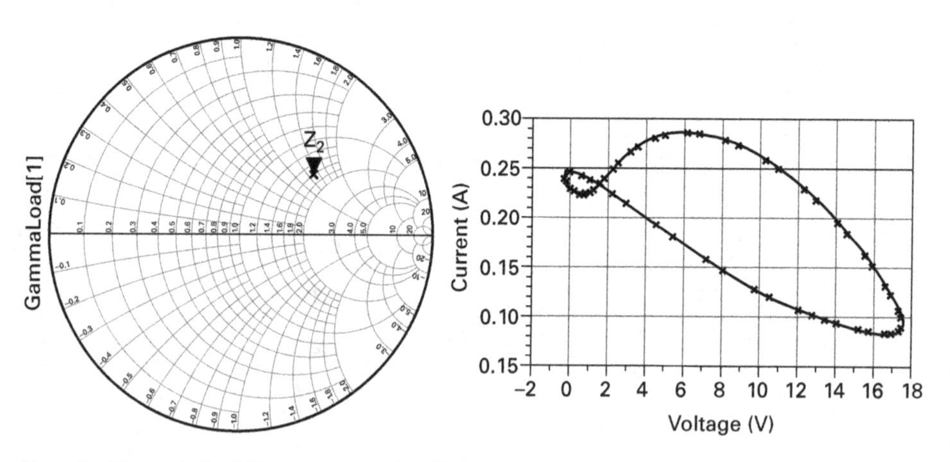

Figure 5.6 Dynamic load-line: measured (symbols) versus simulated (line) at $Z_{2,1} = Z_2$.

The advantage of X-parameters is that the measured characteristics of this FET can be used in a nonlinear circuit design immediately after obtaining the X-parameter model from a measurement process on the hardware DUT.

The device model can be placed in an environment with even moderately large harmonic mismatches, and the resulting DUT behavior is predicted accurately, even without having controlled the harmonic impedances in the X-parameter measurement process. This capability to predict beyond the precise measured harmonic termination conditions is quite analogous to the situation discussed in Chapter 3, where 50 Ω X-parameter measurements could predict well the variation of the DUT behavior with load impedance for a moderate amount of mismatch.

5.5.1 Load-dependent X-parameter model of a GaN HEMT: estimating the effect of independent harmonic impedance tuning

A load-dependent X-parameter model using (5.6) and (5.7) was developed for a 10 W GaN packaged transistor in [1]. The part is a Cree CGH40010 GaN HEMT [5]. A hardware configuration like that of Figure 5.2 was used. A fixed number of nine load states was used in the characterization for X-parameter extraction. The harmonic terminations were uncontrolled during the data acquisition for the X-parameter model; the tuner provides various values of harmonic impedances as it moves from one fundamental complex load state to the next.

Validation measurements were made using a full multi-harmonic load-pull system. Three output tuners were used for the validation measurements to control, independently, the first, second, and third harmonic impedances. The hardware configuration for validation is shown in Figure 5.7. Nine tuner states for each harmonic were used to present the set of independently specified fundamental and harmonic impedances to the device output, for a total of $9^3 = 729$ load states.

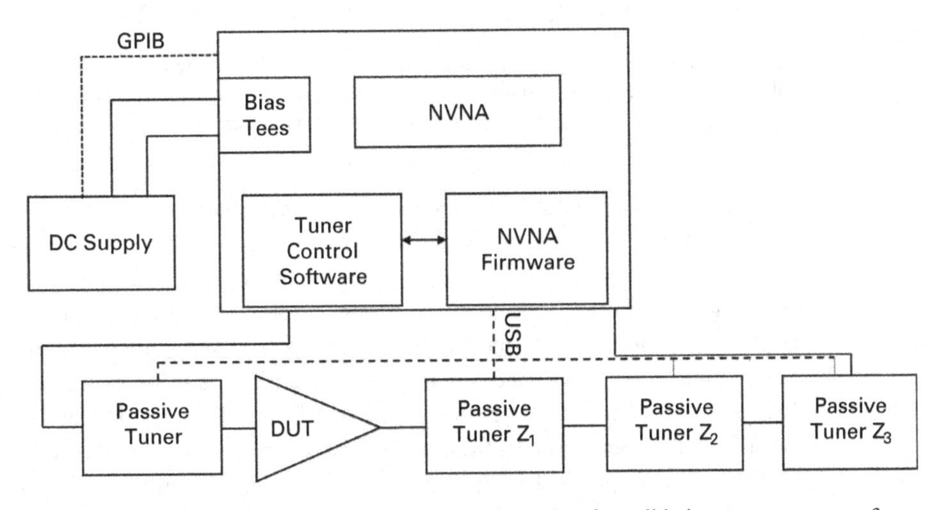

Figure 5.7 Multi-harmonic load-pull hardware configuration for validation measurements of X-parameter model of GaN HEMT.

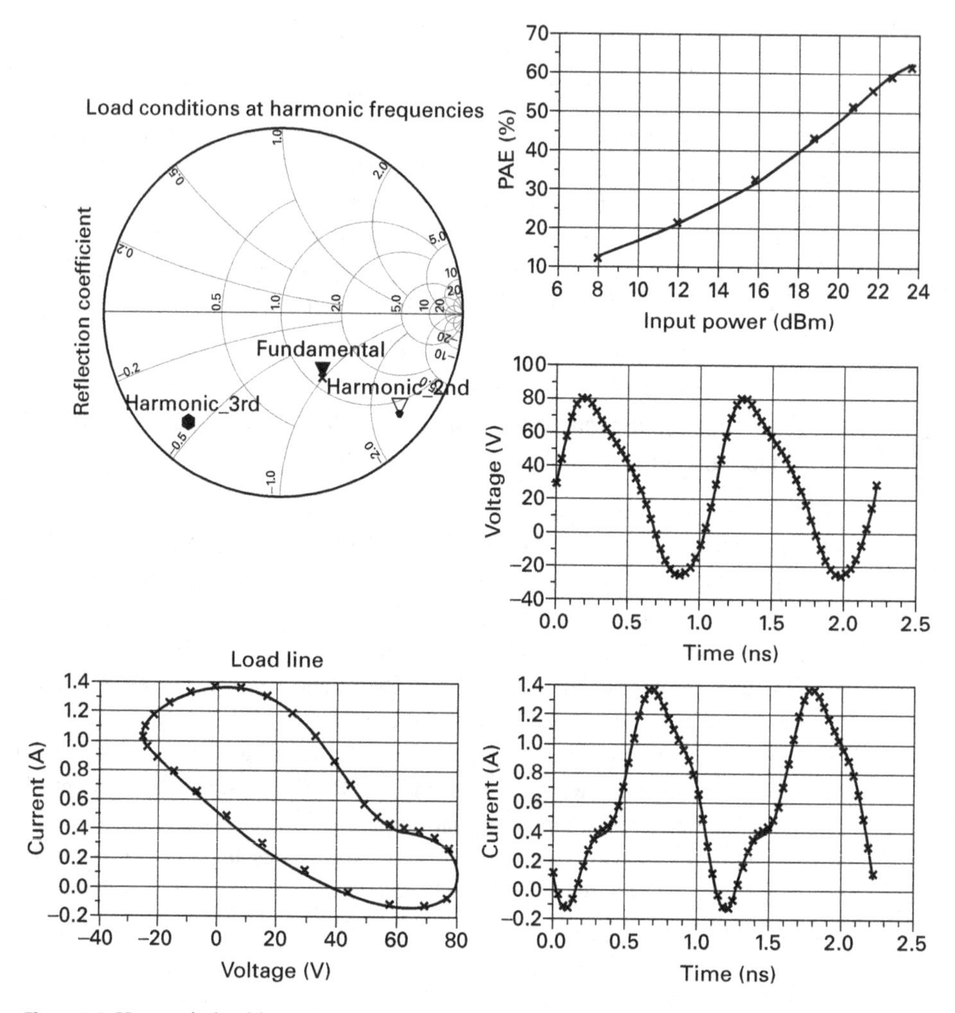

Figure 5.8 Harmonic load impedances set A: validation of X-parameter GaN HEMT model using multi-harmonic load pull. Results from simulations using X-parameter model extracted into controlled fundamental impedance only (line) versus measurements (symbols).

In the validation process, the set of harmonic impedances, controlled in the validation experiments, is taken into the nonlinear simulator and presented to the X-parameter model. The model was constructed by controlling only the fundamental load and extracting the harmonic sensitivity functions at nominally 50 Ω. The model uses the sensitivity functions in (5.6) and (5.7) to predict the effects of tuning the harmonic loads to any value, even outside the Smith chart. Comparison is made of the measured and simulated time-domain voltage and current waveforms, the dynamic load lines, and the power-added efficiency (PAE) from small to large input power for each of the harmonic load impedances specified in the top left of Figure 5.8–Figure 5.11. Figure 5.12 and Figure 5.13 demonstrate the cases when the second and third harmonic impedances are swept around the edge of the Smith chart, meaning that the reflection coefficient has

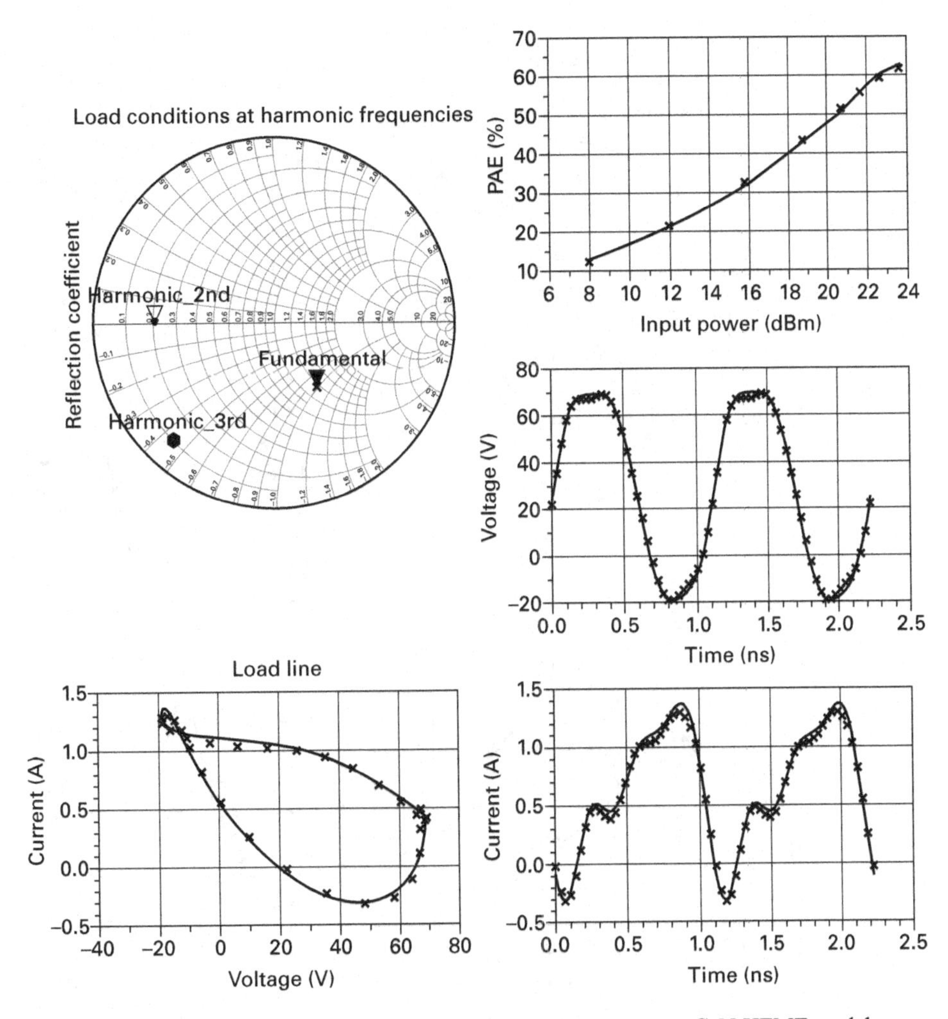

Figure 5.9 Harmonic load impedances set B: validation of X-parameter GaN HEMT model using multi-harmonic load pull. Results from simulations using X-parameter model extracted into controlled fundamental impedance only (line) versus measurements (symbols).

magnitude 1 in these cases. The results demonstrate that the X-parameter model, extracted by controlling only the fundamental impedance value and without controlling harmonic impedance, nevertheless predicts accurately the DUT response to the *independently tuned* (controlled) *fundamental and harmonic impedances* over the entire Smith chart.

The harmonic load-pull characterization required *three* independent load tuners, one each for the fundamental, second, and third harmonics, respectively, at the output. This demonstration is a direct experimental validation of the harmonic superposition principle [3] that led to (5.6) and (5.7). This predictive capability of X-parameters dramatically reduces the hardware complexity, the number of

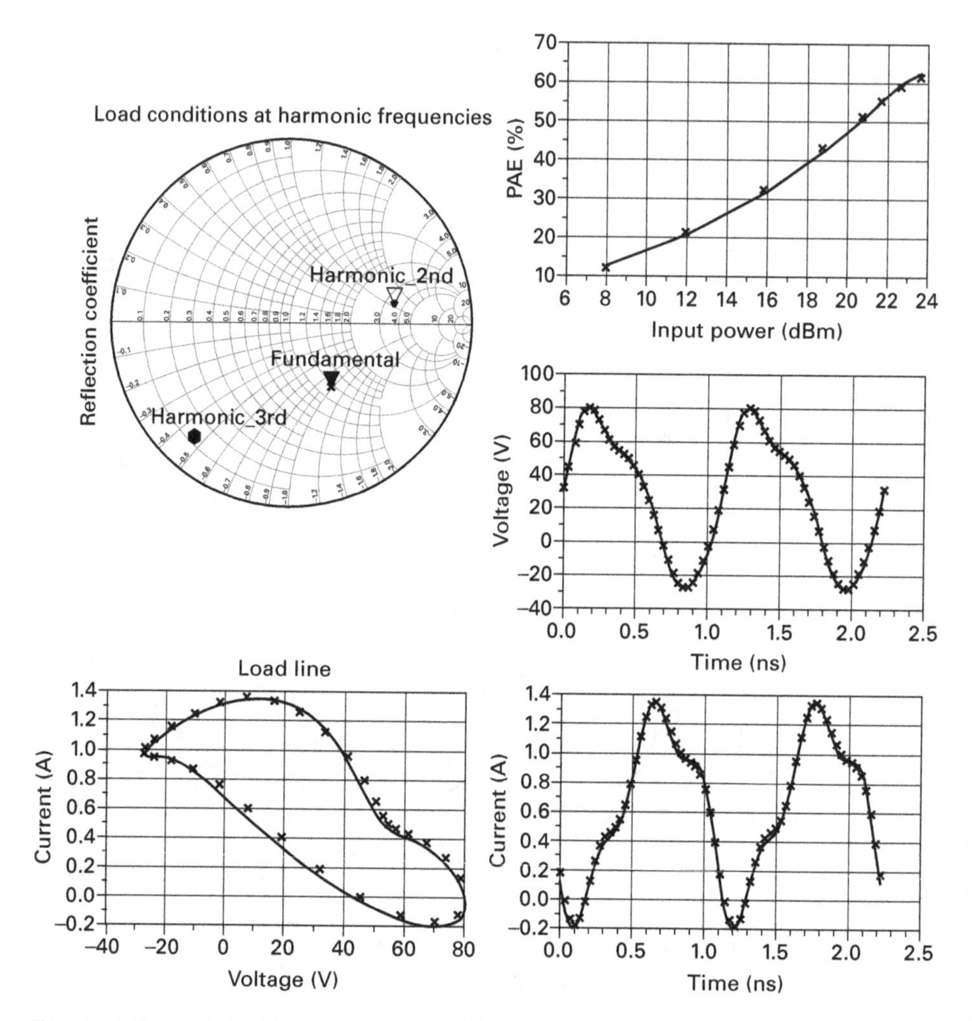

Figure 5.10 Harmonic load impedances set C: validation of X-parameter GaN HEMT model using multi-harmonic load pull. Results from simulations using X-parameter model extracted into controlled fundamental impedance only (line) versus measurements (symbols).

measurements, and the resulting data file size for the design of high-efficiency amplifiers where harmonic terminations may still have significant influence on the DUT behavior. Should the reflection coefficient at a harmonic and the DUT sensitivity to harmonic injection both be very large, the X-parameter framework can treat the harmonic incident waves without the linearity assumption by suitable modification of (5.6) and (5.7).

The ideal number of X-parameter measurements per power and frequency can be computed from (5.14):

$$N_{X\text{-}par} = [(1 \text{ LSOP}) + (\text{two ports}) \cdot (\text{two phases per port})]N_{Z_1} = (1 + 2 \cdot 2)N_{Z_1}. \quad (5.14)$$

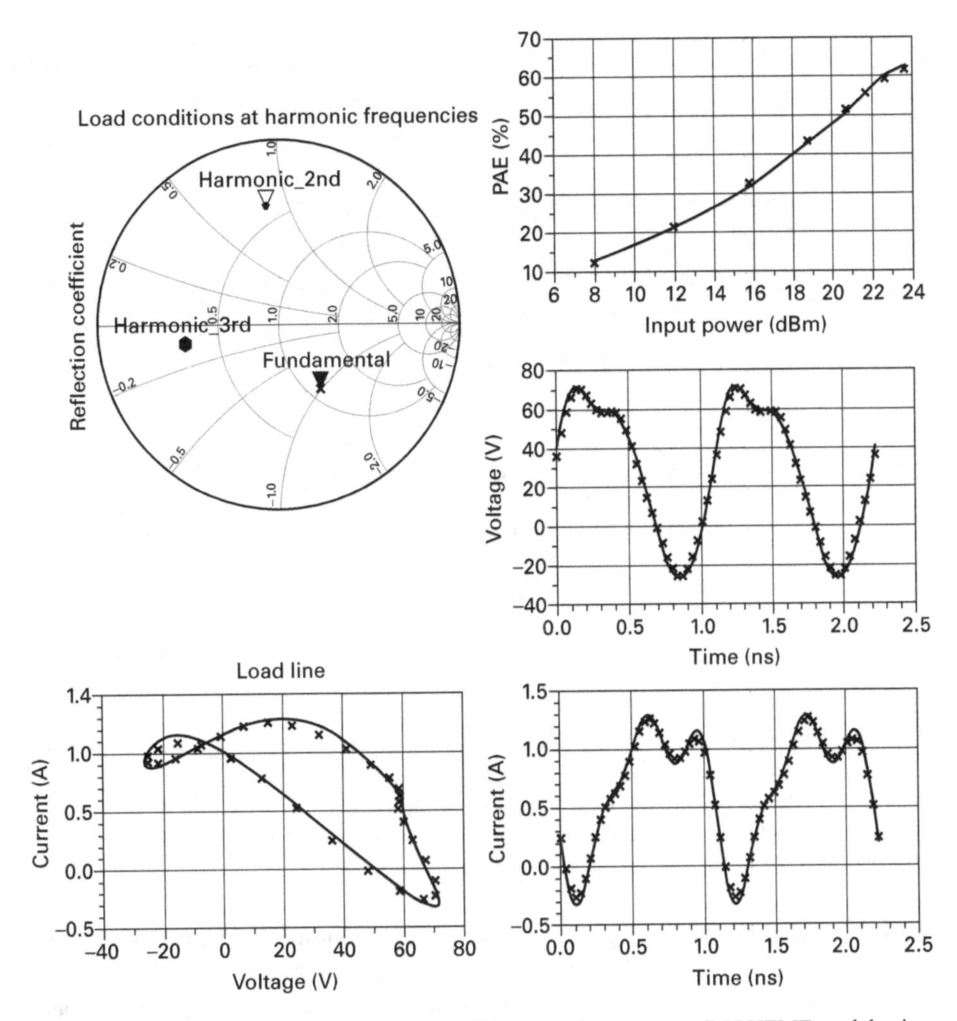

Figure 5.11 Harmonic load impedances set D: validation of X-parameter GaN HEMT model using multi-harmonic load pull. Results from simulations using X-parameter model extracted into controlled fundamental impedance only (line) versus measurements (symbols).

For the harmonic load-pull validation, the number of experiments is given by (5.15):

$$N_{harm\ l\text{-}p} = N_{Z_1} \cdot N_{Z_2} \cdot N_{Z_3}. \tag{5.15}$$

For this example, we have $N_{Z_i} = 9$ for $i = 1, 2, 3$, so only 45 measurements are required for the X-parameter model compared to 729 for the harmonic load-pull measurements. Despite a factor of 16 fewer measurements, the X-parameter model is able to match closely the actual independently tuned harmonic load-pull results. This saves over an order of magnitude of measurement time and file size. It also improves the simulation speed of the resulting model.

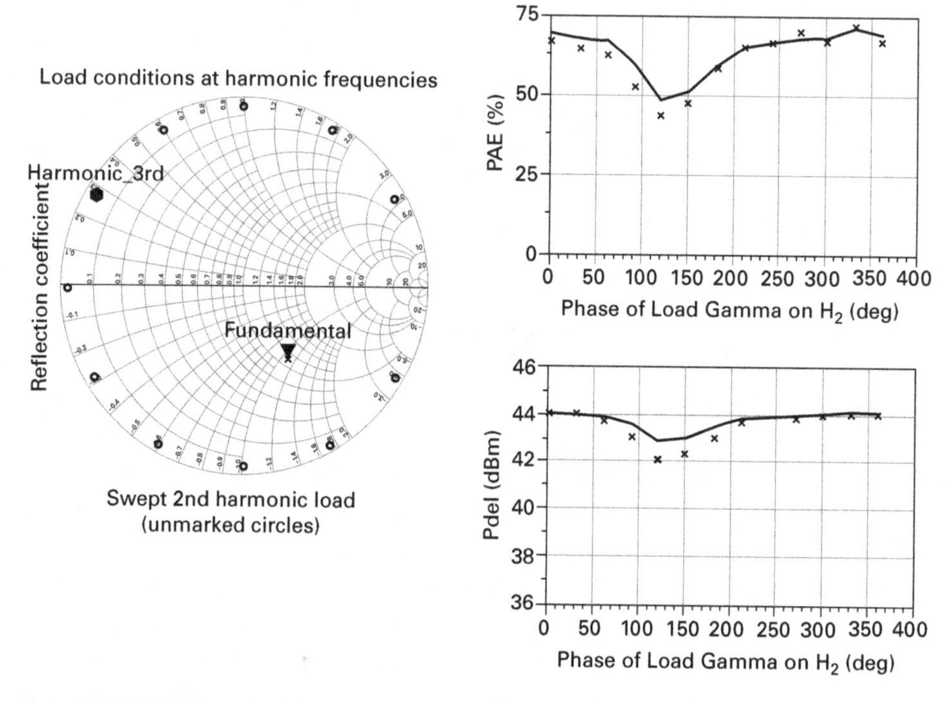

Figure 5.12 PAE and power delivered (Pdel) versus phase of swept second-harmonic (H_2) load impedance at the edge of the Smith chart for fixed fundamental and third-harmonic load impedances. Results from simulations using X-parameter model extracted into controlled fundamental impedance only (line) versus measurements (symbols).

5.5.1.1 Harmonic tuning may not require harmonic load-pull characterization

The above example is important for a number of reasons. The first is that the harmonic terminations presented to a transistor can significantly affect its properties. This is evident in the waveforms shown in Figure 5.8–Figure 5.11 and is a well-known fact in amplifier design [6]. The key here is that just measuring, through the X-parameter formalism, the harmonic sensitivities ($X^{(S)}$ and $X^{(T)}$ functions) as functions of the fundamental load is often sufficient to predict very accurately the effects of independently controlling these harmonic loads, as they vary over the entire Smith chart and even beyond.

5.5.1.2 Source pull is not required

Another important consequence of the X-parameter model is that it will give correct results when placed into an environment with arbitrary impedance at the source, provided that, when the X-parameter model is extracted, the power is swept from low values to the maximum power delivered. This is derived in [1]. This means that it is not necessary to perform an additional fundamental source-pull characterization in order to have the X-parameter model give the correct response when embedded in an arbitrary port 1 impedance environment in the circuit design.

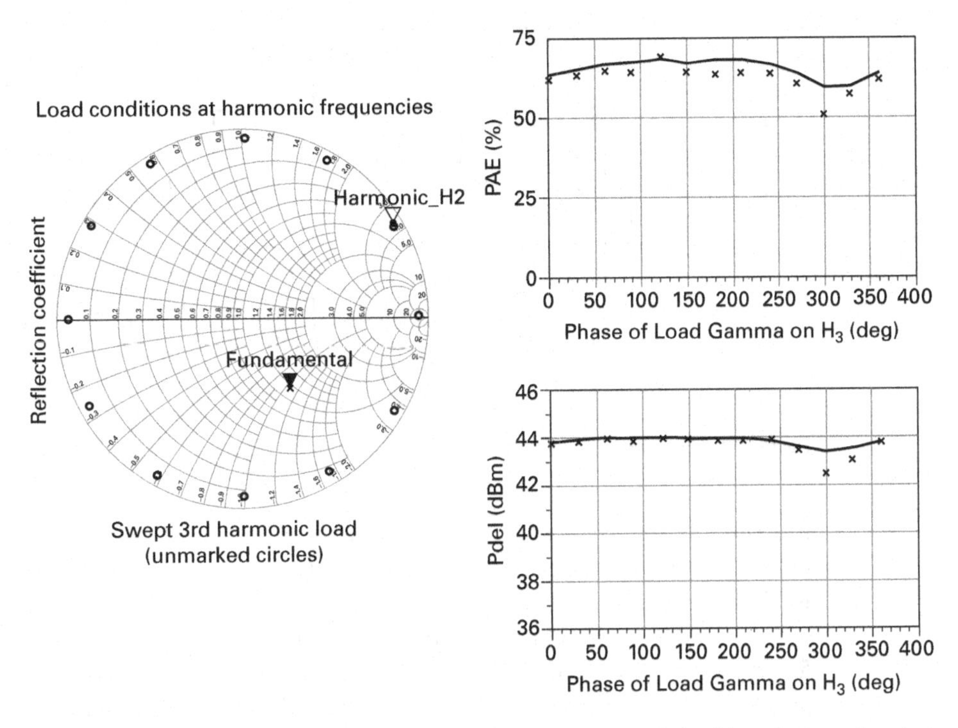

Figure 5.13 PAE and Pdel power versus phase of swept third harmonic load impedance at the edge of the Smith chart for fixed fundamental and second-harmonic load impedances. Results from simulations using X-parameter model extracted into controlled fundamental impedance only (line) versus measurements (symbols).

5.6 Design example: Doherty power amplifier design and validation

This section presents a complete Doherty power amplifier design, entirely within the circuit simulator, based on X-parameter models obtained from NVNA measurements of unmatched GaN power transistors [4]. A key feature of the approach is the application of an active source-injection-based X-parameter measurement setup. This is demonstrated to be necessary for providing sufficient data for an accurate X-parameter model of the transistor used for the auxiliary amplifier over the range it will be exercised in Doherty operation.

The first-pass design success – only one fabrication build was required to meet the design specifications – is confirmed by independent nonlinear measurements of the fabricated Doherty power amplifier.

5.6.1 Doherty power amplifier

The Doherty power amplifier is a high-efficiency architecture based on combining two or more amplifiers, typically operating at different classes, together with coherent input and output phasing, to provide an overall characteristic that is quite efficient, and also

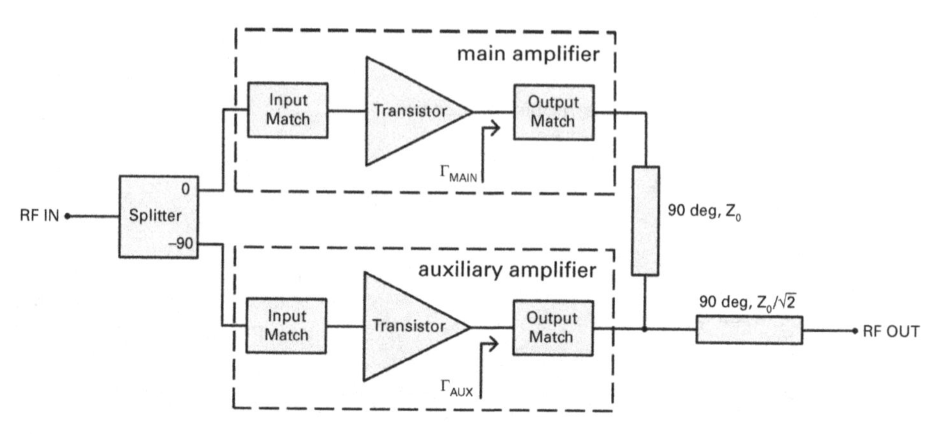

Figure 5.14 Ideal two-way Doherty power amplifier block diagram.

reasonably linear, over a wide range of input power. Details of the operating principles of the Doherty amplifier can be found in [6] and [7]. The design example discussed here is taken from [4].

A simple block diagram for a two-way Doherty power amplifier is given in Figure 5.14. The two amplifiers, with transistors biased for different classes of operation, present loads to one another that are designed to vary as a function of total applied input power. This "mutual load pulling" of the constituent active devices defines a coupled nonlinear problem for the overall performance of the Doherty power amplifier. It cannot be solved by independent scalar load-pull characterizations of each of the individual transistors. The details of the phase relationships of the load pulling are critical for the operation of the overall Doherty design.

The main and auxiliary amplifier matching networks were designed using fundamental and harmonic impedance tuning of the measured transistor X-parameter models in the simulator. Proper splitter and combiner circuitry was also designed and implemented to achieve optimum Doherty output power and power-added efficiency.

While ideal design equations can help with an initial design, [6], an optimized design, based on non-ideal components, requires accurate nonlinear models of both main and auxiliary power amplifiers and their constituent transistors. This design example starts from packaged, unmatched transistors, from which the main and auxiliary amplifiers, including their input- and output-matching networks, must be designed. Transistor X-parameters, measured under active-source-injection conditions, provide a natural measurement-based modeling approach, from which the entire Doherty amplifier can be designed, enabling the details of the mutual interaction between the active components to be solved self-consistently and design parameters to be optimized in the simulator environment.

5.6.2 X-parameter characterization of the transistors

The transistors used in this work are unmatched 10 W high-electron-mobility transistors (HEMT) (CGH40010F from CREE). These transistors are biased in deep class AB and class C operation, for the main and auxiliary amplifiers, respectively.

The transistor used for the main amplifier is biased in an on-state, and can therefore be characterized by X-parameters measured using the NVNA under passive load-pull conditions as described in Section 5.3. However, the transistor used for the auxiliary amplifier is biased "off" in the absence of RF power, and will therefore experience a load outside the Smith chart for the low-power region of operation. This is because, while off, the transistor does not generate any $B_{2,1}$ of its own, yet sees significant $A_{2,1}$ coming from the output of the main amplifier. This means that the nonlinear properties of the transistor in this range cannot be characterized under passive load-pull conditions, and active-source-injection characterization for the X-parameter modeling must be used, as discussed in Section 5.4.3.

5.6.2.1 Nonlinear measurement setup

The measurement system used to characterize the transistor biased for the auxiliary amplifier is shown in Figure 5.15. The figure inset shows the transistor mounted in a fixture for measurement. In this setup, the NVNA employs active source injection to enable device characterization both inside and outside the Smith chart. The auxiliary amplifier transistor will experience extreme load conditions at input power levels below and near transistor turn-on. It is critical for the Doherty amplifier design to capture data in this transition region. It is impossible to acquire such data with measurement setups using passive tuners only. The measurement algorithm controls load conditions by adjusting the amplitude and phase of the RF vector signal generator. The source applying stimulus on port 2 is controlled to achieve specified $A_{2,1}$ values. The X-parameter functions are tabulated versus the complex values of the incident waves at each port, as discussed in Section 5.2.

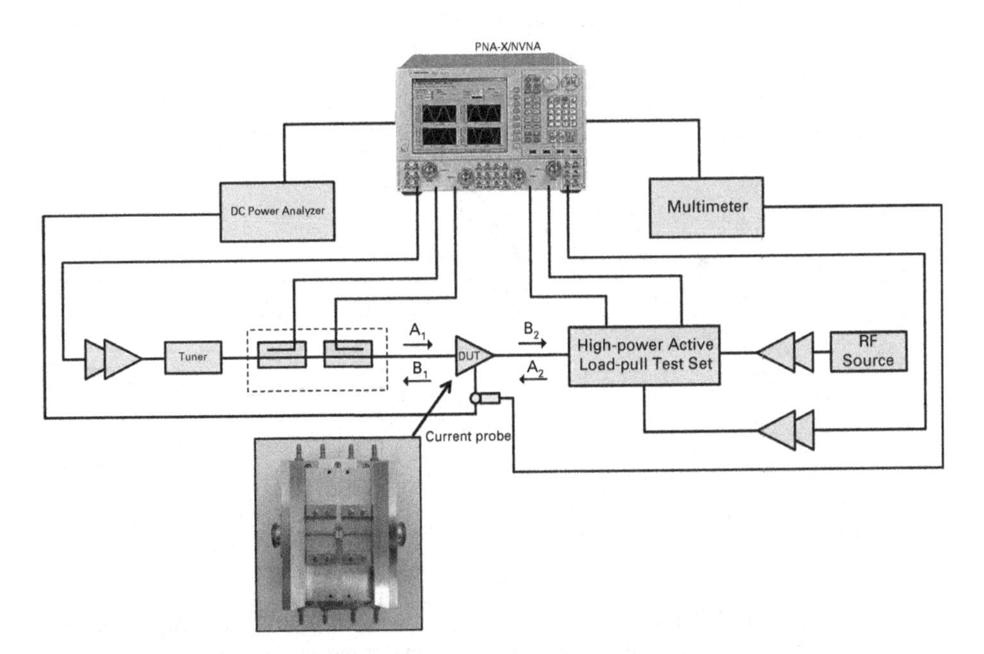

Figure 5.15 Active load-dependent X-parameter measurement setup.

All measurements are performed under pulsed continuous-wave test conditions using a 1.3 GHz carrier frequency with 100 μs pulse width and 10% duty cycle. Pulse generators and RF pulse modulators internal to the PNA-X network analyzer are used to provide the required test conditions. Two DC power analyzer modules provide constant gate and drain DC bias to the transistor under test. The drain current is measured using a current probe, which is placed between the transistor drain terminal and large electrolytic capacitors at the transistor test fixture. Several in-pulse current samples are acquired and averaged by the NVNA to ensure very accurate power-added efficiency (PAE) measurements.

5.6.3 X-parameter model validation

Prior to the full Doherty design, the X-parameter models of the individual transistors are validated extensively. They are compared to DC and small-signal S-parameter measurements over the range of bias conditions. Then they are validated against large-signal measurements, at load conditions included in the extraction data (to verify that the model plays back large-signal measurements correctly) and independent load conditions not used during model extraction (to validate interpolation capabilities within the measured load grid). Example large-signal validation results are shown in Figure 5.16

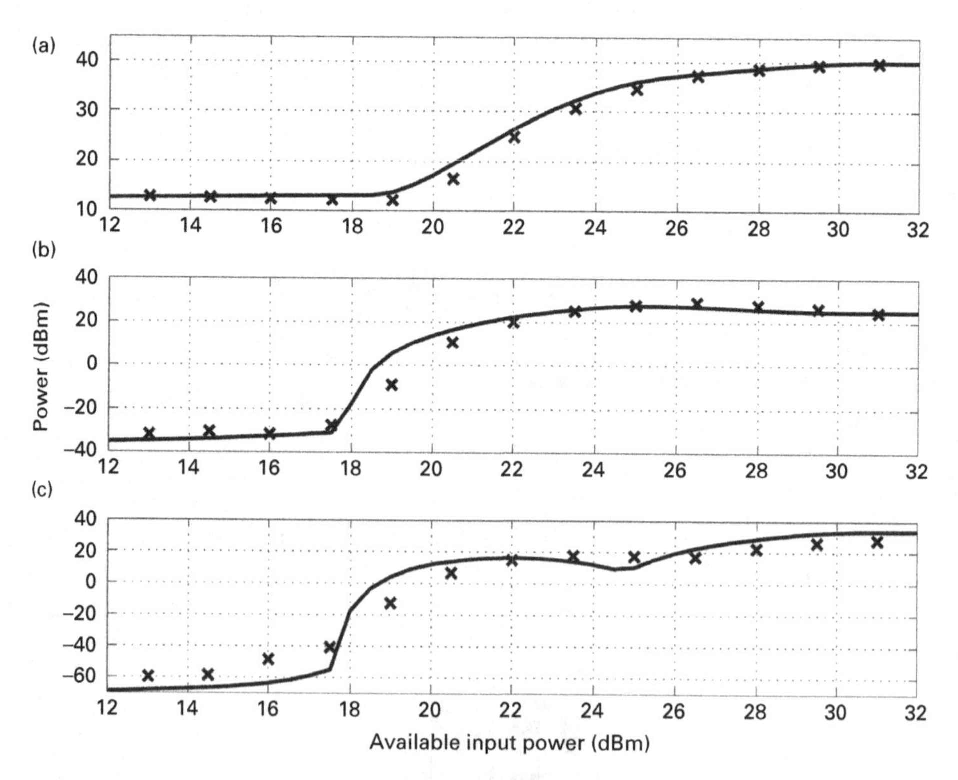

Figure 5.16 Validation results for class C transistor intended for auxiliary amplifier. Simulated using measurement-based X-parameter model (solid lines) and measured (symbols). (a) Fundamental frequency; (b) second harmonic; and (c) third harmonic.

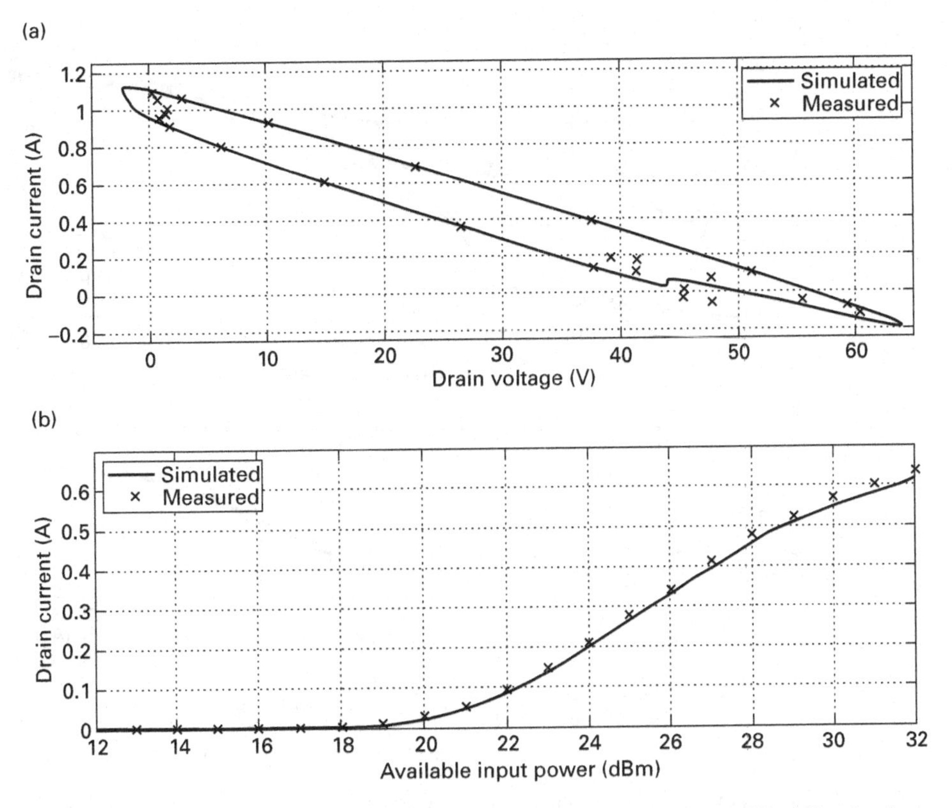

Figure 5.17 Validation results for class C transistor intended for auxiliary amplifier. Simulated using measurement-based X-parameter model (solid lines) and measured (symbols). (a) Dynamic load line; (b) DC-bias current versus input power.

and Figure 5.17. In Figure 5.16, a fixed value of $A_{2,1}$ with magnitude 12 dBm and phase 75 degrees relative to the phase of $A_{1,1}$ is simultaneously incident at port 2 of the transistor. The 12 dBm output power below transistor cutoff (at about +18 dBm input power) in Figure 5.16(a) is therefore due to the nearly complete reflection of this wave from the transistor output impedance.

Figure 5.18 shows measured and simulated fundamental output power versus input power in the presence of an additional $A_{2,1}$ incident wave of varying magnitude and phase angle. The output impedance of the transistor is a highly nonlinear function of transistor input drive. At power levels below turn-on (approximately +18 dBm), the transistor output impedance is very large. When no $A_{2,1}$ is present (lower trace of Figure 5.18(a)), the transistor transmits negligible $B_{2,1}$. For levels of $A_{2,1}$ from 10 dBm to 30 dBm (remaining traces of Figure 5.18(a)), the large $B_{2,1}$ below turn-on is the result of the almost complete reflection of the injected $A_{2,1}$ wave (Figure 5.18(a)). At higher input power levels, above turn-on, the transistor output impedance changes both output reflection magnitude and phase. At certain input power levels and $A_{2,1}$ phase angles, injected and transmitted powers cancel almost completely. These cancellations are hard

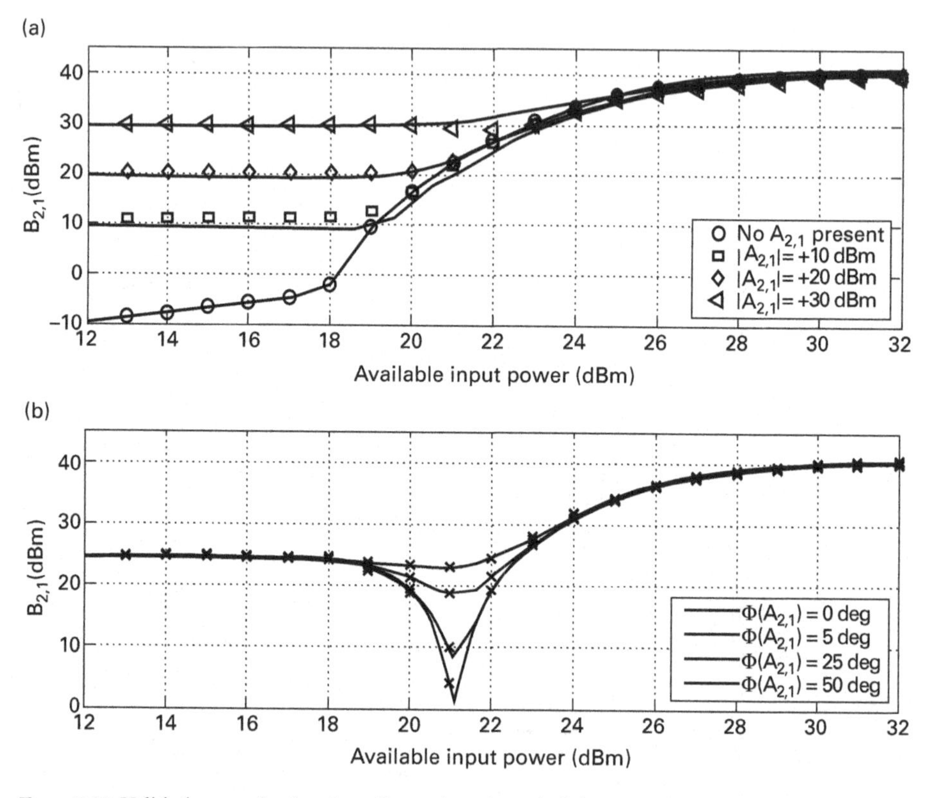

Figure 5.18 Validation results for class C transistor intended for auxiliary amplifier. Simulated using measurement-based X-parameter model (solid lines) and measured (symbols). Fundamental output power versus input power at different $A_{2,1}$ magnitudes (a) and phases (b).

to predict accurately, but are important because they result in highly nonlinear transfer characteristics with input power (Figure 5.18(b)). The X-parameter model (solid line) captures very accurately these measured transfer characteristics (traces with symbols) and thereby enables accurate design of the Doherty output combiner with proper output alignment of the main and auxiliary amplifiers.

In a typical high-efficiency power amplifier design, tuning of the higher-order harmonic loads, at both the input and the output of the active device, is necessary in order to achieve optimum performance. Although the measurement setup depicted in Figure 5.15 controls fundamental terminating impedances only, the X-parameter active-injection-measurement process extracts the DUT harmonic load sensitivity functions, $X^{(S)}$ and $X^{(T)}$, at both the input and output ports, as discussed in Section 5.3. These terms enable the X-parameter model to take into account independent harmonic impedance tuning in the design of the input- and output-transistor-matching networks. The harmonic tuning capabilities of the X-parameter model are therefore essential to achieve an optimal design.

The effect of the $X^{(S)}$ and $X^{(T)}$ terms is implicitly recognized from the comparison between the simulated and measured fundamental load-pull results presented in Figure 5.19. Here, measured results (symbols) are accurately reproduced by the model (lines)

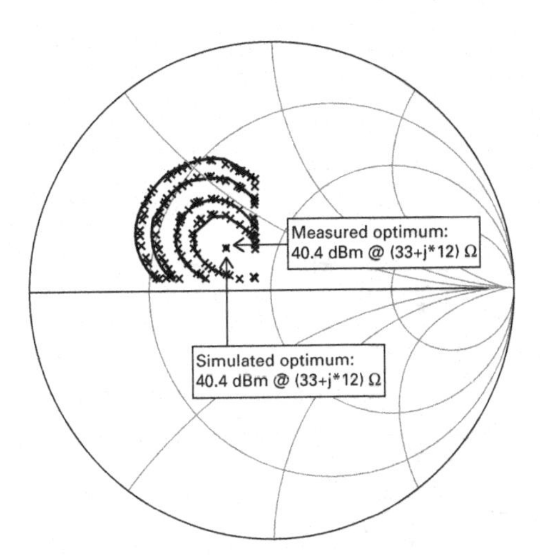

Figure 5.19 Measured (symbols) and simulated (lines) contours for power delivered into the load.

only when the harmonic impedances of the measurement system are embedded properly in the circuit simulator and presented to the X-parameter model. In fact, measured second- and third-harmonic impedances presented to the power transistor drain terminal were relatively far from 50 Ω during the characterization. The X-parameter measurement process calibrates out these uncontrolled harmonic impedances, accounting for them using the harmonic sensitivity terms. The resulting X-parameter model can therefore properly represent the DUT performance when presented with a wide range of harmonic loads in a design, even though the harmonic impedances were not controlled during the X-parameter measurements.

5.6.4 Doherty power amplifier design using X-parameters

The ADS schematic used for the Doherty power amplifier design is shown in Figure 5.20. The X-parameter models for the main and auxiliary amplifiers are at the center.

Harmonic power at the output of the main and auxiliary amplifiers is generally unwanted and should be reflected back to the active devices with proper phase angles. Several matching topologies may fulfill such requirements. Here, low-pass filter structures are employed and realized using microstrip elements on a high-quality Rogers Duroid 6010 laminate.

5.6.4.1 Output match

Design objectives are to maximize output power and drain efficiency while also minimizing harmonic content at the transistor outputs. The fundamental load impedance is thus chosen as a compromise between the power delivered to the load (Pdel) and the PAE. To improve third-harmonic suppression, open-circuited stub elements are designed with electrical lengths corresponding to short circuits at the third harmonic. Final transformation to the transistor drain terminal is achieved using a DC feed with an 80° electrical short at the second harmonic. Impedance transformations are all

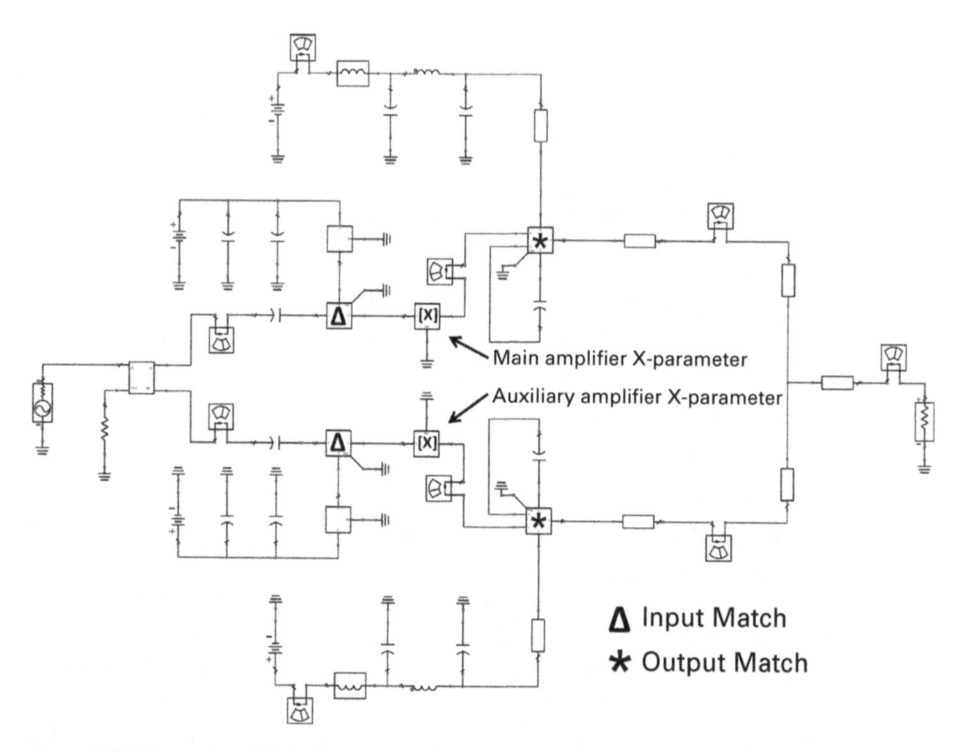

Figure 5.20 Schematic of Doherty design using X-parameter models from measurements of individual power amplifiers.

performed within a 0.6 quality factor circle to reduce element sensitivity to PCB fabrication tolerances. Proper phase alignment of the two amplifier outputs (and power-leakage reduction) is achieved with transmission line transformers.

5.6.4.2 Input match

The purpose of the input-matching network is to transform the source impedance to an appropriate gate impedance that yields optimum power gain and stability. At lower frequencies, the transistor has higher gain and is prone to oscillate. A resonator is added to limit the source reflection coefficient for frequency octaves lower than the fundamental frequency. The DC gate-bias feed is constructed using a high-impedance quarter-wavelength transformer with degraded Q in order to lower the Q of the reflection coefficient presented at the second-harmonic frequency. A branch-line hybrid coupler with meandered series arms drives the two amplifier inputs in quadrature. The meander lines save board space, and additional electromagnetic coupling is accounted for using EM simulations.

5.6.5 Results

The final manufactured Doherty power amplifier is shown in Figure 5.21. Simulated and measured amplifier delivered output power versus available power (Pavs) and drain efficiency versus delivered output power is shown in Figure 5.22. Measured

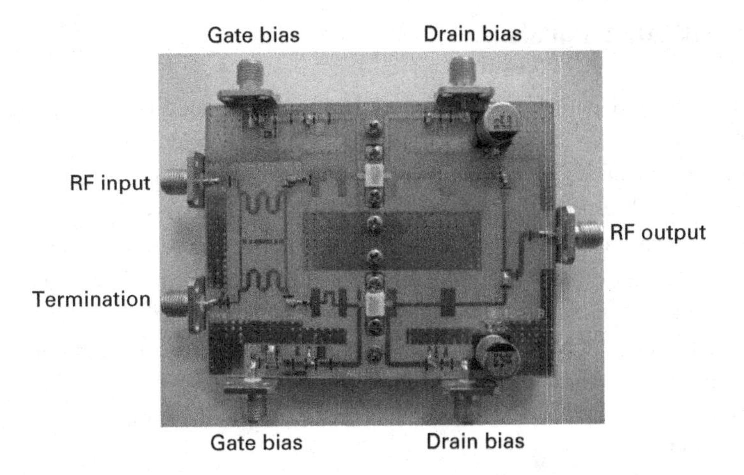

Figure 5.21 Manufactured Doherty power amplifier mounted on an aluminum heat sink.

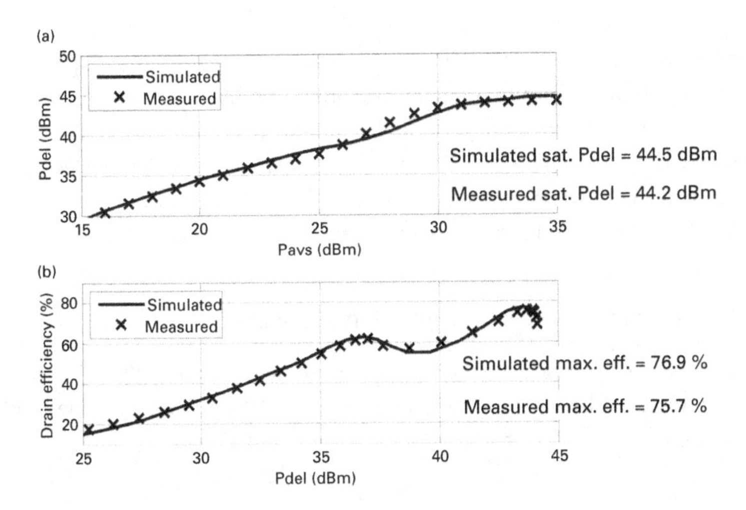

Figure 5.22 Simulated (lines) and measured (symbols) Doherty AM-to-AM and drain efficiency versus delivered power.

saturated output power is +44.2 dBm (simulation predicts +44.5 dBm), and the measured maximum drain efficiency is 75.7% (simulation predicts 76.9%). Overall, the simulation and measurement results agree very well. Measurements of the fabricated power amplifier are within 0.3 dB and 1.2% between simulated and measured saturated output power and maximum drain efficiency, respectively. The characteristic "double-peak" behavior of Figure 5.22(b) is an indication that the design is working in Doherty mode.

5.7 Incommensurate signals

The treatment of X-parameters up to now has been restricted to nonlinear multi-tone maps from signals on a harmonic grid onto the same grid of frequencies. This is sufficient for nonlinear devices being stimulated by a single large tone and generating harmonic spectral responses at multiple ports. It also includes effects of large mismatch at the fundamental frequency, and also at the harmonics, as can happen for waveform engineering of high-efficiency amplifiers by specific harmonic-matching conditions. All signals, including the incident and scattered waves, are necessarily periodic signals with the same period as the inverse of the grid spacing in the frequency domain.

In this portion of the chapter, the treatment is extended to include two incommensurate incident signals. The signals can be incident either on the same port or on different ports of the DUT.

Two sinusoidal signals, with (non-zero) frequencies f_1 and f_2, are incommensurate if the ratio of their frequencies is irrational. Equivalently, f_1 and f_2 are incommensurate if there are no two non-zero integers for which the commensurability condition of Chapter 2 is fulfilled. Mathematically, this is indicated by (5.16):

$$m, n \in \mathbb{Z} \text{ such that } mf_1 + nf_2 = 0 \Rightarrow m = n = 0. \tag{5.16}$$

An example of the incident and scattered waves of a nonlinear 2-port subject to stimulation at the input with two incommensurate frequencies is given in Figure 5.23. Only a finite number of spectral tones in the output spectrum are shown.

Because the output spectrum generated by a nonlinear DUT can be reflected back into the device due to mismatch, or propagate through other linear and nonlinear components, spectral maps defined on all intermodulation products of the large incident signals need to be defined.

5.7.1 Notation for incommensurate two-tone X-parameters

The case of two large fundamental incommensurate tones and their intermodulation products is presented here. The generalization to N large incommensurate fundamental tones is straightforward, but the notation is somewhat tedious.

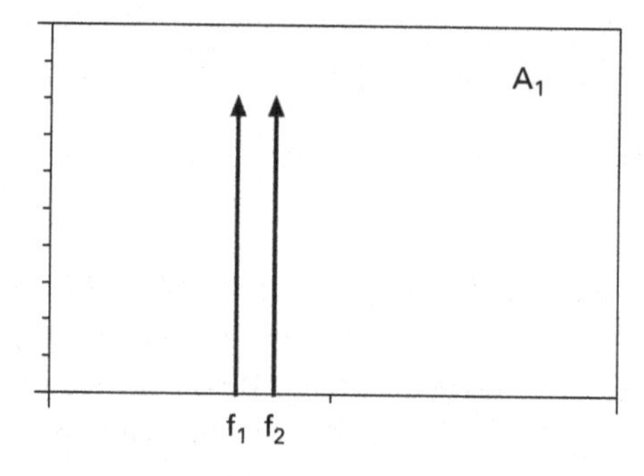

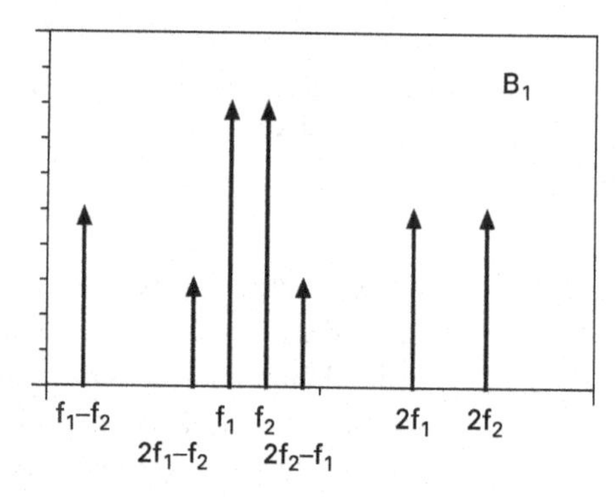

Figure 5.23 Incident and simplified scattered-wave spectra for nonlinear 2-port excited by two large incommensurate tones.

All non-zero signals considered in this section exist on a discrete subset of real intermodulation frequencies, f, given by

$$f = nf_1 + mf_2, \tag{5.17}$$

where n and m are integers. The case where f evaluates to a negative number is treated later. Each wave variable is specified by a complex phasor indexed by a port index, p, a positive integer indicating the port at which the stimulus is applied, and two integers specifying the frequency of the incident signal in terms of contributions from specific orders of each of the two fundamental tones at frequencies f_1 and f_2. This is shown in (5.18) for incident and scattered waves:

$$\begin{aligned} &A_{p,[n,m]}, \\ &B_{p',[n',m']}. \end{aligned} \tag{5.18}$$

Note that there can be more ports than large tones. For example, a mixer may have three ports but there may be only two incident large tones, one at the LO and the other at the RF.

Example: $A_{3,[2,\ -1]}$ is the incident wave at port 3 corresponding to a frequency of $2f_1 - f_2$.

The same convention presented in the preceding chapters is used for DC stimulus and response. If the DC stimulus at port q is a voltage, $DCS_q = V_q$, the DC response is a current, $DCR_q = I_q$. Conversely, if the DC stimulus at port p is a current, $DCS_q = I_q$, the DC response at that port is a voltage, $DCR_q = V_q$.

The complete multi-port multi-tone stimulus includes both the DC and RF stimuli.

Example: Equation (5.19) defines a stimulus with DC voltages applied at ports 1 and 2, a DC current source at port 3, and incident signals applied at each port at different intermodulation frequencies of the underlying large signals:

$$stim = (DCS; RFS) \equiv (V_1, V_2, I_3; A_{1,[0,1]}, A_{3,[1,0]}A_{2,[1,1]}, A_{3,[1,2]}). \tag{5.19}$$

The corresponding response generally has more terms given multiple intermodulation products produced by the DUT, such as that given in (5.20):

$$resp = (DCR; RFR) \equiv (I_1, I_2, V_3; B_{1,[1,0]}, B_{1,[0,1]}, B_{1,[1,1]}, \ldots, B_{1,[N,M]}, B_{2,[1,0]}, B_{2,[0,1]}, \ldots, B_{3,[N,M]}). \tag{5.20}$$

A maximum order of N for harmonics of the first fundamental tone and M harmonics of the second tone is generally specified. This means that *each* of the components in the response, (5.20), is a nonlinear function of *all* of the elements of the stimulus, (5.19). These functions are generally nonlinear, so the general case is complicated.

5.7.2 Time invariance for incommensurate two-tone X-parameters

The concept of cross-frequency phase for commensurate signals was central to implementing time-invariant maps from incident to scattered waves. For incommensurate signals, things are fundamentally different, requiring a different approach.

Cross-frequency phase is an ill-defined concept for incommensurate signals. Each successive zero crossing of the phase of the first signal at frequency f_1 corresponds to a unique value of the phase of the second signal, ϕ_2, at frequency f_2.[1] The infinite sequence of ϕ_2 values is dense on the interval $[0, 2\pi)$, with no finite gaps; roughly speaking, this means that all possible real values of phase are approached arbitrarily closely. A corollary is that it is possible to find, simultaneously, the phase of the first fundamental tone equal to zero and the phase of the second fundamental tone arbitrarily close to zero, just by waiting a long enough time.

An important implication of this fact, related to taking appropriate limits in the space of such signals, is that a two-tone steady-state signal, in the frequency domain, can be uniquely identified by specifying only the magnitudes of the individual signals. This is one fewer degree of freedom compared to the commensurate case considered in the first part of this chapter, where a phase angle was required in addition to the two magnitudes. In an important sense, the signal defined by phasors each with phase 0 will define our reference for enforcing a time-invariant description on the map from inputs to outputs.

Consider now the two fundamental signals as well as signals at any other intermodulation frequency. Each intermodulation frequency is manifestly commensurate with respect to the two fundamental tones. This follows trivially from (5.21) and (5.22). Equation (5.21) defines an equation with integer coefficient among three frequencies that is clearly satisfied when f_3 is given by (5.22):

$$nf_1 + mf_2 - 1f_3 = 0, \tag{5.21}$$

$$f_3 = freq\{A_{p,[n,m]}\} = nf_1 + mf_2. \tag{5.22}$$

The time-invariant process is as follows. Apply independent phase shifts to each of the fundamental tones until each has zero phase. This is shown in (5.23),

$$A_{p,[1,0]} \rightarrow A_{p,[1,0]}P_{[1,0]}^{-1} \quad \text{and} \quad A_{p',[0,1]} \rightarrow A_{p',[0,1]}P_{[0,1]}^{-1}. \tag{5.23}$$

where $P_{[1,0]}$ and $P_{[0,1]}$ are the phases of the two large tones. This translates the large-signal components to the reference state.

Any signal component at an intermodulation frequency labeled $[n,m]$ is commensurate with respect to the fundamental tones, and therefore has a well-defined cross-frequency phase with respect to both of the fundamental tones. The same reasoning of Chapter 2 can now be invoked to deduce that the phase of the incident waves with indices $[n,m]$ must be coherently shifted by the sum of the integral powers of the individual phases of the large tones according to the respective indices. This is shown in (5.24):

$$A_{q,[n,m]} \rightarrow A_{q,[n,m]}P_{[1,0]}^{-n}P_{[0,1]}^{-m}. \tag{5.24}$$

[1] This is not the case for commensurate signals, where there is only a finite number of possible values for the phase of the second signal, ϕ_2 [8].

Since the scattered waves (the response of the nonlinear system) also have components only on the same intermodulation spectrum, we must likewise phase shift the scattered waves by the same terms. This is given in (5.25) for the reference response:

$$B_{q,[n,m]} \to B_{q,[n,m]} P_{[1,0]}^{-n} P_{[0,1]}^{-m}. \tag{5.25}$$

Therefore, under a transformation to the reference state defined by (5.23) and (5.24), where the reference response is given by (5.25), one must have the relationships given in (5.26) for the DUT in order that the equivalent phase shifts of the intermodulation terms preserve their cross-frequency phase. Here the phase factors in (5.25) have been taken to the right-hand side of (5.26):

$$\begin{aligned} B_{q,[n,m]} &= F_{q,[n,m]}\left(DCS, A_{p_1,[1,0]}, A_{p_2,[0,1]}, \ldots, A_{p',[n',m']}\right) \\ &= F_{q,[n,m]}\left(DCS, A_{p_1,[1,0]} P_{[1,0]}^{-1}, A_{p_2,[0,1]} P_{[0,1]}^{-1}, \ldots, A_{p',[n',m']} P_{[1,0]}^{-n'} P_{[0,1]}^{-m'}\right) P_{[1,0]}^{n} P_{[0,1]}^{m}. \end{aligned} \tag{5.26}$$

Not all functions have the transformation properties of (5.26). Only those functions $F(.)$ with the property (5.26) are admissible spectral mappings for a time-invariant DUT being excited by signals at all intermodulation frequencies associated with two fundamental tones at incommensurate frequencies.

Noting that the first two arguments at the reference state are real, the introduction of the X-parameter function is defined according to (5.27):

$$\begin{aligned} &X_{q,[n,m]}^{(FB)}\left(DCS, |A_{p_1,[1,0]}|, |A_{p_2,[0,1]}|, \ldots, A_{p',[n',m']} P_{[1,0]}^{-n'} P_{[0,1]}^{-m'}\right) \\ &= \frac{F_{q,[n,m]}\left(DCS, A_{p_1,[1,0]}, A_{p_2,[0,1]}, \ldots, A_{p',[n',m']}\right)}{P_{[1,0]}^{n} P_{[0,1]}^{m}}. \end{aligned} \tag{5.27a}$$

This means that the DUT X-parameter functions can be defined on a manifold with two fewer real dimensions than the general multi-tone spectral mappings.

Combining the property expressed in (5.26) with the definition in (5.27a) leads to the X-parameter model to be formulated, as shown in (5.27b).

$$B_{q,[n,m]} = X_{q,[n,m]}^{(FB)}\left(DCS, |A_{p_1,[1,0]}|, |A_{p_2,[0,1]}|, \ldots, A_{p',[n',m']} P_{[1,0]}^{-n'} P_{[0,1]}^{-m'}\right) P_{[1,0]}^{n} P_{[0,1]}^{m}. \tag{5.27b}$$

5.7.3 Reference LSOP

The discussion in Section 5.7.2 means it is possible to choose the reference LSOP for the state of a DUT excited by two incommensurate tones to be identified with the phasors of the two signals, each of which corresponds to zero phase. That is, *refLSOPS* depends only on the magnitudes of $A_{p_1,[1,0]}$ and $A_{p_2,[0,1]}$, where p_1 and p_2 are the ports at which the first and second tones are incident, respectively.

5.7.4 Spectral linearization

The nonlinear map (5.27b) depends on many variables, and is therefore very complex to deal with in general. Just as in earlier sections, useful approximations can be developed to simplify things dramatically for practical applications.

It is considered that, out of all the possible signal components represented by $A_{p,[n,m]}$ incident on the DUT, only the fundamental tones, $A_{p_1,[1,0]}$ and $A_{p_2,[0,1]}$, are large, and all other components at intermodulation frequencies are considered small. This is the case, for example, in mixers with nearly matched ports. It is also the case for power amplifiers with ports nearly matched to 50 Ω. In these cases, the rich intermodulation spectra will be reflected, but the terms will be small enough that the spectral linearization process will still be a good approximation to taking their effects into account.

Equation (5.26) is therefore spectrally linearized around the reference LSOPS given by (5.28):

$$refLSOPS = \left(DCS_p,\ |A_{p_1,[1,0]}|,\ |A_{p_2,[0,1]}|, 0, \ldots, 0\right). \tag{5.28}$$

Spectral linearization with respect to all other signals is applied around the *refLSOPS* of (5.28). As all other previous cases, the maps (5.27b) are non-analytic, and so we need to differentiate with respect to the intermodulation phasors and their complex conjugates, independently. The now familiar procedure is applied, and the resulting X-parameter model equations take the form (5.29)–(5.31). The sum is over all ports, q, and all integers n' and m':

$$\begin{aligned} B_{p,[n,m]} &= X^{(F)}_{p,[n,m]}(refLSOPS)P^{n}_{[1,0]}P^{m}_{[0,1]} \\ &+ \sum_{q,n',m'} X^{(S)}_{p,[n,m];q,[n',m']}(refLSOPS)P^{n-n'}_{[1,0]}P^{m-m'}_{[0,1]}A_{q,[n',m']} \\ &+ \sum_{q,n',m'} X^{(T)}_{p,[n,m];q,[n',m']}(refLSOPS)P^{n+n'}_{[1,0]}P^{m+m'}_{[0,1]}A^{*}_{q,[n',m']}, \end{aligned} \tag{5.29}$$

$$I_p = X^{(I)}_{p}(refLSOPS) + \sum_{q,n',m'} \mathrm{Re}\left(X^{(Y)}_{p;q,[n',m']}(refLSOPS)P^{-n'}_{[1,0]}P^{-m'}_{[0,1]}A_{q,[n',m']}\right), \tag{5.30}$$

$$V_p = X^{(V)}_{p}(refLSOPS) + \sum_{q,n',m'} \mathrm{Re}\left(X^{(Z)}_{p;q,[n',m']}(refLSOPS)P^{-n'}_{[1,0]}P^{-m'}_{[0,1]}A_{q,[n',m']}\right). \tag{5.31}$$

The X-parameter sensitivity functions are therefore identified as the partial derivatives of the scattered waves with respect to complex signals at the intermodulation frequencies (and their conjugates) evaluated at the *refLSOPS*:

$$\begin{aligned} X^{(S)}_{p,[n,m];q,[n',m']} &= \left.\frac{\partial B_{p,[n,m]}}{\partial A_{q,[n',m']}}\right|_{refLSOPS} \cdot P^{n'}_{[1,0]}P^{m'}_{[0,1]}, \\ X^{(T)}_{p,[n,m];q,[n',m']} &= \left.\frac{\partial B_{p,[n,m]}}{\partial A^{*}_{q,[n',m']}}\right|_{refLSOPS} \cdot P^{-n'}_{[1,0]}P^{-m'}_{[0,1]}. \end{aligned} \tag{5.32}$$

5.7.4.1 Examples

A two-tone multi-port model (with no explicit DCS dependence), which describes a mixer or amplifier, can be defined by (5.33):

$$
\begin{aligned}
B_{p,[n,m]} = & X^{(F)}{}_{p,[n,m]}\left(|A_{p_1,[1,0]}|,\ |A_{p_2,[0,1]}|\right)P^{n}_{[1,0]}P^{m}_{[0,1]} \\
& +\sum_{q;j,k} X^{(S)}{}_{p,[n,m];q,[j,k]}\left(|A_{p_1,[1,0]}|,\ |A_{p_2,[0,1]}|\right)P^{n-j}_{[1,0]}P^{m-k}_{[0,1]}A_{q,[j,k]} \\
& +\sum_{q;j,k} X^{(T)}{}_{p,[n,m];q,[j,k]}\left(|A_{p_1,[1,0]}|,\ |A_{p_2,[0,1]}|\right)P^{n+j}_{[1,0]}P^{m+k}_{[0,1]}A^{*}{}_{q,[j,k]}.
\end{aligned} \tag{5.33}
$$

For $p_1 = 1$ and $p_2 = 2$, this example describes a mixer with the RF at port 1 and LO at port 2. The IF corresponds to port 3. The *refLSOPS* is specified only by the magnitude of the RF and LO signals (assumed incommensurate).

For both $p_1 = 1$ and $p_2 = 1$, (5.33) describes a power amplifier with a two-tone stimulus at the input port (port 1). In this case, the port index, p, of the scattered B-waves goes from 1 to 2. The *refLSOPS* is specified by the magnitudes of the two input tones.

5.7.5 Discussion

Finite and independent phase shifts are used to shift signals to the reference LSOP and then to shift back the scattered waves appropriately. This is the procedure used to implement time invariance in the incommensurate case. It is noteworthy to consider that the actual time shift necessary to reach the *refLSOP* (where both fundamental signals have zero phase) may be infinite. This fact can be established, rigorously, considering sequences of time shifts and the convergence in function spaces. This level of detail is beyond the scope of this work.

5.7.6 When intermodulation frequencies are negative

X-parameters operate in the frequency domain, mapping the spectrum of the input stimulus to the spectrum of the DUT response. The input and output signals considered in the time domain are real. This means that the spectra in the frequency domain are double sided with conjugate symmetry, and hence contain redundant information by a factor of 2. For efficiency, only half the spectra need to be considered in the equations, and only those X-parameters mapping the chosen spectral components need to be stored in the files. In many applications, it is typical that only positive frequency and DC components are considered. For X-parameters, the frequencies of the signals are determined by the indices n and m and by the values of the fundamental frequencies, f_i, through (5.17). If the frequencies, f_i, are swept, the frequency of the intermodulation term corresponding to a particular set of n and m can become negative. Specifically, for the choice $f_1 = 3$ GHz and for f_2 swept from 2.9 GHz to 3.1 GHz in two steps, the intermodulation term $[1, -1]$ will correspond first to 100 MHz and then to -100 MHz. It is much more convenient to have a consistent indexing scheme independent of the numerical values of the f_i. Therefore, the convention is adopted that the spectral components considered correspond to those sets of n and m where the first non-zero value in the set of these integers is positive. For any two index vectors $[n,m]$ and $[-n,-m]$, each of which corresponds to the same $|f|$, one will be consistent with this convention and the other will be excluded. If the frequency corresponding to the convention on indices is negative, the corresponding positive frequency is considered for the signal and the complex conjugate of the corresponding phasor is taken.

5.7.6.1 Examples

Wave variables and $X^{(F)}$ functions with intermodulation indices [2,3] are selected for processing and storage, whereas those indexed by [−3,5] are not, because, in the latter case, the first non-zero integer is negative. So, in this case, the corresponding intermodulation vector [3,−5] is considered instead. If $3f_1 - 5f_2$ is positive, the complex value of the phasor at that frequency is considered. If $3f_1 - 5f_2$ is negative, the complex conjugate of the corresponding phasor is considered and its contribution attributed to the spectral component with the corresponding positive frequency.

X-parameter functions labeled by two sets of port and intermodulation vectors have rules to remove redundancy that can be deduced in a similar way. The details are not described here, but the following is given as an example:

- the parameter $X^{(S)}_{p,[2,3];q,[0,2]}$ is an allowed (and saved) quantity, but the parameter $X^{(S)}_{p,[-3,2];q,[0,2]}$ is not allowed.

5.7.7 X-parameter models of mixers

An example of the incommensurate two-tone X-parameter is given by the characterization and modeling of a commercial mixer component (Mini-Circuits LAVI-22VH+ (TB-433) double-balanced mixer). A schematic of the component is given in Figure 5.24.

The RF signal is applied at port 1, the LO at port 2. The conversion gain of a mixer with IF matched to a perfect 50 Ω termination is given by the simple X-parameter expression (5.34):

$$conversion\ gain = \frac{|X^{(F)}_{3,[1,-1]}|}{|A_{1,[1,0]}|}. \tag{5.34}$$

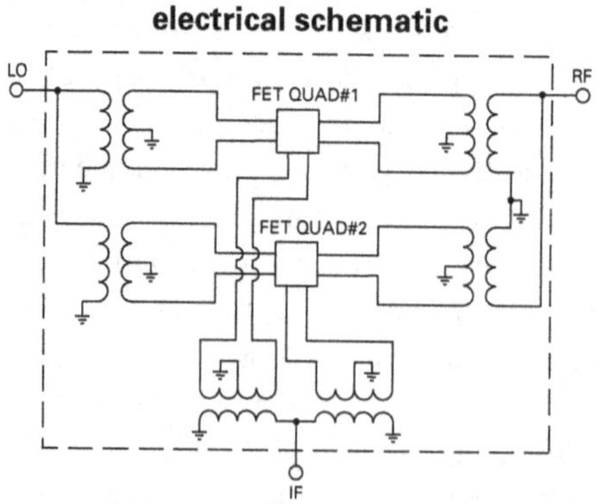

Figure 5.24 Mini-Circuits double-balanced mixer used for three-port X-parameter model. © 2013 Scientific Components Corporation d/b/a Mini-Circuits. Used with permission.

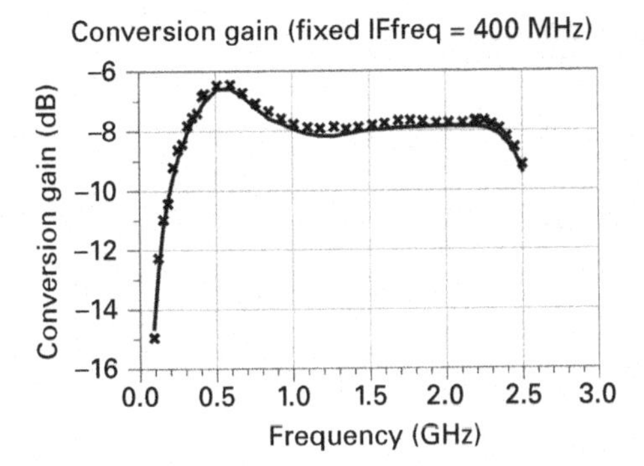

Figure 5.25 Comparison of measurement-based X-parameter mixer model and manufacturer's datasheet for conversion gain (in dB) versus frequency.

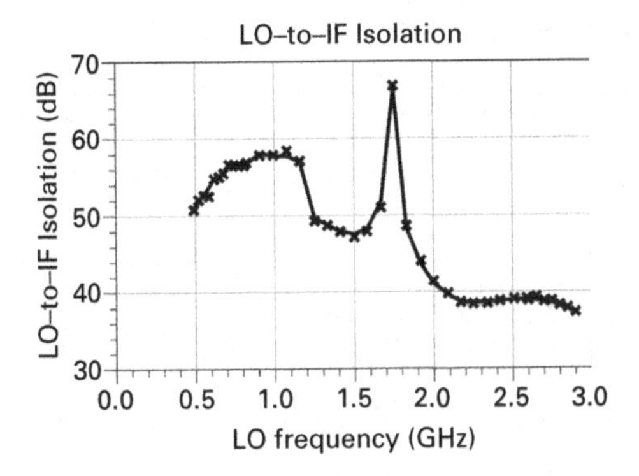

Figure 5.26 Comparison of measurement-based X-parameter mixer model and manufacturer's datasheet for LO–IF isolation (in dB) at the IF port versus frequency.

Here we assume that the IF signal of interest is at the difference of the RF and LO frequency (where LO frequency is lower than the RF frequency).

The X-parameters of the Mini-Circuits mixer were measured. The conversion gain was computed from (5.34) and compared with values from the manufacturer's datasheet. This is shown in Figure 5.25. The LO-to-IF isolation term, at the IF port, can be computed from a similar expression given in (5.35) (see Figure 5.26):

$$\text{LO-to-IF isolation} = \frac{|A_{2,[0,1]}|}{|X^{(F)}_{3,[0,1]}|}. \tag{5.35}$$

The X-parameter model has much more information than just conventional datasheet information. An example is given by the phase of the IF signal, which is a well-defined quantity when normalized to the phase of the RF and LO signals using (5.29). For the Mini-Circuits mixer, this information is shown in Figure 5.27. The expression to compute it is given simply by (5.36):

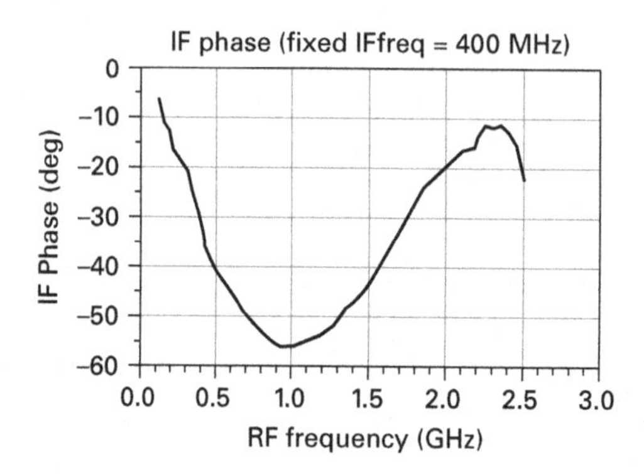

Figure 5.27 Phase of IF signal versus frequency as contained in the X-parameter mixer model.

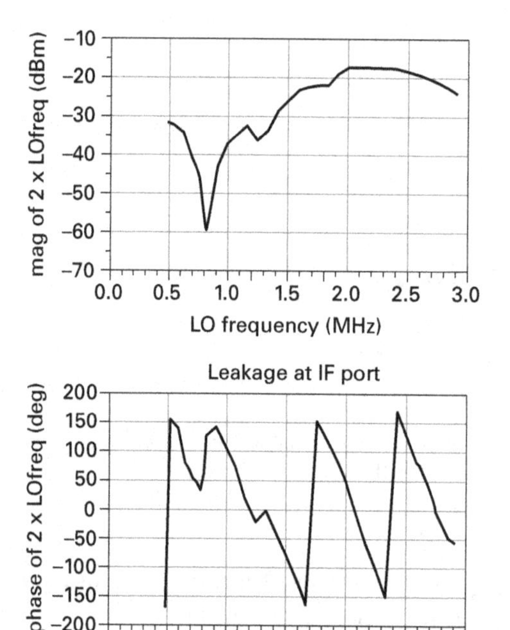

Figure 5.28 Leakage of second-harmonic signal of LO at IF port, in magnitude and phase, versus LO frequency (LO power 21 dBm, fixed IF frequency 400 MHz).

$$\text{IF phase} = phase\left[X^{(F)}_{3,[1,-1]}\right]. \tag{5.36}$$

Additionally, the signal at the second harmonic of the LO signal can be observed at the IF port. This complex quantity, given by $X^{(F)}_{3,[0,2]}$ (normalized in phase to LO and RF at the fundamental), is plotted in Figure 5.28. Many performance parameters of mixers can be expressed in terms of X-parameters. Some examples are given in Table 5.1.

Table 5.1 Mixer behavior in terms of X-parameters

Large RF at port 1, large LO at port 2, IF output at port 3

Mixer terminology	X-parameter expressions or terms	
Conversion gain (difference frequency)	$\frac{\lvert X^{(F)}_{3,[1,-1]}\rvert}{\lvert A_{1,[1,0]}\rvert}$	
LO leakage at IF port for perfectly matched IF	$X^{(F)}_{3,[0,1]}$	
Mismatch terms at IF frequency and IF port	$X^{(S)}_{3,[1,-1];3,[1,-1]}$	$X^{(T)}_{3,[1,-1];3,[1,-1]}$
IF-to-LO isolation terms	$X^{(S)}_{2,[1,-1];3,[1,-1]}$	$X^{(T)}_{2,[1,-1];3,[1,-1]}$
LO-second-harmonic-to-IF isolation	$X^{(S)}_{3,[1,-1];2,[0,2]}$	$X^{(T)}_{3,[1,-1];2,[0,2]}$

5.8 Summary

X-parameters spectrally linearized about an LSOP established by two large incident tones were considered in two important cases. The first case corresponds to two signals incident at different ports at the same fundamental frequency. For power amplifiers, this entails measuring X-parameters as functions of complex-valued load as well as power. This case includes what is sometimes known as "time-domain load pull." It was demonstrated that harmonic sensitivities ($X^{(S)}$ and $X^{(T)}$ terms) measured as functions of the complex load at the fundamental frequency were sufficient to predict accurately the effect of tuning the fundamental and harmonic load impedances independently. This provides nearly the accuracy of harmonic time-domain load pull, but requires only a small fraction of the measurements and time, in a hardware configuration that is much simpler and less expensive. The resulting X-parameter model was shown to produce excellent results for a GaN HEMT amplifier. This approach has been directly applied to more complicated high-efficiency applications, for example the Doherty amplifier in [4], and discussed as a complete design example in Section 5.6.

The second case considered in this chapter corresponds to the case of two signals at incommensurate fundamental frequencies incident at the same or distinct ports of a nonlinear time-invariant DUT. The example provided to illustrate this formalism was a mixer, where RF and LO signals are applied at different ports. The resulting model from NVNA measurements on a commercial mixer showed excellent results compared to the product datasheet, with many more interaction terms provided that could not be obtained by any other means. The same X-parameter two-tone measurement procedure can also be applied to two-port amplifiers, producing X-parameter models where the phases of the intermodulation products are captured along with their magnitudes. This can enable advanced circuit-design techniques, such as derivative superposition [9], in order to cancel distortion terms. Previous to the availability of two-tone X-parameters, this was possible only using nonlinear compact transistor models, the accuracy of which can be questionable.

Exercises

5.1 Prove that two sinusoidal signals, corresponding to the incident waves at ports 1 and 2 at the same fundamental frequency, always have the same relative phase, independent of time shift, given by (5.3).

5.2 Using the ideal X-parameter experiment design of Chapter 3, derive expressions (5.14) and (5.15) for the number of measurements required as functions of the number of independently controlled load impedances.

5.3 Derive (5.29) for the spectral linearization around an incommensurate two-tone LSOP.

References

[1] J. Horn, D. E. Root, and G. Simpson, "GaN device modeling with X-parameters," in *IEEE Compound Semiconductor Integrated Circuit Symp. (CSICS)*, Monterey, CA, Oct. 2010, pp. 1–4.

[2] S. Woodington, R. Saini, D. Williams, J. Lees, J. Benedikt, and P. J. Tasker, "Behavioral model analysis of active harmonic load-pull measurements," in *2010 IEEE MTT-S Int. Microwave Symp. Dig.*, Anaheim, CA, may 2010, pp. 1688–1691.

[3] J. Verspecht, M. V. Bossche, and F. Verbeyst, "Characterizing components under large signal excitation: defining sensible 'large signal S-Parameters'?!," in *49th ARFTG Conf. Dig.*, Denver, CO, 1997, pp. 109–117.

[4] T. S. Nielsen, M. Dieudonné, C. Gillease, and D. E. Root, "Doherty power amplifier design in gallium nitride technology using a nonlinear vector network analyzer and X-parameters," in *IEEE CSICS Dig.*, La Jolla, CA, Oct. 2012, pp. 1–4.

[5] Cree CGH40010 GaN HEMT; available at http://www.cree.com/rf/products/sband-xband-cband/packaged-discrete-transistors/cgh40010.

[6] S. C. Cripps, *RF Power Amplifiers for Wireless Communications*, 2nd edn. Norwood, MA: Artech House, 2006.

[7] W. H. Doherty, "A new high efficiency power amplifier for modulated waves," *Proc. IRE*, vol. **24**, no. 9, pp. 1163–1182, Sept. 1936.

[8] S. Strogatz, *Nonlinear Dynamics and Chaos: With Applications to Physics, Biology, Chemistry, and Engineering*. Reading, MA: Perseus Press, 1994.

[9] D. Webster, J. Scott, and D. Haigh, "Control of circuit distortion by the derivative superposition method," *IEEE Trans. Microw. Guid. Wave Lett.*, vol. **6**, no. 3, pp. 123–125, 1996.

Additional reading

M. S. Hashmi, F. M. Ghannouchi, P. J. Tasker, and K. Rawat, "Highly reflective load-pull," *IEEE Microwave*, vol. **12**, no. 4, pp. 96–107, 2009.

J. Horn, S. Woodington, R. Saini, J. Benedikt, P. J. Tasker, and D. E. Root; "Harmonic load-tuning predictions from X-parameters," in *IEEE PA Symp.*, San Diego, CA, Sept. 2009.

D. E. Root, J. Horn, T. Nielsen, *et al.*, "X-parameters: the emerging paradigm for interoperable characterization, modeling, and design of nonlinear microwave and RF components and systems," in *IEEE Wamicon 2011 Tutorial*, Clearwater, FL, Apr. 2011.

R. S. Saini, S. Woodington, J. Lees, J. Benedikt, and P. J. Tasker, "An intelligence driven active loadpull system," in *IEEE Microwave Measurements Conf. (ARFTG)*, Anaheim, CA, 2010, pp. 1–4.

G. Simpson, J. Horn, D. Gunyan, and D. E. Root, "Load-pull + NVNA = enhanced X-parameters for PA designs with high mismatch and technology-independent large-signal device models," in *IEEE ARFTG Conf.*, Portland, OR, Dec. 2008, pp. 88–91.

P. Tasker, "Practical waveform engineering," *IEEE Microwave*, vol. **10**, no. 7, pp. 65–76, 2009.

J. Verspecht, D. Gunyan, J. Horn, J. Xu, A. Cognata, and D. E. Root, "Multi-tone, multi-port, and dynamic memory enhancements to PHD nonlinear behavioral models from large-signal measurements and simulations," in *2007 IEEE MTT-S Int. Microwave Symp. Dig.*, Honolulu, HI, June 2007, pp. 969–972.

6 Memory

6.1 Introduction

X-parameters were introduced in Chapter 2 as frequency-domain nonlinear mappings. The X-parameter functions are defined on the complex amplitudes of the incident CW signals, returning the complex amplitudes of the scattered CW signals. For fixed power, bias, and load conditions, these input and output amplitudes are constant (in time) complex numbers. As such, X-parameters were defined only for steady-state conditions.[1]

Real applications deal with signals that vary in time and the corresponding time-varying responses of DUTs to such stimuli. In particular, signals used in wireless communications can have independent time-varying values for the amplitudes and phases associated with each of the multiple carriers involved. These time-dependent modulations contain the information associated with the signal. Much of communication systems design is devoted to trying to preserve the information content of these signals while efficiently amplifying, transmitting, and then demodulating them at the receiver to recover the information.

This chapter introduces the notion of dynamics into the signals and into the mappings between input and output stimuli that define the DUT as a nonlinear dynamic system. The envelope domain is introduced to define the class of signals considered. Quasi-static applications of static X-parameters are introduced as methods for estimating the DUT's response to modulated signals, and limitations are discussed. Examples are provided of phenomena that require the abandonment of the static nonlinear system description. This introduces the notion of memory, where the response of the system depends not only on the instantaneous value of the input signal, but also on the history of the input signal. The concept of modulation-induced baseband memory is articulated.

The second part of the chapter is devoted to a detailed exposition of *dynamic X-parameters*. This is a fundamental extension of X-parameters to modulation-induced baseband memory effects. The theory is rigorously developed, starting from a few basic assumptions that can be independently validated. Applications to actual DUT complex stimulus–response characterization and modeling for practical wide-band modulated

[1] This does not include special cases of pulsed X-parameters encountered in Chapter 2.

signals are presented to validate the approach. A significant simplification of the general dynamic X-parameters, valid in the limit of wide-band modulation, is presented under the name *wide-band X-parameters* (X_{WB}).

6.2 Modulated signals: the envelope domain

Modern communication signals can be represented efficiently by a sum of high-frequency carriers with complex amplitudes that vary in time. These signals can be described mathematically by (6.1):

$$A(t) = \mathrm{Re}\left\{\sum_n A_n(t)e^{j\omega_n(t)}\right\}. \tag{6.1}$$

Here, $A_n(t)$ is the time-varying complex amplitude associated with the nth carrier having frequency ω_n. The frequencies may or may not be commensurate. The signal $A(t)$ can have components at each port, but port indices are suppressed for simplicity.

The scattered waves from a nonlinear DUT stimulated with signal $A(t)$ can also be assumed to take a form equivalent to that of (6.1).[2] This is shown in (6.2):

$$B(t) = \mathrm{Re}\left\{\sum_m B_m(t)e^{j\omega_m(t)}\right\}. \tag{6.2}$$

Here, $\{\omega_m\}$ constitutes the set of all intermodulation frequencies produced by the DUT in response to the signal. There can be more spectral components generated by the nonlinear DUT, and therefore more time-varying envelopes, $B_m(t)$, than were present in the incident signal.

Complex envelopes are represented by their magnitudes and phases, each of which varies in time. This is detailed in (6.3) for the incident envelopes, with a similar expression, (6.4), holding for the envelope of each scattered wave:

$$A_n(t) = |A_n(t)|e^{j\phi(A_n(t))}, \tag{6.3}$$

$$B_m(t) = |B_m(t)|e^{j\phi(B_m(t))}. \tag{6.4}$$

The dynamic modeling problem considers how to define the mappings from the $\{A_n(t)\}$ of the incident waves to the $\{B_m(t)\}$ of the scattered waves.

6.3 Quasi-static X-parameter evaluation in the envelope domain

In this section, the case is considered where a static X-parameter model can be evaluated in the envelope domain to calculate the DUT's response to a modulated input signal, under the *quasi-static approximation*. This approximation asserts that, at any time t, the

[2] The case of a chaotic system, where the output spectrum can be qualitatively different, is not considered.

actual output time-dependent envelopes can be approximated by applying the static X-parameter mappings to the values of the input amplitudes at that same instant of time.

To see how this works, one starts with an X-parameter model that was identified from measurements presumed to have been taken under steady-state conditions. For example, consider a simple one-tone X-parameter steady-state mapping defined by (6.5):

$$B_{p,k} = X^{(F)}_{p,k}(|A_{1,1}|)e^{j\phi(A_{1,1})}. \tag{6.5}$$

All mismatch terms are neglected, for simplicity, and therefore only the $X^{(F)}_{p,k}$ terms are considered. The time variation of the input envelope is specified by either an explicit time-dependent *I–Q* signal or one of several modulated source components available in the simulator. The settings for the envelope simulation are chosen to allow adequate time sampling of the envelope corresponding to the input signal and to allow a suitably long period for the simulation.

In the quasi-static approximation, the DUT response, at time t, is computed to be the value of the static X-parameter mapping, as described in (6.5), applied to the value of the input envelope at the same time instant, t. This results in (6.6):

$$B_{p,k}(t) = X^{(F)}_{p,k}\left(|A_{1,1}(t)|\right)e^{j\phi(A_{1,1}(t))}. \tag{6.6}$$

This procedure produces time-varying envelopes for each of the harmonics, indexed by integer k, produced by the DUT in response to the modulated amplitude around the carrier at the fundamental frequency.

The quasi-static approximation is an extrapolation from steady-state conditions to a dynamic (time-varying) condition. It is certainly valid for sufficiently slowly varying $A_n(t)$, since it reduces to the static mapping as the instantaneous envelope amplitudes become constant. It is an accurate approximation to the actual time-dependent response provided that any underlying system dynamics are sufficiently fast that, at any instant of time, the DUT is nearly in its steady-state condition determined by the value of the input amplitude at that same time. That is, the system adiabatically tracks the input from one steady state to the next, parameterized by the time. The quasi-static approximation breaks down as the signal modulation rate increases and becomes comparable to or faster than timescales for which other dynamical effects become observable. In particular, electro-thermal effects, bias-line interactions, and other phenomena come into play for which a more elaborate dynamical description of the DUT is required. This will be discussed in more detail in the following.

6.3.1 Quasi-static two-tone intermodulation distortion from a static one-tone X-parameter model

In the following, the theory is illustrated using an actual amplifier, namely a Mini-Circuits ZFL-11AD+. This amplifier is characterized by performing a one-tone X-parameter model at a carrier frequency of 1750 MHz. The two-tone experiments are performed with a tone spacing of 1200 Hz, whereby the tones are symmetrically placed around the carrier frequency. A time-domain representation of the two-tone input

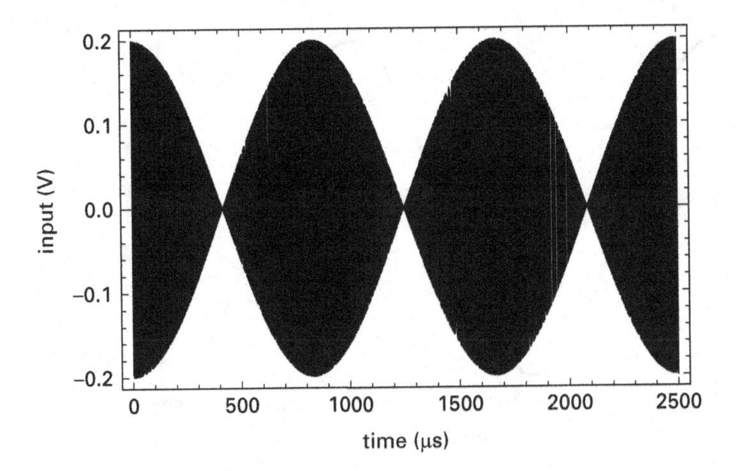

Figure 6.1 Two-tone input signal in the time domain.

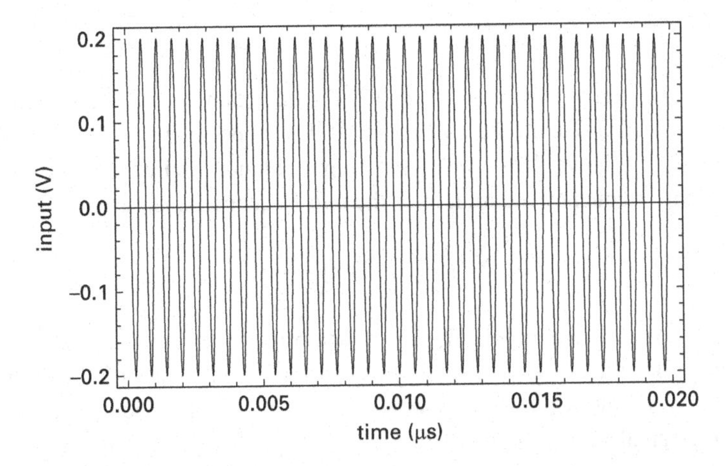

Figure 6.2 Two-tone input signal in the time domain (zoomed in).

signal is shown in Figure 6.1. The carrier oscillation period is so small compared to the millisecond range of the time axis of Figure 6.1 that the signal looks like a solid black area.

The individual oscillations can only be seen on a higher-resolution timescale expressed in nanoseconds rather than milliseconds. This is illustrated in Figure 6.2, which represents a zoom on the first 20 nanoseconds of the 2.5 milliseconds shown in Figure 6.1.

This two-tone sinusoidal signal can be represented as a single carrier with a modulated complex amplitude $A(t)$ according to (6.7). Equal magnitudes, A_1, for the two tones, are considered for simplicity:

$$\begin{aligned} A_1 \cos(\omega_1 t) + A_1 \cos(\omega_2 t) &= 2A_1 \cos\left(\frac{\Delta\omega}{2} t\right) \cos(\omega_0 t) \\ &= \mathrm{Re}\left\{ 2A_1 \cos\left(\frac{\Delta\omega}{2} t\right) e^{j\omega_0 t} \right\} = \mathrm{Re}\{A(t) e^{j\omega_0 t}\}. \end{aligned} \tag{6.7}$$

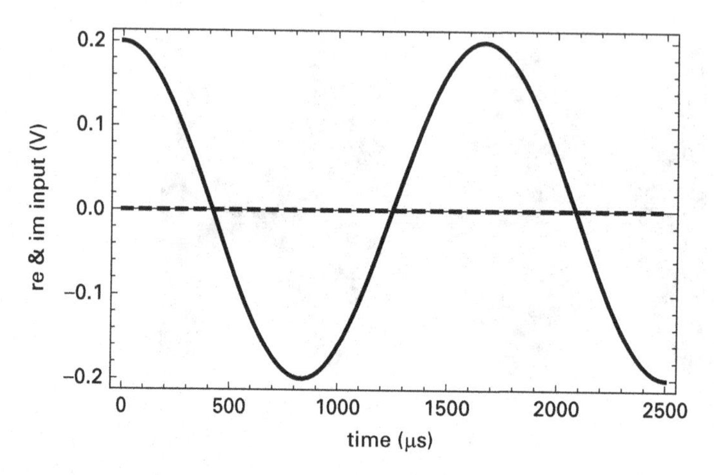

Figure 6.3 Two-tone input signal in the envelope domain: real part (solid line) and imaginary part (dashed line).

Here, $\Delta\omega = \omega_2 - \omega_1$ and $\omega_0 = (\omega_1 + \omega_2)/2$. For our example, ω_1 and ω_2 correspond to (1750 MHz − 600 Hz) and (1750 MHz + 600 Hz), respectively.

From (6.7), it follows that the envelope representation $A(t)$ of any two-tone signal, using as carrier frequency the middle of the two frequencies, ω_0, is simply given by

$$A(t) = 2A_1 cos\left(\frac{\Delta\omega}{2}t\right). \tag{6.8}$$

The real and imaginary parts of $A(t)$ are shown in Figure 6.3.

Consider now the transmission of this two-tone signal through a system that is described by a simple X-parameter model that only takes into account the transmission of the fundamental signal. In other words, only the presence of a term $X^{(F)}_{2,1}(.)$, which is here more simply denoted by $X^F(.)$, is considered in the following. The amplifier response $B(t)$ using the quasi-static approximation is calculated according to (6.6), whereby $A(t)$ is given by (6.8). The result is given below:

$$B(t) = X^F\left(|A(t)|\right)e^{j\phi(A(t))} \tag{6.9}$$

with

$$|A(t)| = 2A_1\left|\cos\left(\frac{\Delta\omega}{2}t\right)\right|, \tag{6.10}$$

$$\phi(A(t)) = \arccos\left(\operatorname{sgn}\left(\cos\left(\frac{\Delta\omega}{2}t\right)\right)\right). \tag{6.11}$$

The amplitude $|A(t)|$ and phase $\phi(A(t))$ are shown in Figure 6.4 and Figure 6.5, respectively.

Note that (6.11) is simply a way to express that $\phi(A(t))$ toggles between 0° and 180°, whereby $e^{j\phi(A(t))}$ toggles between +1 and −1, depending on the sign of the function

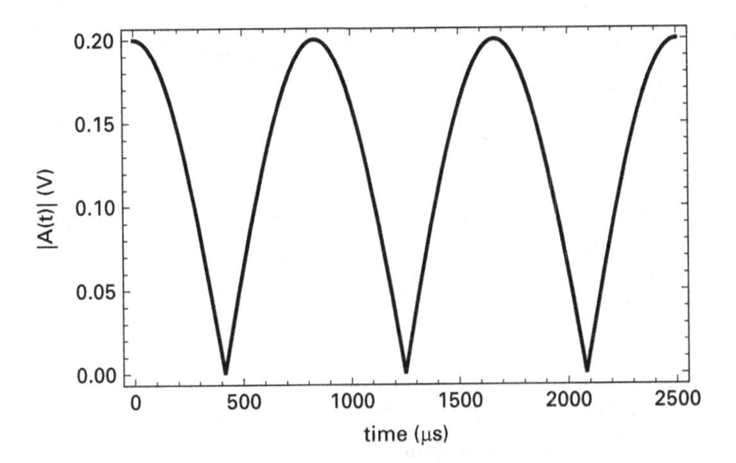

Figure 6.4 Input signal envelope amplitude.

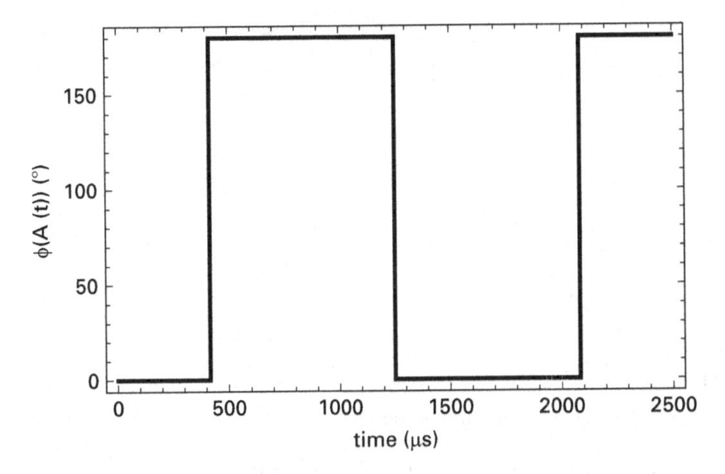

Figure 6.5 Input signal envelope phase.

$\cos(\Delta\omega t/2)$. The amplitude and phase of the X-parameter function $X^{(F)}(.)$ of our amplifier example are shown in Figure 6.6 and Figure 6.7, respectively.

Note that Figure 6.6 corresponds to a classic AM-to-AM characteristic, whereas Figure 6.7 corresponds to a classic AM-to-PM characteristic. The corresponding compression characteristic is shown in Figure 6.8.

Equations (6.9)–(6.11) are used to evaluate and analyze the DUT response. For the amplifier example considered, the real and imaginary parts of the two-tone response $B(t)$ are shown in Figure 6.9.

Note that the output signal $B(t)$, regardless of the shape of the nonlinear function $X^{(F)}(.)$, has exactly the same period as the function $\cos(\Delta\omega t/2)$, namely a period equal to $4\pi/\Delta\omega$. The periodic output envelope $B(t)$ can therefore be expanded into a complex Fourier series, with harmonic frequencies equal to an integer times $\Delta\omega/2$, as shown in (6.12):

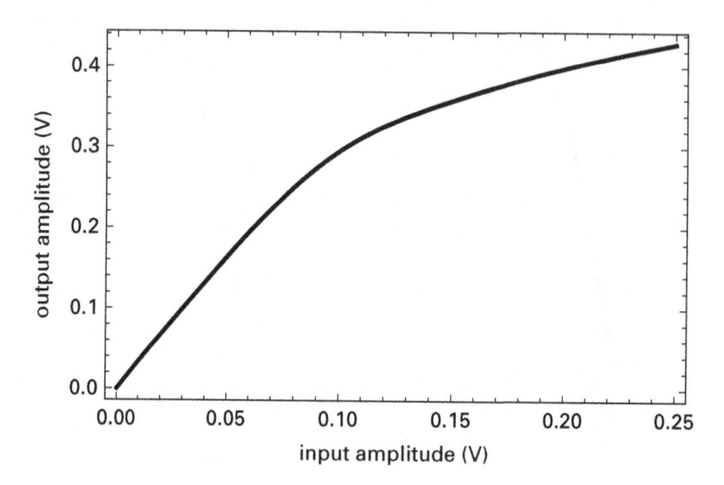

Figure 6.6 Amplitude of $X^{(F)}(.)$ versus input amplitude (AM-to-AM).

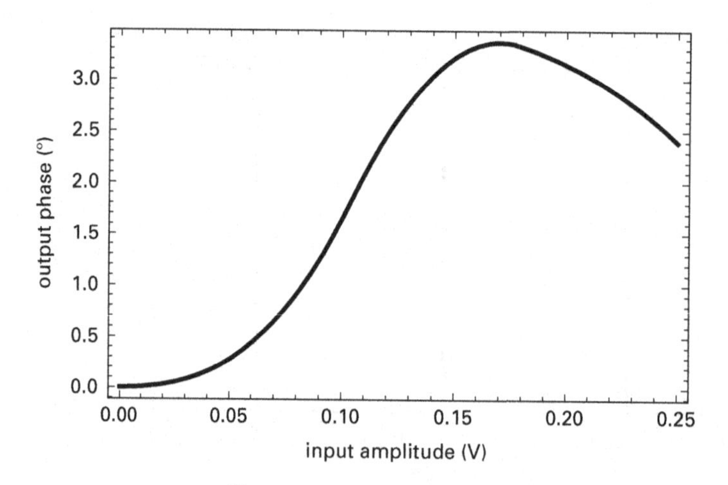

Figure 6.7 Phase of $X^{(F)}(.)$ versus input amplitude (AM-to-PM).

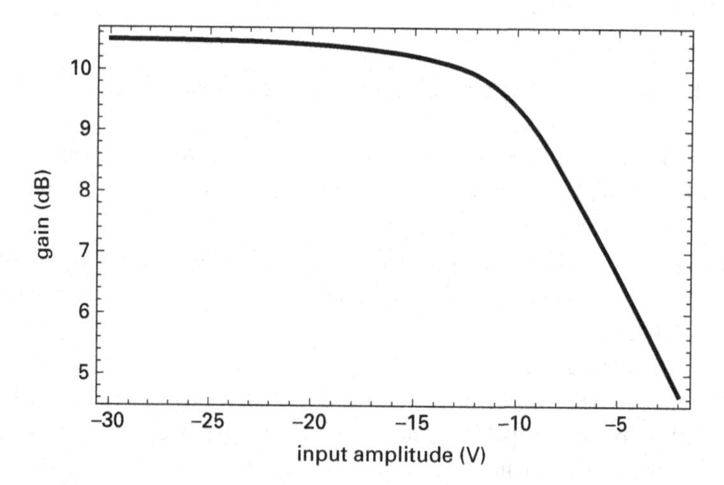

Figure 6.8 Compression characteristic.

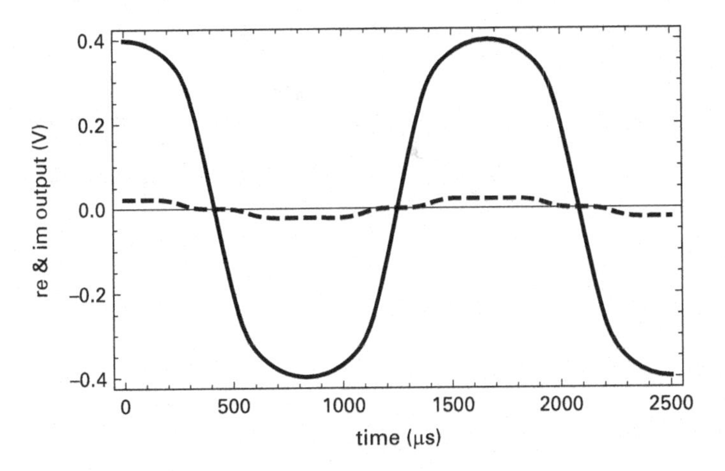

Figure 6.9 Calculated output signal in the envelope domain.

$$B(t) = \sum_{k=-\infty}^{+\infty} B_k e^{jk\frac{\Delta\omega}{2}t}. \tag{6.12}$$

The coefficients of this series, B_k, have some interesting properties. Based on the symmetry conditions

$$B\left(\frac{2\pi}{\Delta\omega} - t\right) = -B(t) \tag{6.13}$$

and

$$B(-t) = B(t), \tag{6.14}$$

which can readily be verified using (6.9)–(6.11), the resulting series B_k will always be a symmetric odd harmonic series. In other words, $B_k = 0$ if k is an even integer, and

$$B_k = B_{-k}. \tag{6.15}$$

These odd components B_k, having harmonic indices 1, 3, 5,... or −1, −3, −5,..., correspond to the upper and lower intermodulation products, respectively, generated by the two-tone signal. Note that the spacing between the harmonic frequencies, as one would expect for a two-tone signal, equals $\Delta\omega$. The lower third-order intermodulation (IM3) product corresponds to $k = -3$, the lower fifth-order intermodulation (IM5) corresponds to $k = -5$, the upper IM3 corresponds to $k = +3$, the upper IM5 corresponds to $k = +5$, etc.

By making use of the symmetry conditions (6.13) and (6.14) and the definition of the complex Fourier series, the following simple relationship between the one-tone X-parameter function $X^{(F)}(.)$ and the quasi-statically generated intermodulation products B_k results:

$$B_k = \frac{2}{\pi} \int_0^{\pi/2} X^{(F)}(2A_1 \cos(\theta)) \cos(k\theta) d\theta, \tag{6.16}$$

only valid, of course, for k equal to an odd integer.

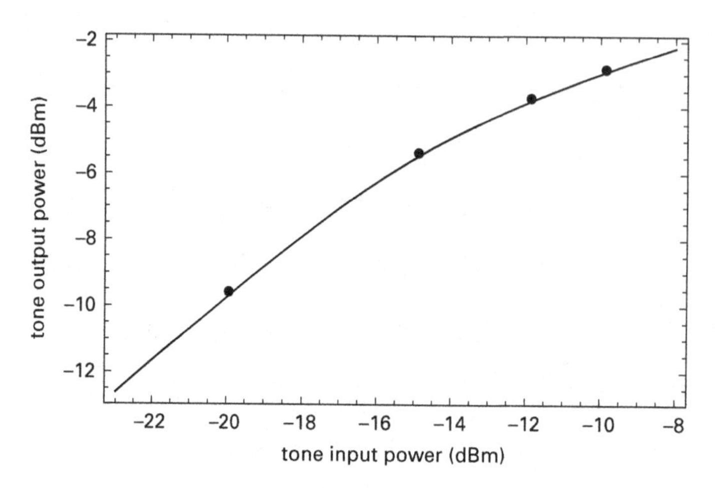

Figure 6.10 Measured and calculated tone output power.

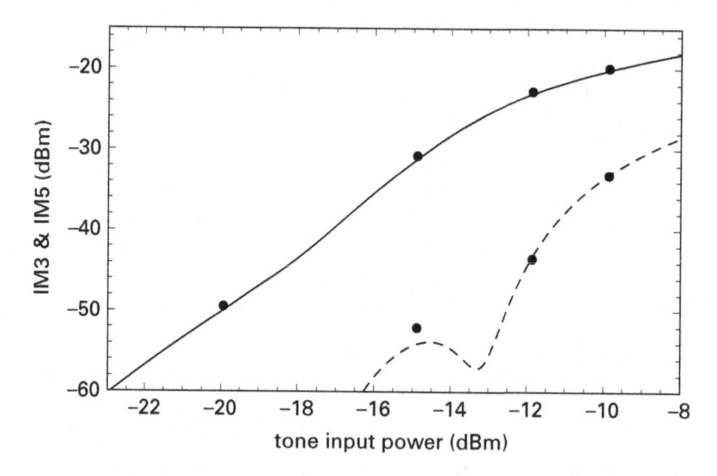

Figure 6.11 Measured (symbols) and calculated IM3 (solid line) and IM5 (dashed line).

Experimental validation of the described quasi-static approach is provided by comparing amplitude values of the intermodulation products B_k as calculated by using (6.16) with actual measured amplitude values, for several different input power levels. Figure 6.10 shows the calculated output power, per tone, as a function of the input power, per tone (solid line), together with actual measured data acquired from the example two-tone experiments (dots).

Figure 6.11 shows the calculated values of B_3 and B_5, which correspond to IM3 and IM5, respectively, together with the measured data points. The measured and calculated values correspond very well, thereby proving the validity of the quasi-static approach.

Equation (6.16) reveals another important characteristic of quasi-statically generated intermodulation products, namely that B_k only depends on A_1 and $X^{(F)}(.)$, but is

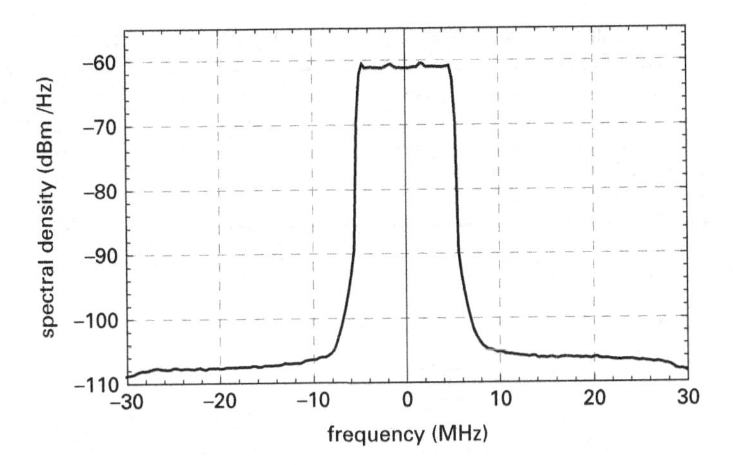

Figure 6.12 Input signal spectrum.

independent of $\Delta\omega$. In other words, the intermodulation products are independent of the spacing between the two tones of the input signal.

6.3.2 ACPR estimations using quasi-static approach

In a similar manner to the case of the sinusoidal amplitude modulation considered above, it is possible to estimate other nonlinear figures of merit (FOMs), such as the adjacent channel power ratio (ACPR) and others common to digital communications circuits, from a simple static X-parameter model in an envelope analysis. For digitally modulated signals, it is easiest to use built-in modulated sources available in the simulator. The simulator simply evaluates the static X-parameter model at each sampled time according to (6.6). This is illustrated using an actual device, namely a 25 W GaN monolithic microwave integrated circuit (MMIC) power amplifier (CREE CMPA2560025F). The device is first characterized by a static X-parameter model. This is equivalent to measuring the AM-to-AM AM-to-PM characteristic of the device. Next, a long-term-evolution (LTE) input signal with a 2.5 GHz carrier frequency is applied to the device. The spectrum of the input signal is shown in Figure 6.12.

The output signal is predicted by applying the static X-parameter model to the measured input envelope. Then the spectrum of the modeled output signal is compared to the actual measured output spectrum. The result is shown in Figure 6.13. Note the significant spectral re-growth in both the measured and the modeled output spectra.

The relevant FOM is computed from the simulated output spectrum according to the specific protocol appropriate to the modulation format. Examples of quasi-statically estimated ACPR from X-parameters for a real amplifier and independent experimental validation are provided in Table 6.1.

Note that there is a good correspondence between the measurements and the values derived from the static X-parameter model.

Table 6.1 ACPR estimated from static X-parameters versus independent measurements

	Output power (dBm)	Lower sideband ACPR (dB)	Upper sideband ACPR (dB)
X-parameter model	37.9	18.6	18.7
Measurements	38.0	19.0	19.1

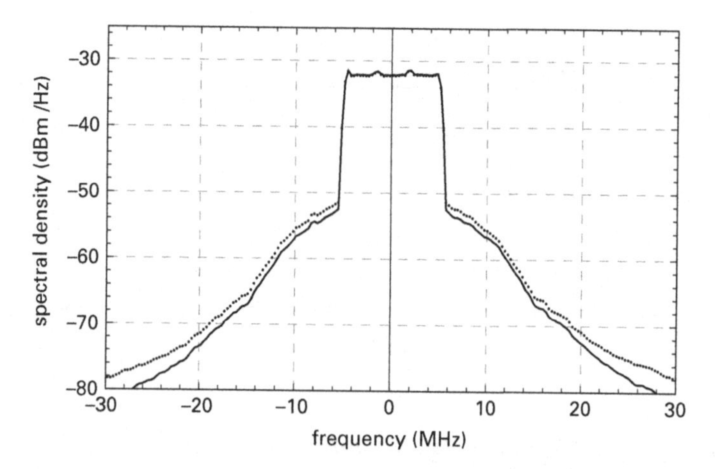

Figure 6.13 Output spectra: static X-parameter model (dotted line) and measurements (solid line).

6.3.3 Limitations of quasi-static approach

As explained in Section 6.3.1 the quasi-static evaluation of a static X-parameter model will produce an intermodulation spectrum with identical levels for upper and lower sidebands from a two-tone input signal. Moreover, the simulated levels are independent of the modulation rate – the frequency separation, Δf, between the two incident tones. That is, the distortion spectrum shows no "bandwidth dependence." In the envelope domain, it can be shown that the time-varying output envelopes are symmetric with respect to their peaks.

In the limit of slowly varying input envelopes, (6.6) reduces to evaluating a set of independent steady-state mappings at different power levels. That is, (6.6) reduces to (6.5), provided the latter was characterized over the full range of amplitudes covered by the time-varying input signal of (6.6). Simulations for modulated signals become exact as the modulation rate, or signal bandwidth (BW), approaches zero (the narrow-band limit).

For the two-tone case, it is possible actually to measure true steady-state X-parameters as functions of an LSOP that depends on both large input tones, using the approach discussed in Chapter 5. Two-tone X-parameters provide exact intermodulation spectra, in magnitude, phase, and their dependence on the frequency separation of the tones.

It should not be surprising that full information about the DUT response to two steady-state tones cannot be obtained from measurements taken with an LSOP set by only a single incident CW signal. It is clear there is more information in a set of two-tone measurements than in a single-tone measurement.

At high modulation bandwidths, the actual time-dependent scattered waves are no longer accurately computable from (6.6). That is, the quasi-static approximation of going from (6.5) to (6.6) becomes less valid. Any FOMs, such as third-order intercept (IP3) or ACPR, derived from the scattered waves simulated under the quasi-static approximation will therefore not be in complete quantitative agreement with the actual performance characteristics of the DUT.

6.3.4 Advantages of quasi-static X-parameters for digital modulation

Key nonlinear FOMs are usually scalars (numbers). For example, IP3, ACPR, and (error vector magnitude) EVM are numbers, typically computed from measurements made with a spectrum analyzer. The FOMs of particular components, such as individual stages of a multi-stage amplifier, cannot generally be used to infer the overall FOM for the composite system. That is, just knowing the FOMs of individual parts is not enough to design a nonlinear RF system for lowest ACPR.

X-parameters, on the other hand, enable at least a quantitative estimate of any nonlinear FOM for the full system just from knowledge of the X-parameters of the constitutive components. The X-parameters of the composite system follow from the nonlinear algebraic composition of the component X-parameters according to the general treatment in Chapter 2. The quasi-static approximation can be used easily in the envelope domain to evaluate the overall system response to modulated signals of various kinds.

Improved estimates of DUT behavior in response to wide-band modulated signals require a more careful treatment of dynamical behavior that begins in the following section.

6.4 Manifestations of memory

Before defining memory and dealing with it precisely, the topic is introduced with a set of frequently observed manifestations, or *signatures*, of memory. These can be considered evidence of underlying dynamical interactions that cause the DUT to behave differently from quasi-static inferences. None of these manifestations of memory are sufficient on their own to "characterize" the dynamics in a complete way, but rather they are evidence of the existence of dynamics for which a general treatment is required.

Real amplifiers exhibit behaviors more complicated than those predicted by the techniques of Section 6.2 for the case of rapidly varying input envelopes. For example, the intermodulation spectrum of a power amplifier in response to a large-amplitude two-tone excitation will show asymmetric upper and lower levels, with levels depending strongly on bandwidth (frequency separation of the tones). These

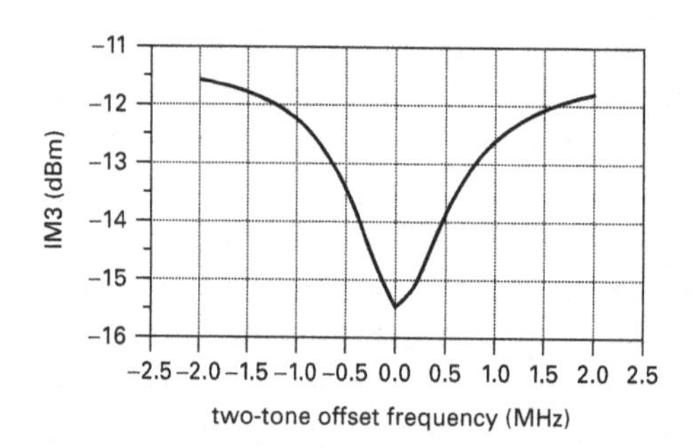

Figure 6.14 Frequency-dependent and asymmetric intermodulation spectrum.

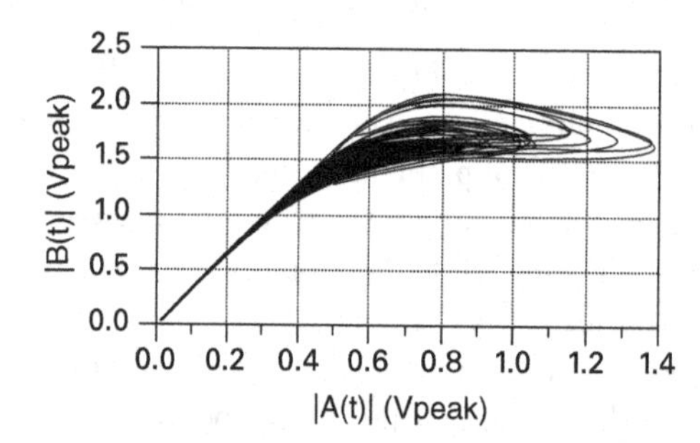

Figure 6.15 Multi-valued instantaneous AM-to-AM amplifier characteristics.

effects are induced by nonlinear behavior and are especially visible under high levels of compression. Some simulated results are shown in Figure 6.14.

For this simulation, a constant input amplitude of 6 dBm per tone is chosen, whereby the two-tone spacing is swept from nearly 0 Hz up to 4 MHz. The carrier frequency for the experiment is fixed at 2.6 GHz. The device is a radio-frequency integrated circuit (RFIC) amplifier, whereby the nonlinear memory effects are caused by the presence of a 6 nH inductor in series with the DC power supply. Note that the x-axis scale indicates the "offset frequency" of each of the two tones, thereby corresponding to half the tone spacing. The power at the output per tone is about 9 dBm. The magnitude of intermodulation distortion depends significantly on the separation frequency of the two excitation tones. Also evident is that the lower frequency sideband, at the maximum separation frequency, is about 1 dBm larger than the upper sideband. Another manifestation of memory is that the instantaneous AM-to-AM characteristic exhibits *multi-valuedness*. These effects are illustrated by performing a simulation whereby one uses bandwidth-limited noise as the input signal. The instantaneous output envelope amplitude $|B(t)|$ versus the instantaneous input envelope amplitude $|A(t)|$ is depicted in Figure 6.15.

It is evident that the output power is not uniquely determined by the input power at the same time instant. Rather, the output depends on the history of the input signal.

6.5 Causes of memory

Primary causes of memory effects include dynamic self-heating, bias-modulation effects, charge-storage (junction capacitances) and transit-time effects in active devices (transistors and diodes), and trapping effects in some semiconductor devices (e.g. drain lag in GaN FETs).

6.5.1 Self-heating

Self-heating involves the self-consistent coupling of electrical power dissipation in the device (transistor or amplifier) to a temperature change that in turn modulates the gain and therefore changes the output waveforms. Typically, self-heating in a transistor is modeled by two coupled equivalent circuits, one electrical and one thermal. The electrical circuit has current sources and capacitors, the values of which depend on voltages in the electric circuit and also the dynamic junction temperature, T_j, of the thermal circuit. The thermal circuit calculates the junction temperature, T_j, as a response to electrical power dissipation computed in terms of the voltages and currents of the electrical circuit, and thermal resistance and thermal capacitance parameters in the thermal circuit [13].

A hallmark of dynamic-thermal effects is that they vary at timescales typically many orders of magnitude slower than those of the RF signals. As the modulation rate becomes comparable to thermal-relaxation timescales, significant dynamical effects from the mutual electrical and thermal coupling become observable.

6.5.2 Bias modulation

Bias modulation occurs when a nonlinear device produces energy at baseband frequencies by intermodulation of higher-frequency RF components. The low-frequency signal components thus produced can pass through the bias lines, modulating the voltages and currents, causing them to vary in time, which in turn modulates the RF-scattering properties of the DUT. This is depicted in Figure 6.16 for the case of a pulsed RF excitation.

The figure illustrates how an inductor that is present in series with the DC power supply will cause memory effects. Consider an RFIC amplifier whereby one suddenly applies a significant amount of RF power at time zero, with no RF being present before time zero. As such, the envelope representation of the input signal $A_1(t)$ looks like a perfect step. The amplitude after the step will be noted as A_0. So what is the envelope representation of the corresponding output signal $B_2(t)$? As will be explained in the following, the $B_2(t)$ will typically look like a step with an exponential relaxation part on top, as shown in Figure 6.16. This can be explained as follows. In general, $B_2(t)$ will be a function, $F(.)$, of both $A_1(t)$ and the bias voltage. This is expressed as follows:

$$B_2(t) = F(A_1(t), V_{bias}(t)). \tag{6.17}$$

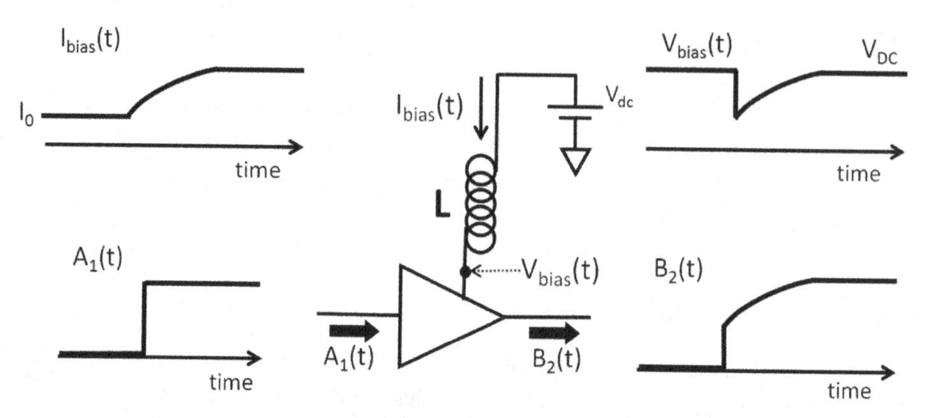

Figure 6.16 Illustration of biasing effects.

The bias voltage $V_{bias}(t)$ depends dynamically on the bias current $I_{bias}(t)$, expressed by

$$V_{bias}(t) = V_{DC} - L\,\frac{dI_{bias}(t)}{dt}. \tag{6.18}$$

Finally, the bias current $I_{bias}(t)$ is a function $I(.)$ of both $A_1(t)$ and $V_{bias}(t)$:

$$I_{bias}(t) = I\big(A_1(t), V_{bias}(t)\big). \tag{6.19}$$

So what happens if one applies a step RF input? Before the step is applied, the bias current has reached a steady-state value, I_0, and the bias voltage equals V_{DC}. The value of I_0 is given by

$$I_0 = I(0, V_{DC}). \tag{6.20}$$

Right after the step, at an infinitesimal time ε, the bias current $I_{bias}(\varepsilon)$ will still be equal to I_0 as the bias current through the inductor can only vary at a finite rate, as shown by (6.18). The bias voltage right after the step, $V_{bias}(\varepsilon)$, will be given by solving (6.21) as an implicit equation:

$$I_0 = I\big(A_0, V_{bias}(\varepsilon)\big). \tag{6.21}$$

For most circuits, $V_{bias}(\varepsilon)$ will be smaller than V_{DC}. The output amplitude right after the step, $B_2(\varepsilon)$, is calculated using (6.17):

$$B_2(\varepsilon) = F\big(A_0, V_{bias}(\varepsilon)\big). \tag{6.22}$$

Immediately after the step, the bias current $I_{bias}(t)$ starts to increase at a rate given by (6.23):

$$\frac{dI_{bias}(\varepsilon)}{dt} = \frac{1}{L}\big(V_{DC} - V_{bias}(\varepsilon)\big). \tag{6.23}$$

As time goes by, $I_{bias}(t)$ and $V_{bias}(t)$ will gradually, in a relaxation-like manner, approach their steady-state values corresponding to a CW excitation signal with amplitude A_0. The CW steady-state solutions are given by

$$\lim_{t \to \infty} V_{bias}(t) = V_{DC}, \tag{6.24}$$

$$\lim_{t \to \infty} I_{bias}(t) = I(A_0, V_{DC}) \tag{6.25}$$

and

$$\lim_{t \to \infty} B_2(t) = F(A_0, V_{DC}). \tag{6.26}$$

The shapes of the $V_{bias}(t)$, $I_{bias}(t)$, and $B_2(t)$ are qualitatively depicted in Figure 6.16.

A natural question then arises as to how this relates to the validity of the quasi-static X-parameter approach, as explained in Section 6.3. Suppose that such an inductor is present – under what circumstances will one be able to use the quasi-static X-parameter approach? The answer is present in (6.23), which shows that the derivative of the bias current is proportional to the reciprocal of the inductance L. In other words, the smaller the value of L, the faster all transitions occur. If any transitions happen faster than any noticeable variations of the input envelope amplitude, the resulting behavior will still look like a quasi-static map. The conclusion is that the quasi-static approach will still work for small enough L or for slow enough modulation.

But if that is not the case, what are our options? If one has direct access to the voltage bias node shown in Figure 6.16, one can still use CW X-parameters. The procedure is to measure directly the functions $F(.)$ and $I(.)$, whereby one not only sweeps the RF amplitude, A, but also V_{bias}, and whereby one measures the dependences of the RF output, $B(t)$, and the bias current, $I_{bias}(t)$, not only on the RF amplitude, but also on the bias voltage. Note that the relationships (6.17) and (6.19) remain valid under a CW RF excitation, so the functions $F(.)$ and $I(.)$ can be measured under CW operating conditions. In practice, this can be done, for example, by using an Agilent NVNA. The Agilent NVNA controls the biasing instruments and can be made to sweep the applied DC bias over a range of values large enough to cover any set of conditions expected to be encountered in the actual application. The DC currents (voltages), as well as their sensitivities to power and load, are simultaneously measured during the X-parameter extraction. They are represented by the $X^{(I)}$ and $X^{(Y)}$ functions that depend on power, bias, and RF frequency.

The extracted model will have an external port for the bias. The bias-modulation effects can be simulated by adding a circuit model of the bias network between this pin and the battery. This network can be a simple inductor, an L-R network, or more complicated L-R-C networks. Now the model can be simulated with a modulated signal. The low-frequency components generated by the device will cause the node voltage at the X-parameter bias pin to vary in time. This is another application of the quasi-static approximation, where the previously fixed voltages were once just parameters specifying the mapping from incident A's to scattered B's, but have now become time-varying variables that modulate the mapping. An example of such a circuit, and the results, are given in Figure 6.17.

The circuit that is used to illustrate the principle is an RFIC amplifier that needs two bias voltages. A first bias supply enters through the pin marked "v3ic" and a second bias supply enters the RFIC through the RF output (as such, a DC block is needed). The first bias does not draw any current and is not contributing to any memory effects. The

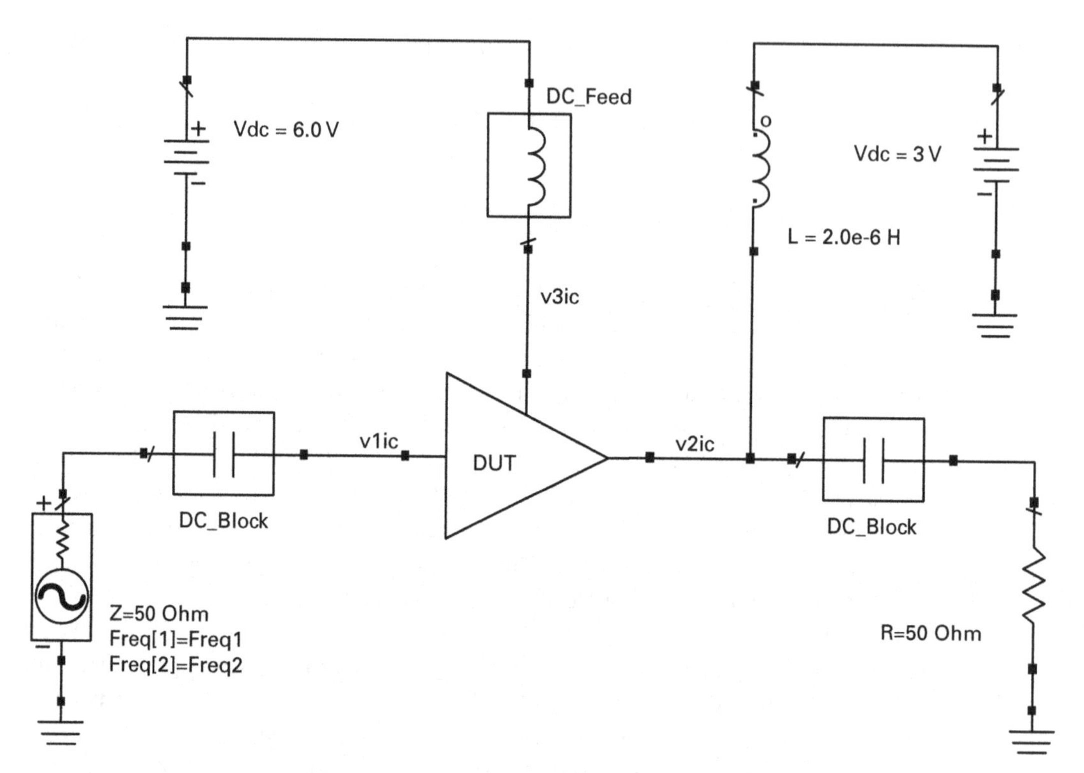

Figure 6.17 Schematic used for simulating bias-line-induced memory.

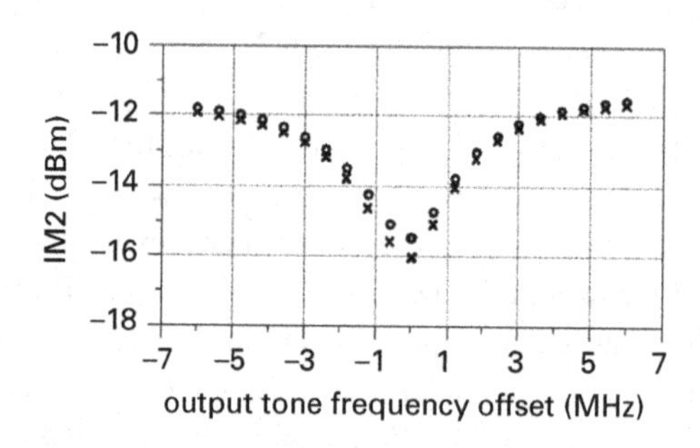

Figure 6.18 IM3: comparison between circuit schematic and X-parameter (o = circuit schematic, x = X-parameter model).

second bias draws all of the supply current. For the purpose of demonstration, a series 2 μH inductor is connected between the bias pin and the 3 V DC supply. As a first step, through simulation, an X-parameter model is extracted that contains the dependency of the bias current and the RF output on the bias voltage (pin "v2ic"). Next, a set of two-tone simulations is performed, whereby the simulation is performed on the original circuit schematic as well as on a schematic in which the RFIC has been replaced by the extracted X-parameter model. The results for IM3 and the output power per tone are shown in Figure 6.18 and Figure 6.19, respectively. It is evident that the method of

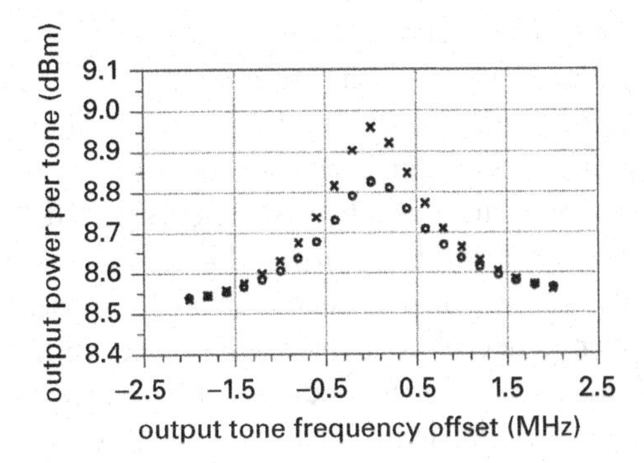

Figure 6.19 Two-tone output power: comparison between circuit schematic and X-parameter (o = circuit schematic, x = X-parameter model).

using bias-dependent X-parameters can produce the tone-separation dependence as well as the (slight) asymmetry of the magnitudes of the intermodulation spectral products.

6.6 Importance of memory

Memory effects make it much more difficult to quantify nonlinearity, since the standard figures of merit, such as IM3 and ACPR, depend sensitively on the modulation bandwidth and format. In particular, these FOMs cannot be a-priori calculated just from knowledge of static X-parameters (e.g. AM-to-AM and AM-to-PM characteristics). It is therefore more difficult to design circuits and systems trading off efficiency for linearity with nonlinear components demonstrating significant memory effects.

Efficiency is of such paramount importance that the power amplifier market is ready to sacrifice open-loop linearity, preferring to linearize these nonlinear devices externally, primarily with digital pre-distortion (DPD) and other techniques. Pre-distortion in the presence of strong memory effects is one of the major problems facing the power amplifier industry today. Control, elimination, or (ultimately) exploitation of memory effects would enable a dramatic improvement in performance from existing technologies. In all these cases, a good model of memory is a key component of a solution.

6.6.1 Modulation-induced baseband memory and carrier memory

Two distinct kinds of memory effects are significant for power amplifiers. They will be referred to as carrier-memory effects and modulation-induced baseband-memory effects.

Carrier memory is caused by dynamical effects that physically occur at the carrier frequency, typically at RF and microwave frequencies. Group-velocity effects and gain variation with frequency fall into this category. Simply sweeping the frequency of a single RF tone across the band of interest for an amplifier and noting the frequency response indicates RF carrier memory. Such effects have to be taken into account when one wants to construct a model that remains accurate across a bandwidth that is significant when compared to the RF carrier frequency. The addition of RF carrier memory to X-parameters is ongoing; it is not described here [14]. Note that this kind of memory does not require any nonlinear effects. It shows up for a linear filter as well as for an amplifier under small-signal operating conditions if there is a significant variation of the scattering parameters across the modulation bandwidth. For this reason, this kind of memory is often also called linear memory. It is sometimes called "short-term memory," referring to the fact that the timescale of this kind of memory effect is typically in the nanosecond range, comparable to the period of the RF carrier frequency. When a high-Q resonating filter structure is involved, however, carrier memory can introduce long time transients. Under such conditions, the terminology "short-term memory" is misleading. Therefore, the term "short-term memory" is not used in this book.

Modulation-induced baseband memory is of a different nature. It is caused by dynamical effects that physically happen in the baseband, at timescales many orders of magnitude slower than the RF stimuli, such as microseconds, milliseconds, or even seconds. Examples include the time variation of the bias settings, dynamic self-heating, and trapping effects. Although the physical effects happen relatively slowly, their influence shows up at the high-frequency carrier through the process of modulation. Dynamic temperature variations and dynamic bias settings may slowly modulate the compression and AM-to-PM characteristics of the amplifier. Such modulation-induced baseband-memory effects can become significant even at relatively small modulation bandwidths, for example a few kilohertz or megahertz. A general way to treat such memory effects in the framework of X-parameters is described in the following sections. Modulation-induced baseband memory is sometimes called "long-term memory," complementary to the term "short-term memory." As previously explained, this terminology is misleading as some carrier-memory effects may actually correspond to transitions that have longer settling times than some modulation-induced baseband-memory effects.

6.6.2 Dynamic X-parameters

6.6.2.1 Overview

The dynamic X-parameter method [3] is an original approach taken to extend X-parameters to include modulation-induced baseband-memory effects. Starting from a few basic assumptions, which can be validated by independent experiments, a full theory is developed by rigorous derivation. The approach results in a new term subsuming all long-term memory effects that simply adds to the existing static X-parameter expressions developed previously. The core of the new term is a "memory kernel" function that is uniquely identified from a new type of complex envelope transient measurement that can be measured on an NVNA or vector signal analyzer (VSA). The approach is extremely powerful in that within the original assumptions the resulting model is valid for any type of modulated signal, independent of format, power level, or details of the probability density function (PDF) defining the statistics of the signal. That is, the resulting behavioral model is highly transportable [5], [8]. The dynamic X-parameter method generically abstracts the intrinsic dynamics and nonlinearities of the DUT. It can therefore be directly used to predict accurately the DUT response to a wide variety of modulated signals, including those with high peak-to-average ratios (PARs) and wide modulation bandwidths, without the need to re-extract the model for different formats or average power conditions. Independent experimental validation of the model, applied to a real power amplifier subjected to very different types of modulated signals, confirms the theory. The model and its identification has a compelling and intuitive conceptual interpretation as well.

6.6.2.2 Motivation

The hysteretic behavior of the instantaneous gain depicted in Figure 6.15 suggests the starting point for the approach. The fact that the output RF amplitude is multiple valued means there must be additional independent variables that need be specified besides the RF input amplitude. Moreover, the dependence on these auxiliary variables must be identified by the modeling process.

For simplicity, only the case of a unilateral, perfectly matched, input–output map around a single carrier is considered here. The approach can be easily extended to multiple ports with mismatch, harmonics, and intermods.

6.6.2.3 Assumptions

The general time-dependent complex envelope of the scattered wave, $B(t)$, is assumed to be a nonlinear function of the input envelope, $A(t)$, and an arbitrary number of unknown real-valued dynamical variables, $h_i(t)$, for $i = 1,2,\ldots$. These "hidden" variables represent underlying physical processes such as time-varying temperature, bias voltages and currents, or trapping states in semiconductors. The unknown constitutive relation mapping these time-varying inputs to the output is given by

$$B(t) = X\Big(|A(t)|, h_1(t), \ldots, h_N(t)\Big)e^{j\phi(A(t))}. \tag{6.27}$$

The first assumption is that only the magnitude of $A(t)$ enters the nonlinear function in (6.27), with the phase an overall multiplicative factor. This was demonstrated to be a good approximation in [7]. Recall that it is rigorously true that the phase is a purely multiplicative factor in the static one-tone case, as derived in Chapter 2.

The second assumption is that the values of the hidden variables never depart greatly from their values under steady-state conditions. That is, $\Delta h_i(t)$, defined in (6.28), is small:

$$\Delta h_i(t) = h_i(t) - s_i\Big(|A(t)|\Big). \tag{6.28}$$

Here, $s_i(|A(t)|$ is the quasi-statically mapped value of $h_i(t)$, namely the value that $h_i(t)$ would have under steady-state conditions corresponding to an input amplitude magnitude of value $|A(t)|$ (see the discussion of Section 6.3). Small $\Delta h_i(t)$ means (6.27) can be linearized around the steady-state values of the hidden variables, simplifying the description and subsequent analysis considerably.

Substituting for $h_i(t)$ using (6.28) in (6.27), the function X is linearized around its value, with all hidden variables taking their (quasi-) static values, to first order in $\Delta h_i(t)$, to obtain the expression in (6.29):

$$B(t)=X\left(|A(t)|,s_1\left(|A(t)|\right)+\Delta h_1(t),\ldots,s_N\left(|A(t)|\right)+\Delta h_N(t)\right)e^{j\phi(A(t))}$$

$$\approx\left(X\left(|A(t)|,s_1\left(|A(t)|\right),\ldots,s_N\left(|A(t)|\right)\right)+\sum_{i=1}^{N}\frac{\partial X}{\partial h_i}\bigg|_{\left\{|A(t)|,s_1\left(|A(t)|\right),\ldots,s_N\left(|A(t)|\right)\right\}}\Delta h_i(t)\right)e^{j\phi(A(t))}$$

$$\equiv\left(X^{(F)}\left(|A(t)|\right)+\sum_{i=1}^{N}D_i\left(|A(t)|\right)\Delta h_i(t)\right)e^{j\phi(A(t))}. \tag{6.29}$$

The first term on the right-hand side of (6.29) is identified as the familiar static X-parameter function, $X^{(F)}$, because, under steady-state conditions, $\Delta h_i(t) = 0$ for each hidden variable, and this corresponds to the quasi-static case.

Specific assumptions about the explicit form of the hidden variables are now introduced in order to evaluate the new terms that appear in addition to the static X-parameter function in (6.29).

Each hidden variable is expressed according to (6.30):

$$h_i(t) = \int_0^\infty P_i(|A(t-\tau)|)k_i(\tau)d\tau. \tag{6.30}$$

The function $P_i(.)$ can be interpreted as a source term, describing how the device's nonlinear dependence on the input amplitude excites the hidden variable. The function $k_i(.)$ describes the dynamics of the process, namely how the source term at some time in the past affects the present value of the hidden variable. As such, $k_i(t)$ represents an impulse response of a linear filter governing the relaxation of the particular phenomenon represented by the ith hidden variable. Equation (6.30) is a convolution representing the causal response of the hidden variable to the source term generated by the input-signal envelope at all points in the past, according to the impulse response governing that dynamical process.

The fact that $P_i(.)$ depends on the magnitude of the complex amplitude describes the generic phenomena of conversion between the first zone signal and the baseband auxiliary variables. This type of "feedback" architecture has been considered in [10] and [11].

The term $\Delta h_i(t)$ appearing in (6.29) can now be evaluated using (6.28) and (6.30). For a constant input value of A (a true CW signal), the value of the hidden variables can be calculated from (6.30) to be

$$s_i(A) = P_i(|A|)\int_0^\infty k_i(\tau)d\tau. \tag{6.31}$$

The quasi-static value of the hidden variable is therefore given by

$$s_i(|A(t)|) = P_i(|A(t)|)\int_0^\infty k_i(\tau)d\tau. \tag{6.32}$$

From Equations (6.28), (6.30), and (6.32), (6.29) is re-expressed as (6.33), where the order of summation and integration have been interchanged:

$$
\begin{aligned}
B(t) &\approx \left(X^{(F)}\left(|A(t)|\right) + \int_0^{\infty} \sum_{i=1}^{N} D_i\left(|A(t)|\right)\left(P_i\left(|A(t-\tau)|\right) - P_i\left(|A(t)|\right)\right) k_i(\tau) \right) e^{j\phi(A(t))} \\
&= \left(X^{(F)}\left(|A(t)|\right) + \int_0^{\infty} G\left(|A(t)|, |A(t-\tau)|, \tau\right) d\tau \right) e^{j\phi(A(t))}.
\end{aligned}
\tag{6.33}
$$

In (6.33), the convenient notation $G(x,y,t)$ has been introduced, defining the *memory kernel* as follows:

$$
G(x,y,t) = \sum_{i=1}^{N} D_i(x)\left(P_i(y) - P_i(x)\right)k_i(t). \tag{6.34}
$$

It follows immediately from (6.34) that, for all values of x and t, the memory kernel vanishes when the first two arguments are equal. That is,

$$
G(x,x,t) = 0. \tag{6.35}
$$

Condition (6.35) corresponds to the case that the present and past values of the amplitude are the same, which is the condition defining the steady state. In steady state, the memory contribution must vanish since the entire contribution is already accounted for by the $X^{(F)}$ term in (6.33).

6.6.2.4 Discussion

The memory kernel, $G(x,y,t)$, depends nonlinearly on the instantaneous amplitude of the input signal, the past value of the input-signal envelope, and explicitly on the time at which the past value of the input occurred.

The remarkable fact, to be demonstrated below, is that the memory kernel can be identified uniquely from a specific set of nonlinear measurements. Once the memory kernel is "measured," it can be substituted into (6.33), and the contributions to the scattered waves from *all hidden variables* can be accounted for, no matter how many variables are assumed to be present in (6.27), in particular without the need to identify the functions P_i and k_i separately for each variable.

6.6.3 Identification of the memory kernel: conceptual motivation

Under a one-tone steady-state excitation at microwave frequencies, physical variables such as temperature assume specific constant values. That is, the temperature does not vary in time with the frequency of the RF signal, but rather assumes a fixed value based on the power dissipation averaged over a much longer timescale. It can be said that temperature has a "low-pass" dynamical behavior.[3] Steady-state measurements at different power levels (different values of A) therefore generally correspond to different values of DUT temperature. Care must be taken to vary the amplitude slowly enough so

[3] The particular value of the temperature depends on the power-added efficiency of the device at that power level and the effective thermal resistance.

that sampled measurements at different power levels will reflect the characteristics of the system at steady state.[4] Alternatively, the power can be stepped from one value to the next, but sufficient time must be allocated after each transition, such that the system reaches its steady-state value before making an RF measurement of the output. That is, each step in amplitude causes a thermal transient, where the RF and thermal variables are initially decoupled. Eventually the thermal transient dies away and the system returns to the new steady state corresponding to the value to which the RF amplitude was stepped.

The foregoing considerations motivate the procedure for identifying the memory kernel. Each hidden variable can be made to depart from its steady-state value by stepping the value of the input-signal amplitude. The step transition is related to the source term, and the time-dependent relaxation is related to the linear filter term of (6.30), respectively. In general, the response to such steps between steady-state values depends on both the initial state and the final state of the transition.

6.6.4 Step response of the memory kernel

A stepped-input complex-amplitude signal defined by (6.36) is now applied:

$$A(t) = \begin{cases} A_1, (t \leq 0), \\ A_2, (t > 0). \end{cases} \tag{6.36}$$

Substituting (6.36) into (6.33), the step response, denoted by $B^{step}(A_1, A_2, t)$, is evaluated for $t > 0$:

$$B^{step}(A_1, A_2, t) = \left(X^{(F)}(|A_2|) + \int_0^t G(|A_2|, |A_2|, \tau)d\tau + \int_t^\infty G(|A_2|, |A_1|, \tau)d\tau \right) e^{j\phi(A_2)}. \tag{6.37}$$

The second term on the right-hand side of (6.37) vanishes due to (6.35). The memory kernel can be isolated from (6.37) by taking the time derivative of both sides. The result is given by

$$G(A_2, A_1, t) = -\frac{d}{dt} B^{step}(A_1, A_2, t) e^{-j\phi(A_2)}. \tag{6.38}$$

It is apparent from (6.38) that the memory kernel is completely identified from the set of complex-amplitude step measurements from initial states A_1 to final states A_2. The time derivative converts the step response to an impulse response. Since G must be defined for all values of its arguments, the required measurements must sample transitions from all possible initial values of A to all possible final values. That is, if the range of amplitudes from $A = 0$ to $A = A^{max}$ is sampled at N discrete values, there are N^2 transitions that must be considered. For each transition, the time-dependent response, in magnitude and phase, must be sampled for all times from immediately after the step to times long enough for the DUT to have relaxed to its steady-state value.

[4] This type of variation is known as *adiabatic*.

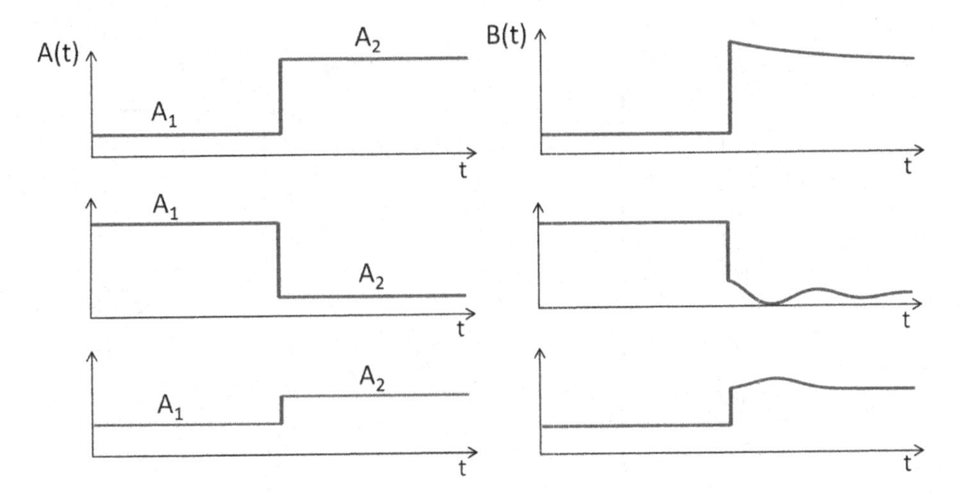

Figure 6.20 Step inputs and responses.

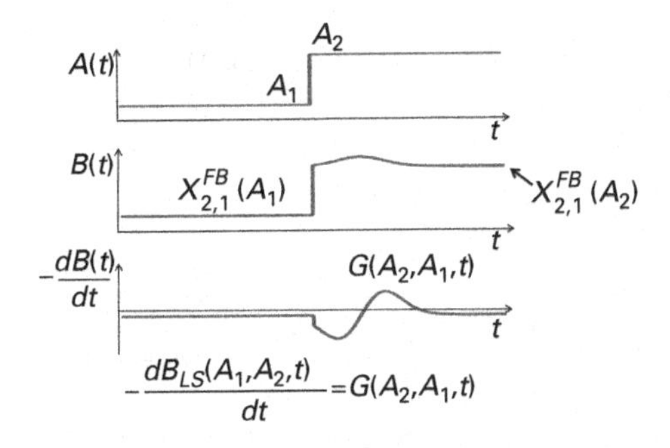

Figure 6.21 Identification of memory kernel from step responses.

The method is presented graphically in Figure 6.20 and Figure 6.21 for different steps and corresponding responses.

6.6.5 Application to real amplifier

The methodology described in Section 6.6.4 was applied to a Mini-Circuits ZFL11AD+ amplifier [4]. The carrier frequency used was 1.75 GHz. Fifteen different values of amplitudes from $0.01V_{peak}$ to $0.2V_{peak}$ were selected for the set of step envelope measurements. Figure 6.22 depicts input steps from two different initial values of A_1, a small value (top left) and a larger value (bottom left), to all other levels. The right side of the figure shows the magnitudes of the time-dependent responses corresponding to transitions to each of the final values of A_2 from the low initial value (top right) and from the higher initial value (bottom right).

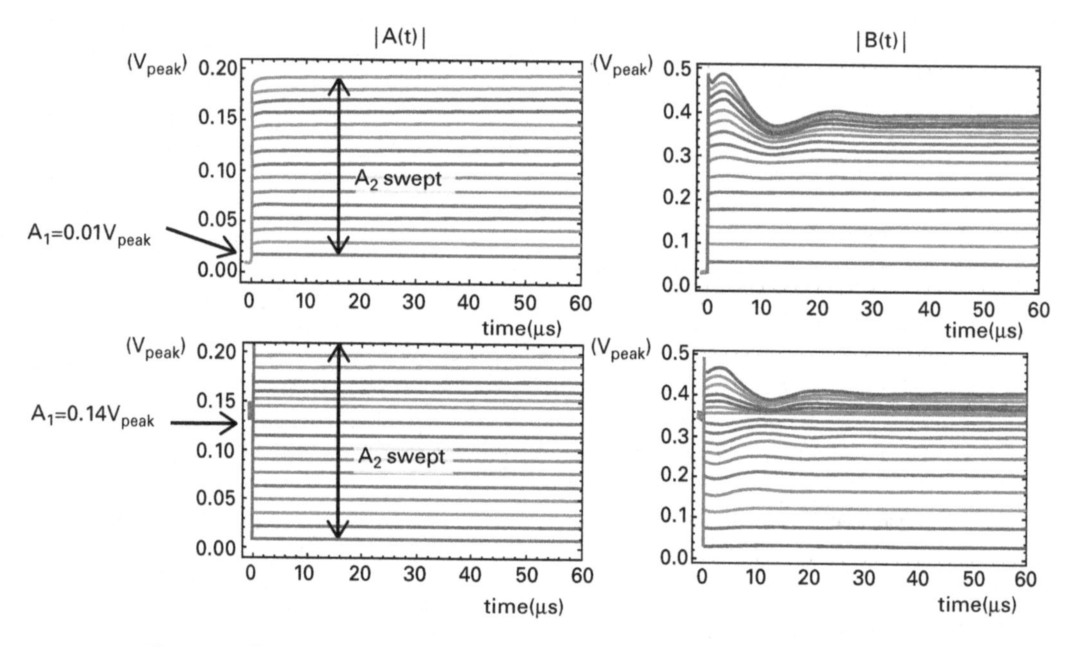

Figure 6.22 Step-response magnitudes from two initial amplitudes for Mini-Circuits ZFL11AD+ amplifier.

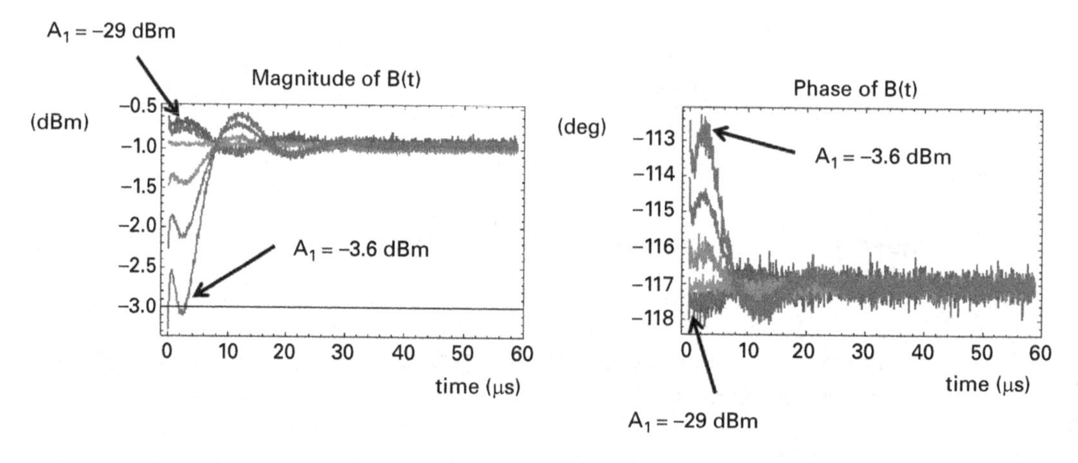

Figure 6.23 Magnitude and phase responses from various initial states to a fixed final step value of $A_2 = -10.6$ dBm.

The magnitude response and phase response corresponding to transitions to a fixed final value of $A_2 = -10.6$ dBm from a variety of initial values are shown in Figure 6.23.

The steady-state X-parameter unilateral transfer function at the fundamental frequency can be obtained from the long-term response of the step, or alternatively from the static X-parameter measurements described in Chapter 2. For the Mini-Circuits amplifier of this example, the results are presented in Figure 6.24.

The memory kernel, $G(A_2, A_1, t)$, is plotted versus A_2 and A_1 at a fixed time (50 ns) in Figure 6.25.

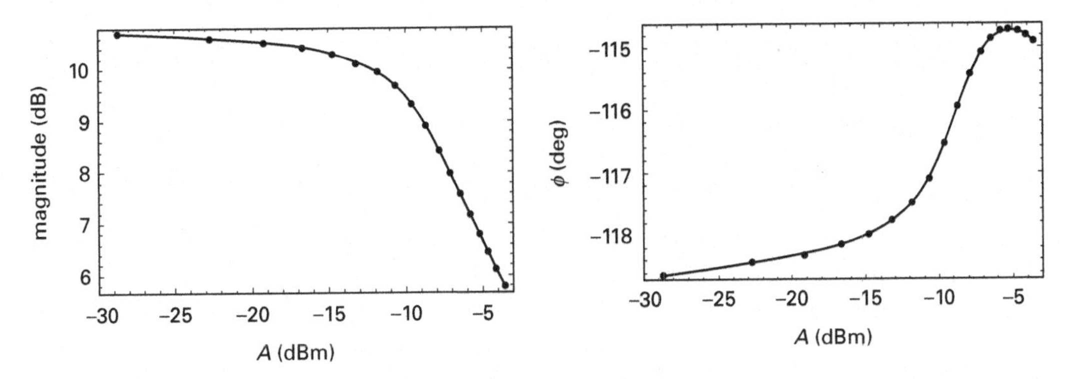

Figure 6.24 Static X-parameter gain function for Mini-Circuits ZFL11AD+ amplifier.

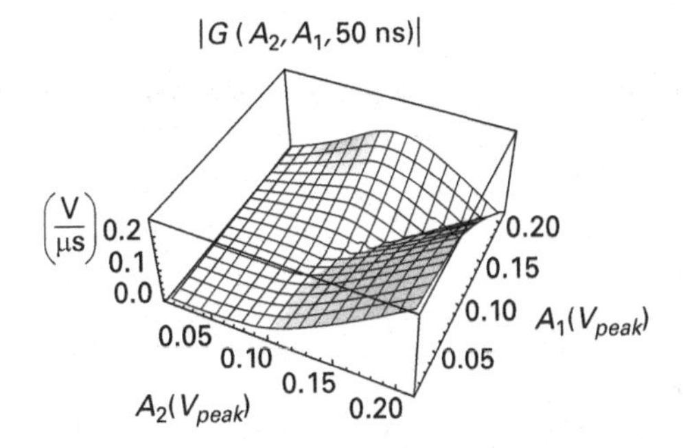

Figure 6.25 Memory kernel at fixed $t = 50\,\text{ns}$ for Mini-Circuits ZFL11AD+ amplifier.

The magnitude of the memory kernel as a function of time for a particular step transition is shown in Figure 6.26.

6.6.6 Validation of memory model

The memory model thus obtained from the step envelope measurements is implemented in the Agilent Advanced Design System (ADS) simulator.

The model is validated by comparing the actual measured DUT response to simulations using the model stimulated with the same signals. That is, the actual measured signals are brought into the simulator to excite the model of the DUT for a more accurate comparison. Any difference between measured response and simulated response is therefore due entirely to the model, and not due to slight differences between the ideal and actual modulated stimuli. The modulated signals are applied using an Agilent signal generator. The response measurements for validation are made on the Agilent NVNA. The measured response data are taken back into ADS to make the comparison with the simulated performance.

Envelope-domain validation results are shown in Figure 6.27, Figure 6.28, and Figure 6.29. Figure 6.27 shows the amplitude of the input envelopes for two-tone experiments at a frequency spacing of 19.2 kHz and for four different power levels. In

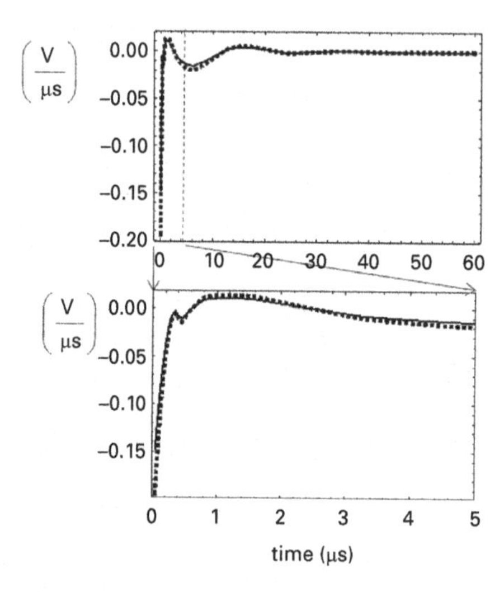

Figure 6.26 Memory kernel (V/μs) versus time (μs) for a step transition from low (0.01) to high (0.2) input amplitude. Real part (solid line); imaginary part (dotted line). The lower plot is a zoomed-in area of the upper plot.

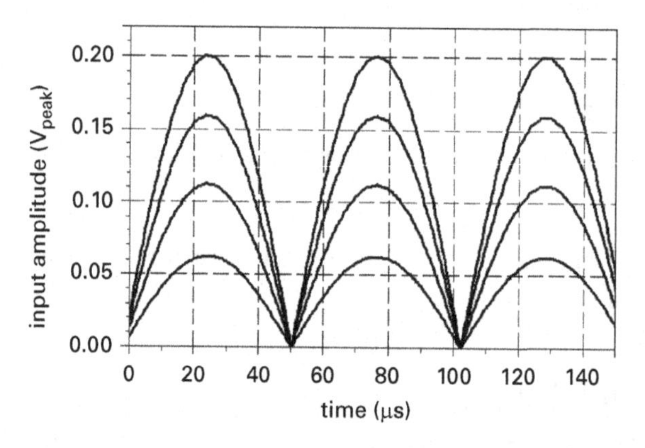

Figure 6.27 Dynamic X-parameter model validation: measured input envelope amplitudes for a two-tone frequency difference of 19.2 kHz.

Figure 6.28, the measured and simulated output waveforms are shown. The output envelopes corresponding to higher power levels are clearly not symmetric about their peak values, despite the fact that the input amplitudes are symmetric. In Figure 6.29, the amplitude of the error is shown between the measured and the modeled output envelopes. The simulated and measured output amplitudes agree very well. The memory model is therefore able to go far beyond the quasi-static application of the static X-parameter model discussed earlier. The improvement in accuracy is purely attributable to the second term in (6.33), involving the memory kernel.

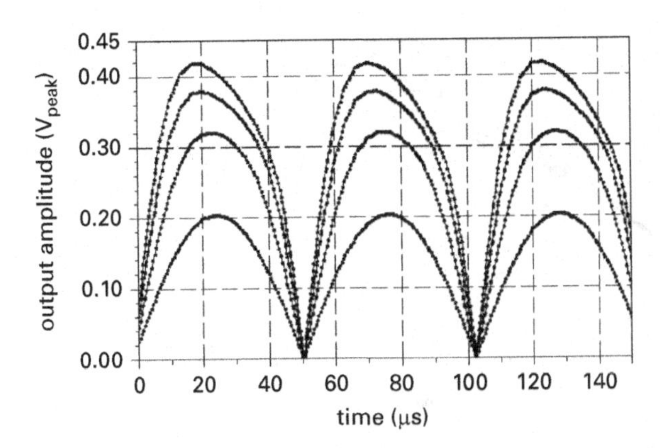

Figure 6.28 Measured and modeled output envelope amplitudes (measurement: solid line, model: dots).

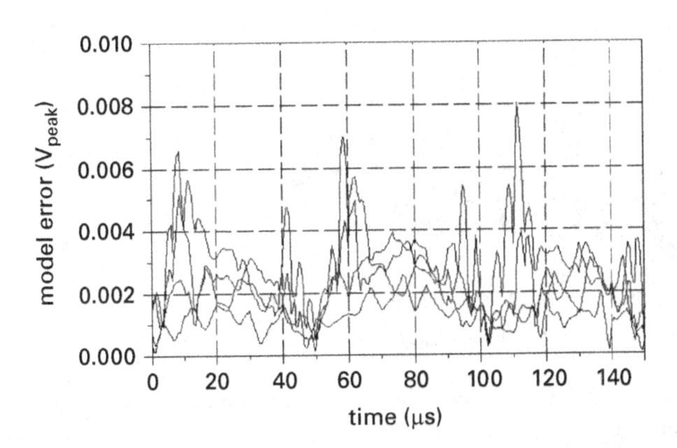

Figure 6.29 Amplitude of difference between measured and modeled output envelopes versus time.

Figure 6.30 shows the measured amplitudes of the output envelopes versus the instantaneous input amplitude; this is the so-called "dynamic compression characteristic." For the higher input-power levels, one can clearly see the looping of the characteristics. Figure 6.31 shows the differences between the model and the measurements (versus instantaneous input amplitude). One concludes that the dynamic X-parameter model accurately models the looping.

The memory model can also predict the detailed dependence of DUT gain and intermodulation distortion characteristics on modulation bandwidth. This is demonstrated in Figures 6.32–6.34. In Figure 6.32, the measured and simulated gains of the lower and upper signals are compared as a function of tone spacing over a wide range of frequencies. The sharp resonances and asymmetric values are well predicted.

In Figures 6.33 and 6.34, a comparison is presented between the measured and simulated frequency dependences of the two third-order intermodulation sidebands.

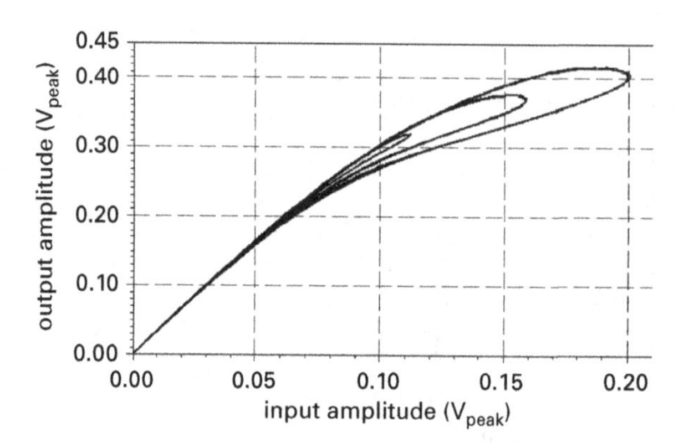

Figure 6.30 Measured dynamic compression characteristics at four different power levels.

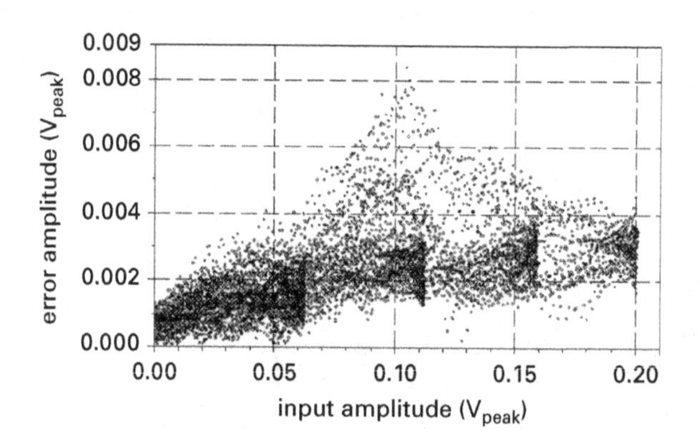

Figure 6.31 Amplitude of difference between measured and modeled output envelopes versus instantaneous input amplitude.

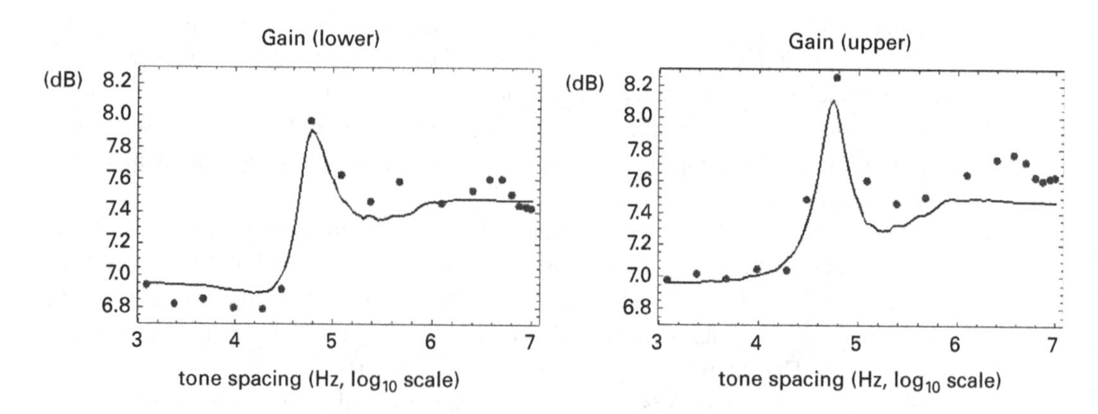

Figure 6.32 Gain versus tone spacing: measured and simulated with dynamic X-parameters.

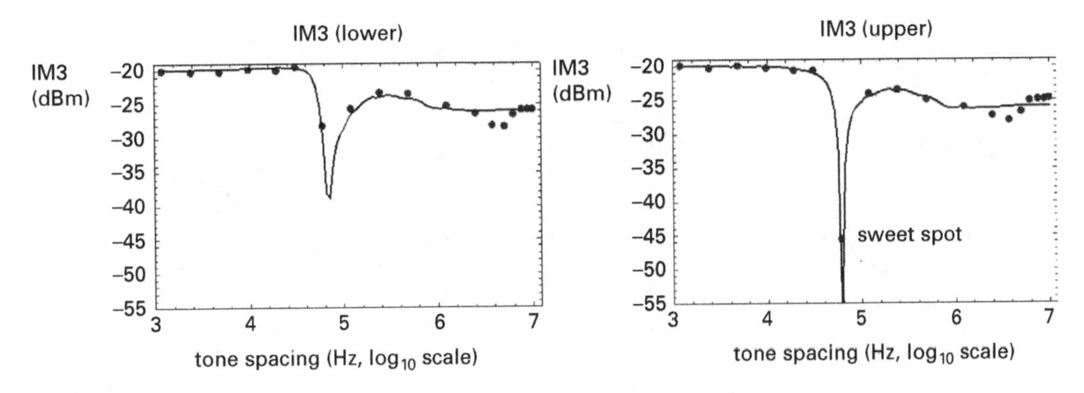

Figure 6.33 Magnitude of IM3 versus tone spacing: measured and simulated with dynamic X-parameters.

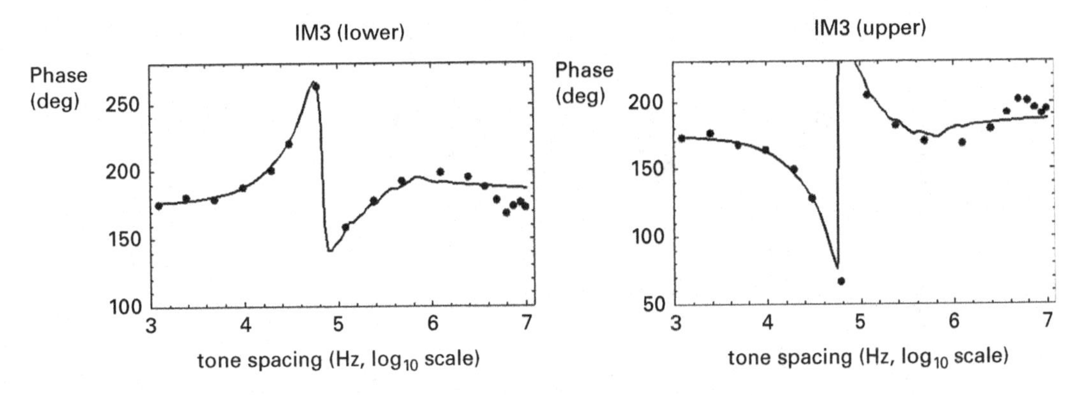

Figure 6.34 Phase of IM3 versus tone spacing: measured and simulated with dynamic X-parameters.

The model is able to predict very accurately both the magnitudes and the phases of the intermodulation products, including the sharp resonance.

The "IMD sweet spot" – an operating condition where the device behaves more linearly than at other nearby conditions, is clearly identified by the memory model. Another demonstration of this is given in Figure 6.35. The 60 kHz tone spacing corresponds to a peak in the gain and a minimum in the power of the third-order distortion product. The instantaneous gain, while still showing considerable hysteresis, is evidently more linear than the characteristics at 120 kHz tone spacing that compresses more with incident power.

Another stringent test of the memory model is how well it can predict the measured DUT characteristics in response to a wide-band digitally modulated signal. The same dynamic X-parameter model obtained for the Mini-Circuits amplifier was used to predict the modulated output envelope and spectral re-growth in response to a wide-band code-division multiple-access (WCDMA) signal. Comparisons of the dynamic X-parameter model simulation with measurements are shown in Figures 6.36 and 6.37. Also compared are the predictions of the static (CW) X-parameter model. The CDMA signal

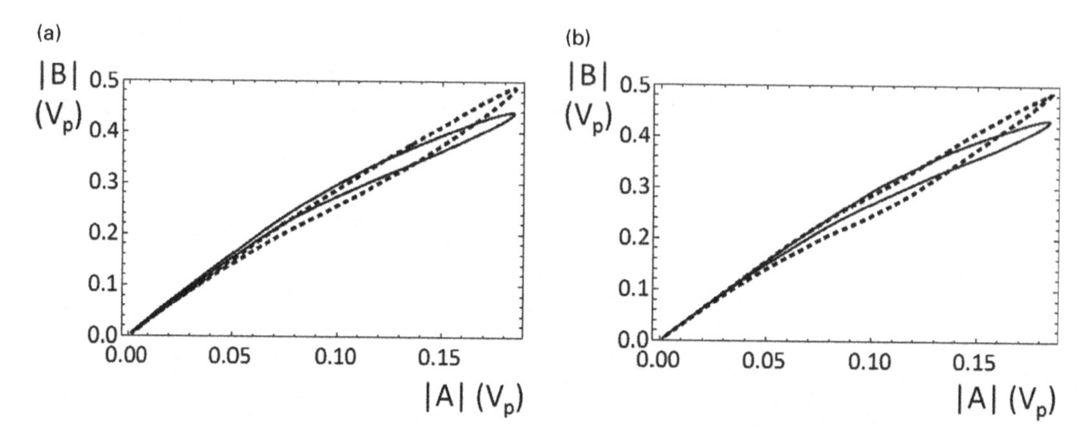

Figure 6.35 Measured (a) and modeled (b) dynamic compression characteristics (120 kHz tone spacing: solid line; 60 kHz tone spacing: dashed line).

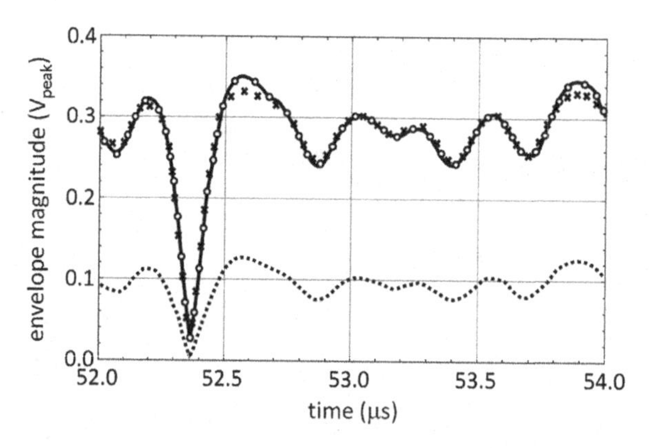

Figure 6.36 Validation of dynamic X-parameter and quasi-static X-parameter model simulations for CDMA input signal. Waveforms: solid line, measured; dotted line, input signal; x, static X-parameters; o, dynamic X-parameters.

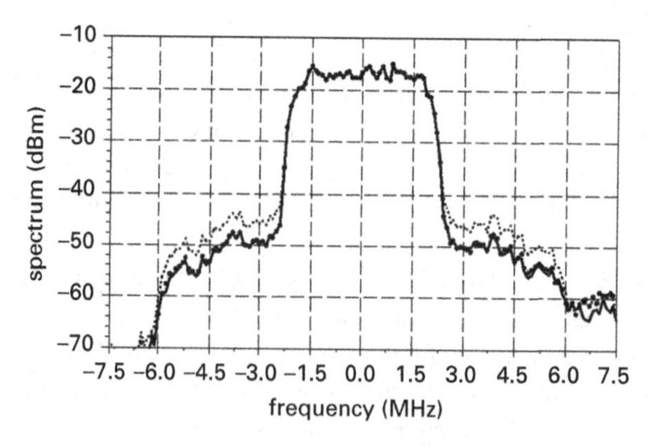

Figure 6.37 Validation of dynamic X-parameter and quasi-static X-parameter model simulations for CDMA input signal. Spectral re-growth: solid line, measured; •, dynamic X-parameters, dotted line, static X-parameters.

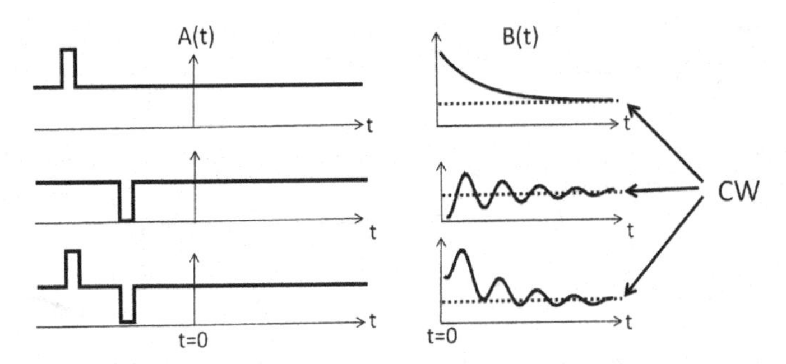

Figure 6.38 Interpretation of dynamic X-parameters in terms of superposition of memory effects.

was generated and uploaded to an Agilent ESG, and the waveforms were measured on the NVNA in envelope mode.

6.6.6.1 Discussion

It should be noted that both the two-tone signal (sinusoidally modulated carrier) and the WCDMA signals are completely different types of signals compared to the complex step excitations used to derive the memory kernel. Nevertheless, the dynamic X-parameter model is able to reproduce very accurately the actual characteristics of the measured DUT responses to each signal. In the case of the two-tone signal, the model is able to resolve the narrow resonance in gain and distortion with frequency separation of the tones. For the digitally modulated signal, the model does an excellent job describing the DUT output envelope time-dependent waveforms in the envelope domain and spectral re-growth in the frequency domain. In both cases, the simulations with the memory model are in much better agreement with the detailed measurements than the predictions of the CW X-parameter model evaluated in the quasi-static approximation. This is a powerful validation of the claim of transportability of the memory model.

6.6.7 Interpretation of dynamic X-parameters

An interesting and useful interpretation of the dynamic X-parameter theory can be made in terms of a principle of superposition of "nonlinear impulse responses." It has been demonstrated that step responses – or the impulse response given their time derivative – generate a departure from the steady state that eventually relaxes. The linearity assumption around the steady state is equivalent to the notion that the departure from steady state induced by a sequence of steps (impulses) is the superposition of the departures from steady state induced by each impulse separately. This is graphically illustrated in Figure 6.38. It is possible to check the applicability of dynamic X-parameter theory by seeing if the DUT's response to multiple pulses is indeed the superposition of the responses to individual pulses.

A key difference between these sets of impulses and other similar looking theories, such as [6] or [11], is that these impulses depend nonlinearly on both the initial and final

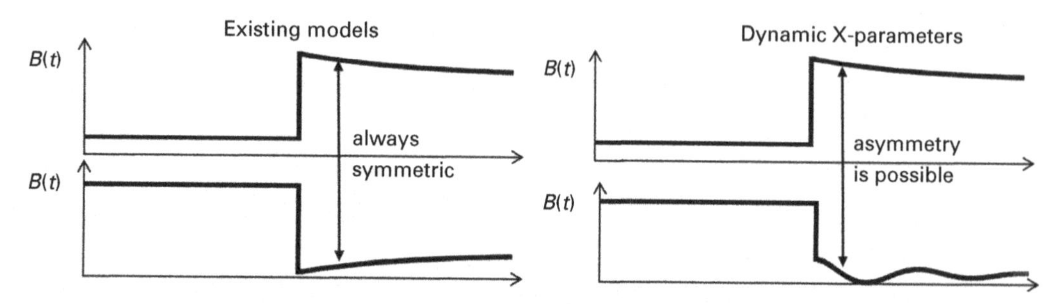

Figure 6.39 Complementary step responses are non-physical consequences of some "impulse response" envelope models, but not dynamic X-parameters.

values for the steps. The form of the resulting expression for $B(t)$, given by (6.33), is much more general. In particular, the turn-on and turn-off transients in a real power amplifier are completely different from each other. An actual turn-on transient will occur rapidly but not settle into its steady-state value for, say, microseconds until the device temperature reaches its steady-state value. When turned off, however, the RF amplitude becomes zero on timescales shorter than the RF frequency (limited only by transit time and stored charge in transistors or group delay considerations in amplifiers). These characteristics are accurately predicted by dynamic X-parameters. Other approaches, such as the nonlinear complex impulse response model (NCIRM), and more classical models, including the parallel Hammerstein model, can be shown to produce symmetric turn-on and turn-off transients. A schematic representation of this is shown in Figure 6.39. A mathematical proof of this statement for the parallel Hammerstein model is presented in Appendix C. Note that the calculation for an NCIRM model can be performed in a similar way.

6.6.8 Wide-band X-parameters (X_{WB})

6.6.8.1 Motivation

In the limit of narrow-band modulation, the system becomes arbitrarily close to its steady-state value and the contribution from the memory kernel vanishes. The device can be described in the narrow-band limit by the static (CW) X-parameters of Chapter 2. In many practical cases, the opposite limit, namely *wide-band modulation*, can also result in a simplification of the general dynamic X-parameter formalism [4].

The indication that this is possible is given by the following observations. A two-tone stimulus at various modulation rates (offset frequencies) is applied to the Mini-Circuits amplifier in Figure 6.40.

At 1.2 kHz offset (Figure 6.40(a)), the instantaneous AM-to-AM characteristic is essentially static. There is no hysteresis, which was a key signature of memory effects. At 30 kHz (Figure 6.40(b)), things are quite different. There is substantial hysteresis, indicative of significant dynamic memory effects. At a still higher frequency offset of 8.75 MHz (Figure 6.40(c)), the behavior qualitatively changes again. The looping is nearly gone and the device appears approximately static. That is, the AM-to-AM

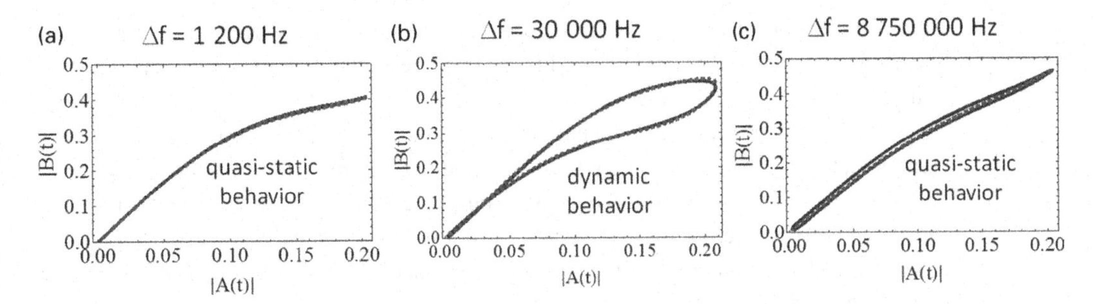

Figure 6.40 Nonlinear behavior as a function of signal bandwidth for Mini-Circuits ZFL-11AD+ amplifier.

characteristic is essentially single valued again. However, there is still a substantial difference between Figures 6.40(a) and (c). At large frequency offset, the effective static characteristics are quite different from the actual static characteristics. In particular, there is effectively less compression at high offset frequencies – so the system appears more linear than could be inferred from the actual static characteristics.

This can be understood in terms of timescales associated with hidden variables. At very low modulation rates, the hidden variables track the RF input-amplitude variation. At intermediate modulation rates, resonances in the dynamics of the hidden variables may be excited. The hidden variables depart from their quasi-static values, but their values vary in time, causing the hysteresis observed in Figure 6.40(b). Finally, at very high modulation rates, the hidden variables can no longer respond to the rapidly varying input envelope signal. Thermal variables have a low-pass characteristic as described in Section 6.6.3. Here the hidden variables assume constant values given by a suitable average over the amplitude variation.

An example of these considerations is furnished by a self-heating transistor. Static AM-to-AM and AM-to-PM characteristics are traced out by sampled measurements at different power levels. Measurements at each power level are made after waiting for the DUT to equilibrate thermally. The actual DUT "junction" temperature[5] is different at every different CW power level. For class B and class C amplifiers, the device typically gets hotter as the RF power increases.[6] The gain of the amplifier typically decreases with increasing temperature. If, instead, one ramped up the input power at a rate faster than the typical thermal response, one would get a higher value for the gain at the same instantaneous power, since the temperature would be lower than at the steady-state condition. One therefore expects more compression from static AM-to-AM and AM-to-PM measurements than from signals characterized by rapidly varying instantaneous power levels. In particular, for a two-tone input signal, with tones separated more widely than the inverse thermal relaxation time, the instantaneous AM-to-AM and AM-to-PM characteristics are essentially iso-thermal. The effective temperature in this case is somewhere between the low and high power values for the static case. From

[5] Temperature is defined everywhere in space and time. For simplicity, and according to standard practice, it is approximated by a single value that may vary in time.

[6] This is not true for class A amplifiers, which become cooler with increased applied power.

these simple considerations, the qualitative characteristics of Figure 6.40(a) and (c) can be explained. In particular, the wide-band DUT response is more linear than can be inferred from static characteristics.

Starting from the dynamic X-parameter formalism, one can derive the following simplification appropriate in the wide-band modulation limit. If the amplitude values in the past were always varying rapidly, their effects can only be manifested in the scattered waves according to constant time-averaged values. That means that the memory kernel appearing in (6.33) can be replaced by its time-averaged value, as indicated in (6.39):

$$\int_0^{\infty} G(|A(t)|,\ |A(t-\tau)|,\tau)d\tau \approx \int_0^{\infty}\left(\int_0^{\infty} G(|A(t)|,a,\tau)\ d\tau\right)\rho(a)da. \tag{6.39}$$

The average can be calculated by integrating the possible amplitudes over their probability distribution, $\rho(a)$, as indicated by the last equality of (6.39).[7] The advantage of the latter expression is that it is valid also for *stochastic* information signals that are defined only in terms of their PDFs.[8] Further discussion of this point is given in Appendix D.

The inner integral in (6.39) is evaluated by substituting for G the time derivative of the envelope response to the step excitations using (6.38), where the fact is used that, at infinite time, the step response from any initial condition to $A(t)$ is precisely the static X-parameter steady-state function $X^{(F)}(|A(t)|)$:

$$\begin{aligned}\int_0^{\infty} G(|A(t)|,a,\tau)d\tau &= -\int_0^{\infty}\frac{d}{d\tau}B^{step}(a,|A(t)|,\tau)d\tau \\ &= -B^{step}(a,|A(t)|,\infty)+B^{step}(a,|A(t)|,0) \\ &= -X^{(F)}(|A(t)|)+B^{step}(a,|A(t)|,0).\end{aligned} \tag{6.40}$$

Substituting (6.40) into (6.39), and adding the result back into (6.33), the final expression for wide-band X-parameters is obtained as follows:

$$B(t) \approx \left(\int_0^{\infty} X_{WB}(|A(t)|,a)\rho(a)da\right)\cdot \exp\left(j\phi(A(t))\right), \tag{6.41}$$

whereby the bivariate model kernel $X_{WB}(.)$ is defined as follows:

$$X_{WB}(x,y) = B^{step}(y,x,0). \tag{6.42}$$

The above equation implies that only the output value at time zero, immediately after a step, determines the behavior of the system under wide-band modulated excitation signals. A pictorial representation is given in Figure 6.41.

[7] A lower-case a is used here to avoid confusion with the instantaneous amplitude denoted by A.
[8] This is justified by the ergodic principle, namely that the time average is equal to the statistical average.

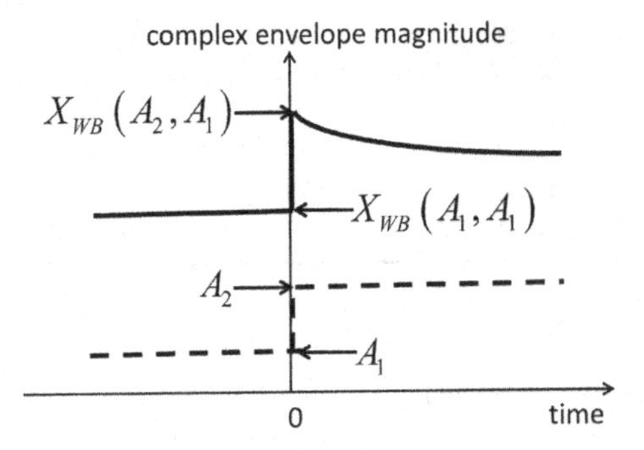

Figure 6.41 Wide-band X-parameters can be identified from the initial step response.

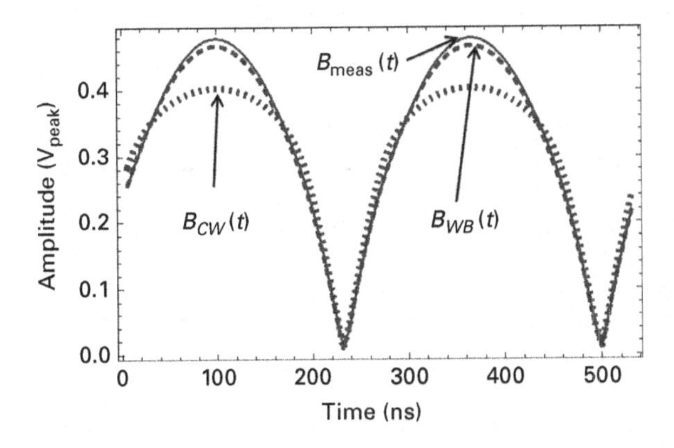

Figure 6.42 Validation of wide-band X-parameters for a two-tone signal.

6.6.8.2 Results for wide-band X-parameters

Figure 6.42 and Figure 6.43 demonstrate the results for wide-band X-parameters on the Mini-Circuits amplifier. A two-tone experiment is performed with a tone spacing of 3.75 MHz. Wide-band X-parameters, denoted with the "WB" subscript, are compared to the NVNA measured data and the static X-parameters, which are here denoted with the "CW" subscript (for continuous wave). Figure 6.42 shows the fundamental output time-varying envelope of the amplifier produced in response to a two-tone input signal. Figure 6.43 shows the gain compression and magnitude of the third-order intermodulation products versus input tone power.

Tables 6.2 to 6.4 contain the measured and modeled two-tone gain, IM3 products, and IM5 products for the four input power levels.

Except for the lowest power level IM5 product, which is near the noise floor of the measurements, the wide-band X-parameter model predicts gain, IM3, and IM5 products with an error smaller than 1 dB; the errors of the static model are significantly higher.

Table 6.2 Comparison of measured and modeled gain

Input power (dBm per tone)	−20	−15	−12	−10
Measured gain (dB)	10.41	9.84	8.85	7.97
WB X-parameter gain (dB)	10.39	9.71	8.60	7.65
CW model gain (dB)	10.35	9.52	8.19	7.07

Table 6.3 Comparison of measured and modeled IM3

Input power (dBm per tone)	−20	−15	−12	−10
Measured IM3 (dBm)	−53.0	−35.5	−28.9	−28.0
WB X-parameter IM3 (dBm)	−52.4	−35.1	−28.5	−27.2
CW model IM3 (dBm)	−48.0	−31.0	−23.1	−20.1

Table 6.4 Comparison of measured and modeled IM5

Input power (dBm per tone)	−20	−15	−12	−10
Measured IM5 (dBm)	−82.0	−68.4	−42.8	−36.1
WB X-parameter IM5 (dBm)	−71.2	−67.5	−42.8	−36.9
CW model IM5 (dBm)	−67.4	−53.8	−44.0	−33.5

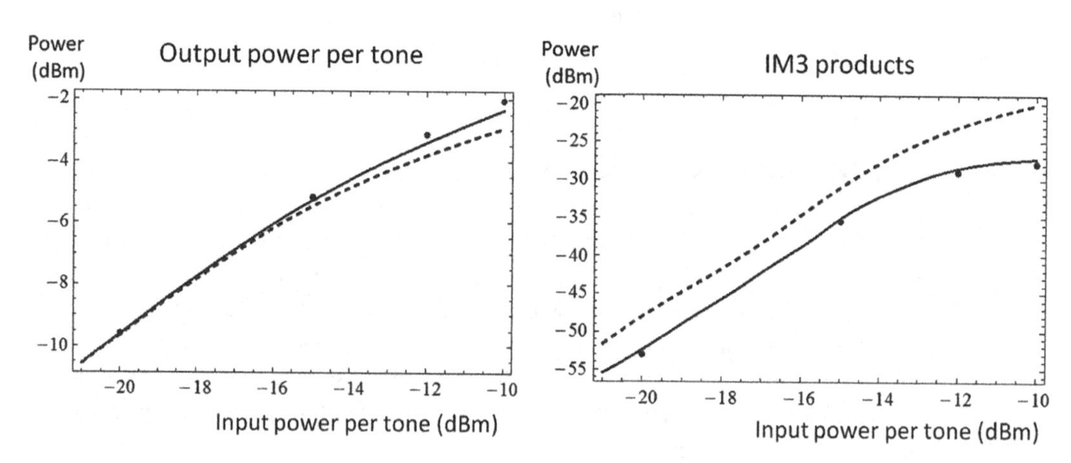

Figure 6.43 Validation of wide-band X-parameter: gain compression and intermodulation (measured, dots; WB model, solid line; CW model, dashed line).

It is observed that the wide-band X-parameters define an *effective quasi-static* model that is different from the actual static (CW) X-parameters. There is no hysteresis in the predicted instantaneous AM-to-AM, for instance. The predictions with wide-band X-parameters are much closer to the actual measured characteristics of the DUT than those predicted by the CW X-parameters. The results are quite consistent with the qualitative reasoning presented in Section 6.6.8.1. In particular, the device appears to be "more linear" for wider bandwidths based on its less compressed AM-to-AM characteristics and also its decreased intermodulation distortion values.

6.6.8.3 Discussion

Wide-band X-parameters still require stepped complex amplitude step responses to characterize the bivariate kernel appearing in (6.41). However, this represents a drastic simplification of the trivariate kernel of the full dynamic X-parameters required for (6.38). This results in many fewer measurement points, since only the initial time point after the transition needs to be measured, rather than the entire time record of the response all the way to steady state.

It is also evident from (6.41) that this effective static model requires knowledge only of the bivariate step response, and is therefore valid for any (fast) modulated signal. It is independent of format, PDF, average power, or other details of the signals.

The accuracy of the approach is expected to be well suited for continuously modulated very wide-band signals, as is characteristic of digitally modulated communications signals. Even if some of the energy in such broad-band signals happens to fall near a narrow dynamical resonance, it represents a tiny fraction of the total energy of the signal and is not expected to require the full dynamic X-parameter approach.

Of course, in general, the dynamic X-parameter theory is a superset of both the CW and WB X-parameters, and reduces to them in the appropriate static and wide-band limits, respectively.

References

[1] M. Rudolph, C. Fager, and D. E. Root, Eds., *Nonlinear Transistor Model Parameter Extraction Techniques*. Cambridge: Cambridge University Press.

[2] A. Soury and E. Ngoya, "Handling long-term memory effects in X-parameter model," in *IEEE Int. Microwave Symp. Dig.*, Montreal, Canada, June 2012, pp. 1–3.

[3] J. Verspecht, J. Horn, L. Betts, C. Gillease, and D. E. Root, "Extension of X-parameters to include long-term dynamic memory effects," in *2009 IEEE MTT-S Int. Microwave Symp. Dig.*, Boston, MA, June 2009, pp. 741–744.

[4] J. Wood, D. E. Root, and N. B. Tufillaro, "A behavioral modeling approach to nonlinear model-order reduction for RF/microwave ICs and systems," *IEEE Trans. Microw. Theory Tech.*, vol. **52**, no. 9, part 2, pp. 2274–2284, Sept. 2004.

[5] D. E. Root, J. Wood, and N. Tufillaro, "New techniques for nonlinear behavioral modeling of microwave/RF ICs from simulation and nonlinear microwave measurements," in *40th ACM/IEEE Design Automation Conf. Proc.*, Anaheim, CA, June 2003, pp. 85–90.

[6] J. Xu, M. M. Iwamoto, J. Horn, and D. E. Root, "Large-signal FET model with multiple time scale dynamics from nonlinear vector network analyzer data," in *IEEE MTT-S Int. Microwave Symp. Dig.*, Anaheim, CA, May 2010, pp. 417–420.

[7] J. C. Pedro and S. A. Maas, "A comparative overview of microwave and wireless power-amplifier behavioral modeling approaches," *IEEE Trans. Microw. Theory Tech.*, vol. **53**, no. 4, Apr. 2005.

[8] A. Soury, E. Ngoya, and J. M. Nebus, "A new behavioral model taking into account nonlinear memory effects and transient behaviors in wideband SSPAs," in *2002 IEEE MTT-S Conf. Dig.*, Seattle, WA, June 2002, pp. 853–856.

[9] J. Verspecht, J. Horn, T. Nielsen, and D. E. Root, "Extension of X-parameters to include long-term memory effects," in *IEEE ARFTG Conference*, Clearwater Beach, FL, Dec. 2010.

[10] J. Wood and D. E. Root, *Fundamentals of Nonlinear Behavioral Modeling for RF and Microwave Design*. Norwood, MA: Artech House, 2005, chap. 3.

Additional reading

P. Roblin, D. E. Root, J. Verspecht, Y. Ko, and J. P. Teyssier, "New trends for the nonlinear measurement and modeling of high-power RF transistors and amplifiers with memory effects," *IEEE Trans. Microw. Theory Tech.*, vol. **60**, no. 6, part 2, pp. 1964–1978, 2012.

D. E. Root, M. Iwamoto, and J. Wood, "Device modeling for III-V semiconductors: an overview," *IEEE CSIC Symp.*, Monterey, CA, 2004, pp. 279–282.

D. E. Root, D. Sharrit, and J. Verspecht, "Nonlinear behavioral models with memory: formulation, identification, and implementation," *IEEE MTT-S Int. Microwave Symp. Workshop (WSL) on Memory Effects in Power Amplifiers*, San Francisco, CA, June 2006.

J. Verspecht, D. Gunyan, J. Horn, J. Xu, A. Cognata, and D.E. Root, "Multi-tone, multi-port, and dynamic memory enhancements to PHD nonlinear behavioral models from large-signal measurements and simulations," in *IEEE MTT-S Int. Microwave Symp. Dig.*, Honolulu, HI, June 2007, pp. 969–972.

Appendix A: Notations and general definitions

A.1 Sets

$\{x_1, x_2, x_3\}$ a set of elements (curly brackets)
$\mathbb{N}$ the set of natural numbers: $\{1, 2, \ldots\}$
$\mathbb{N}^*$ the set of natural numbers, including zero: $\{0, 1, 2, \ldots\}$

$$\mathbb{N}^* = \mathbb{N} \cup \{0\} \tag{A.1}$$

$\mathbb{Z}$ the set of integer numbers: $\{\ldots, -2, -1, 0, 1, 2, \ldots\}$
$\mathbb{R}$ the set of real numbers
$\mathbb{R}_+$ the set of real positive numbers
$\mathbb{R}_-$ the set of real negative numbers
$\mathbb{C}$ the set of complex numbers
z^* complex conjugate

A.2 Vectors and matrices

A vector or a matrix is identified by a square bracket. The square bracket may be omitted in the text when the ideas remain clear. Sometimes the omission of the square brackets might be mentioned in the text, for clarity.

A vector of n elements (square brackets):

$$[x] = \begin{bmatrix} x_1 \\ x_2 \\ \vdots \\ x_n \end{bmatrix}. \tag{A.2}$$

A matrix of $n \times m$ elements (square brackets):

$$[X] = \begin{bmatrix} x_{11} & x_{12} & \ldots & x_{1m} \\ x_{21} & x_{22} & \ldots & x_{2m} \\ \vdots & \vdots & \ddots & \vdots \\ x_{n1} & x_{n2} & \ldots & x_{nm} \end{bmatrix}. \tag{A.3}$$

Transposed matrix:

$$[X]^T = \begin{bmatrix} x_{11} & x_{12} & \dots & x_{1m} \\ x_{21} & x_{22} & \dots & x_{2m} \\ \vdots & \vdots & \ddots & \vdots \\ x_{n1} & x_{n2} & \dots & x_{nm} \end{bmatrix}^T = \begin{bmatrix} x_{11} & x_{21} & \dots & x_{n1} \\ x_{12} & x_{22} & \dots & x_{n2} \\ \vdots & \vdots & \ddots & \vdots \\ x_{1m} & x_{2m} & \dots & x_{nm} \end{bmatrix}. \quad (A.4)$$

Unity matrix:

$$[I] = \begin{bmatrix} 1 & 0 & \dots & 0 \\ 0 & 1 & \dots & 0 \\ \vdots & \vdots & \ddots & \vdots \\ 0 & 0 & \dots & 1 \end{bmatrix}. \quad (A.5)$$

A.3 Signal representations

A.3.1 Time-domain representation (real signal)

$$\begin{aligned} s(t) &= S^{(r)} \cos(\omega t + \phi), \\ \omega &= 2\pi f, \end{aligned} \quad (A.6)$$

where

$S^{(r)}$ = (real) amplitude
ω = angular frequency
f = frequency
ϕ = initial phase
$\omega t + \phi$ = (instantaneous) phase

A.3.2 Complex representation (complex envelope signal)

$$\begin{aligned} s(t) &= \mathrm{Re}\{Se^{j\omega t}\}, \\ j &= \sqrt{-1}, \end{aligned} \quad (A.7)$$

where S is the complex amplitude:

$$S = S^{(r)} e^{j \cdot \phi}. \quad (A.8)$$

It is common in the industry to use an identifier for the complex amplitude (S_c, for example). In this text the preference is to use an identifier for the real amplitude ($S^{(r)}$) because it is the complex amplitude that is used in the majority of cases presented.

The complex amplitude, S, is typically interpreted as a vector in the complex plane, also known as a phasor. The RMS value of the complex amplitude, shown in (A.9), is used often in engineering practice:

$$S^{(RMS)} = \frac{S}{\sqrt{2}}. \quad (A.9)$$

In this text, its vector interpretation is called the RMS vector or RMS phasor.

A.4 Fourier analysis

Although other conventions exist and are in use, the convention presented here is used in this book and is commonly used in the electrical engineering field.

The Fourier series is expressed as shown in (A.10):

$$x(t) = \frac{c_0}{2} + \sum_{n=1}^{\infty} c_n \cos(n\omega t) + \sum_{n=1}^{\infty} s_n \sin(n\omega t), \tag{A.10}$$

where the Fourier coefficients are determined as follows:

$$\begin{aligned} c_0 &= \frac{2}{T}\int_0^T x(t)dt, \\ c_n &= \frac{2}{T}\int_0^T x(t)\cos(n\omega t)dt, \\ s_n &= \frac{2}{T}\int_0^T x(t)\sin(n\omega t)dt. \end{aligned} \tag{A.11}$$

A more compact form can be obtained using the notations in (A.12):

$$\begin{aligned} A_0 &= c_0, \\ A_n &= \sqrt{c_n^2 + s_n^2}, \\ \phi_n &= \begin{cases} \operatorname{atan}\left(\frac{-s_n}{c_n}\right) & \begin{cases} s_n > 0 \text{ and } c_n \geq 0 \\ s_n < 0 \text{ and } c_n \geq 0 \\ s_n = 0 \text{ and } c_n > 0, \end{cases} \\ \operatorname{atan}\left(\frac{-s_n}{c_n}\right) + \pi & \begin{cases} s_n < 0 \text{ and } c_n \leq 0 \\ s_n > 0 \text{ and } c_n \leq 0 \\ s_n = 0 \text{ and } c_n < 0, \end{cases} \\ 0, & s_n = 0 \text{ and } c_n = 0. \end{cases} \end{aligned} \tag{A.12}$$

The coefficient A_n is the amplitude and ϕ_n is the phase of the spectral component on the frequency $n\omega$.

The Fourier series can then be expressed as shown in (A.13):

$$x(t) = \frac{A_0}{2} + \sum_{n=1}^{\infty} A_n \cos(n\omega t + \phi_n). \tag{A.13}$$

The complex amplitude is defined using (A.14):

$$A_{nc} = A_n e^{j\phi_n}. \tag{A.14}$$

The complex form of the Fourier series using only positive frequencies is shown in (A.15):

$$x(t) = \frac{A_0}{2} + \sum_{n=1}^{\infty} \operatorname{Re}\left(A_{nc} e^{jn\omega t}\right). \tag{A.15}$$

The complex form of the Fourier series using both positive and negative frequencies is shown in (A.16):

$$x(t) = \frac{1}{2}\sum_{n=-\infty}^{\infty} A_{nc}e^{jn\omega t}. \tag{A.16}$$

Using (A.12), the complex amplitude, A_{nc}, is expressed as shown in (A.17):

$$A_{nc} = c_n - js_n. \tag{A.17}$$

Considering (A.11), the complex amplitude, A_{nc}, can be determined directly from the original function, $x(t)$, as shown in (A.18):

$$A_{nc} = \frac{2}{T}\int_0^T x(t)e^{-jn\omega t}\, dt. \tag{A.18}$$

The spectrum of a non-periodic signal is continuous and is determined by the Fourier transform, $\mathfrak{F}\{.\}$, as shown in (A.19):

$$X(\omega) = \mathfrak{F}\left\{x(t)\right\} = \int_{-\infty}^{\infty} x(t)e^{-j\omega t}\, dt. \tag{A.19}$$

The time-domain signal can be determined from its frequency-domain spectrum using the inverse Fourier transform, as shown in (A.20):

$$x(t) = \mathfrak{F}^{-1}\left\{X(\omega)\right\} = \frac{1}{2\pi}\int_{-\infty}^{\infty} X(\omega)e^{j\omega t}\, d\omega. \tag{A.20}$$

A.5 Wave definitions

A.5.1 Generalized power waves

Given a network port, as shown in Figure A.1, the generalized power waves are defined as a function of port voltage and port current.

The voltage, current, and waves in Figure A.1 are sinusoidal signals with the same frequency, and they are represented in the frequency domain by their complex amplitudes.

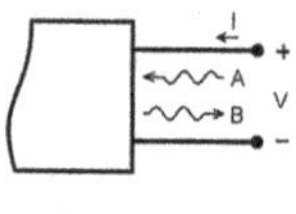

Figure A.1 A port of a network, with port voltage, current, and waves.

Considering a real and positive impedance for the port, $Z_0 \in \mathbb{R}_+$, the generalized power waves are defined by (A.21) [1], where A is the incident wave and B is the scattered wave:

$$\begin{aligned} A &= \frac{V + Z_0 I}{2\sqrt{Z_0}} \\ B &= \frac{V - Z_0 I}{2\sqrt{Z_0}} \end{aligned} \quad Z_0 \in \mathbb{R}_+. \tag{A.21}$$

The unit of measurement for generalized power waves is $\sqrt{\text{watt}}$, which results from (A.21).

The reference impedance is real and positive for the vast majority of practical situations. Almost all modern network analyzers use transmission lines, of characteristic impedance 50 Ω, which set the reference impedances for the measurement process.

A more general notation is used in this text than in [1], in the sense that V and I in (A.21) are not restricted to their RMS (complex) values, but rather they may have either peak (complex amplitude) or RMS (complex) values depending on the purpose for which the mathematical expression is used:

$$A^{(RMS)} = \frac{A^{(pk)}}{\sqrt{2}} \in \mathbb{C}. \tag{A.22}$$

For example, when power calculations are conducted, the RMS value must be used, as shown in (A.23) for the power of wave A:

$$\begin{aligned} A^{(RMS)} &= \frac{V^{(RMS)} + Z_0 I^{(RMS)}}{2\sqrt{Z_0}}, \\ \text{power of } A\text{-wave} &= \left|A^{(RMS)}\right|^2. \end{aligned} \tag{A.23}$$

When a time-domain representation is needed, the peak value must be used, as shown in (A.24) for the incident wave, A:

$$\begin{aligned} A^{(pk)} &= \frac{V^{(pk)} + Z_0 I^{(pk)}}{2\sqrt{Z_0}}, \\ a(t) &= \mathrm{Re}\left\{A^{(pk)} e^{j\cdot\omega\cdot t}\right\}. \end{aligned} \tag{A.24}$$

The time-domain representation of the generalized power waves is a relatively new concept, and is needed for a better understanding of the cross-frequency phase relationships between various sinusoidal signals used in the analysis of the behavior of nonlinear systems.

It can be easily proven that the time-domain waves are directly related to the time-domain representation of the port voltage and port current, as shown in (A.25):

$$\begin{aligned} a(t) &= \frac{v(t) + Z_0 i(t)}{2\sqrt{Z_0}} \\ b(t) &= \frac{v(t) - Z_0 i(t)}{2\sqrt{Z_0}} \end{aligned} \quad Z_0 \in \mathbb{R}_+. \tag{A.25}$$

The relationship between frequency and time domains, established by the Fourier transform, is thus extended to the generalized power waves. The complex amplitudes represent the coefficients of the Fourier series of a time-periodic wave, $b(t)$, as shown in (A.26):

$$B_k^{(pk)} = \frac{2}{T}\int_0^T b(t)e^{-jk\omega t}\,dt, \qquad \omega = \frac{2\pi}{T}. \tag{A.26}$$

The reciprocal conversion shown in (A.27) determines the time-domain wave when its frequency-domain representation is known (the complex amplitudes of its harmonics):

$$b(t) = \mathrm{Re}\left(\sum_{k=1}^{N} B_k^{(pk)} e^{jk\omega t}\right). \tag{A.27}$$

For convenience of notation, the "RMS" identifier is omitted in this text when possible. As a result, unless otherwise noted, the upper-case letters represent RMS values of the complex amplitudes. The "pk" identifier for peak values is used explicitly, where needed.

A.5.2 Voltage waves

An alternative set of waves is sometimes used in the practice of RF/MW measurement and design. These are voltage waves, which are defined as shown in (A.28), as a function of port voltage and current using the reference impedance, which is considered a real and positive number in this book:

$$\begin{aligned} A^{(v)} &= \frac{V + Z_0 I}{2} \\ B^{(v)} &= \frac{V - Z_0 I}{2} \end{aligned} \qquad Z_0 \in \mathbb{R}_+. \tag{A.28}$$

The identifier v is used here to distinguish clearly between the voltage waves $A^{(v)}$ and $B^{(v)}$ and the generalized power waves A and B.

The unit of measurement for voltage waves is the volt, which results from (A.28).

All the other comments presented for generalized power waves are also valid for the voltage waves.

A.6 Linear network matrix descriptions

A more detailed presentation of this topic can be found in [2].

An N-port network has $N + 1$ pins. One of the pins is selected as reference, and each of the remaining N pins forms a port with respect to the reference pin.

All port voltages, currents, incident waves, and scattered waves form corresponding vectors, as shown in (A.29):

$$[V] = \begin{bmatrix} V_1 \\ V_2 \\ \vdots \\ V_N \end{bmatrix}, \quad [I] = \begin{bmatrix} I_1 \\ I_2 \\ \vdots \\ I_N \end{bmatrix}, \quad [A] = \begin{bmatrix} A_1 \\ A_2 \\ \vdots \\ A_N \end{bmatrix}, \quad [B] = \begin{bmatrix} B_1 \\ B_2 \\ \vdots \\ B_N \end{bmatrix}. \tag{A.29}$$

A.6.1 S-parameters

The S-parameters relate scattered waves to the incident waves, as shown in

$$[B] = [S][A]. \tag{A.30}$$

A.6.2 Z-parameters

The Z-parameters relate voltages to currents, as shown in

$$[V] = [Z][I]. \tag{A.31}$$

A.6.3 Y-parameters

The Y-parameters relate currents to voltages, as shown in

$$[I] = [Y][V]. \tag{A.32}$$

References

[1] K. Kurokawa, "Power waves and the scattering matrix," *IEEE Trans. Microw. Theory Tech.*, vol. **13**, no. 2, Mar. 1965.

[2] G. Gonzalez, *Microwave Transistor Amplifiers*, 2nd edn. Englewood Cliffs, NJ: Prentice Hall, 1984.

Appendix B: X-parameters and Volterra theory

B.1 Introduction

This appendix describes the theoretical derivation of the properties of the X-parameter functions for a small $A_{1,1}$ amplitude. The theory is based on the assumption that the modeled behavior can be described by the Volterra theory in the limit for the amplitude of $A_{1,1}$ going to zero. The theory shows that the frequency-domain version of the Volterra theory can be considered as a McLaurin series polynomial approximation of the X-parameters.

B.2 Mathematical notation and problem definition

For clarity, the mathematical notation that is used is recalled in the following:

- $A_{p,h}$ the spectral component with harmonic index "h" of the incident wave at port "p"
- $B_{p,h}$ the spectral component with harmonic index "h" of the scattered wave at port "p"
- Y^* refers to the complex conjugate of the phasor "Y"
- P the unity length phasor having the same phase as $A_{1,1}$, defined as

$$P = \exp\left(j\phi(A_{1,1})\right). \tag{B.1}$$

The symbols $X_{r,m}^{(FB)}(.)$, $X_{r,m;s,n}^{(S)}(.)$, and $X_{r,m;s,n}^{(T)}(.)$ refer to the different kinds of X-parameter functions. They are defined by the following X-parameter equation (for simplicity it is assumed that the LSOP is defined solely by $|A_{1,1}|$):

$$B_{r,m} = X_{r,m}^{(FB)}\left(|A_{1,1}|\right) P^m + \sum_{(s,n)\neq(1,1)} \left(X_{r,m;s,n}^{(S)}\left(|A_{1,1}|\right) P^{m-n} A_{s,n} + X_{r,m;s,n}^{(T)}\left(|A_{1,1}|\right) P^{m+n} A_{s,n}^*\right). \tag{B.2}$$

Note that $X_{r,m}^{(FB)}(.)$ refers to the large-signal part of $B_{r,m}$, which is a function of the amplitude of $A_{1,1}$ only, $X_{r,m;s,n}^{(S)}(.)$ refers to the coefficient associated with the contribution to $B_{r,m}$ of the small-signal component $A_{s,n}$, and $X_{r,m;s,n}^{(T)}(.)$ refers to the coefficient associated with the contribution to $B_{r,m}$ of the conjugate of the small-signal component $A_{s,n}$. In order to simplify the analysis, the DC waves will first be excluded. The degenerate DC case will be analyzed in Section B.5.

The problem that is solved is the behavior of the functions $X^{(FB)}_{r,m}(.)$, $X^{(S)}_{r,m;s,n}(.)$, and $X^{(T)}_{r,m;s,n}(.)$ in the limit of the amplitude of $A_{1,1}$ going to zero. In other words, the goal is to find the properties of the McLaurin series for these functions versus $A_{1,1}$.

B.3 Application of the Volterra theory

In order to find the characteristics of the scattering functions for small amplitudes of $A_{1,1}$ it is assumed that the behavior can be described by the Volterra theory in the limit for the amplitude of $A_{1,1}$ going to zero. In the Volterra theory the output spectra are written as a series of convolutions of the input spectra [1][2]. As a consequence, if the input spectra are sets of discrete tones, where each tone is represented by an input phasor, the output spectra will also comprise sets of discrete tones. The frequencies of the tones present in the output are integer linear combinations of the frequencies of the input tones, and each output phasor can be written as a polynomial function of the input phasors and the conjugate of the input phasors. The order of the input phasors in each term of the polynomial that represents an output phasor will be such that the time-invariance principle holds. Expressed in the time domain, the time-invariance principle states that the application of an arbitrary delay in the input signal has to result in exactly the same delay for the output signal. Expressed in the frequency domain, the time-invariance principle states that the application of any linear phase shift to the input phasors has to result in the same linear phase shift to the output phasors. The above will be illustrated by the following example.

Consider that there are two tones present as the input signal. The input signal is in this case represented by a phasor x having a frequency f_x and a phasor y having a frequency f_y. According to the Volterra theory, the output spectrum is discrete and is represented by a set of phasors $z_{[mn]}$, with two integer indices "m" and "n" denoting the output frequency, whereby the output phasors $z_{[mn]}$ have a frequency equal to $mf_x + nf_y$. Also according to the Volterra theory, $z_{[mn]}$ can be written as the following polynomial in x, y, x^*, and y^*:

$$z_{[mn]} = \sum_{\substack{\forall p,q,r,s \geq 0 \\ (p-q-m)f_x + (r-s-n)f_y = 0}} H_{pqrs} x^p x^{*q} y^r y^{*s}. \tag{B.3}$$

In (B.3), the summation goes over a constrained set of positive integers p, q, r, and s that satisfy the time-invariance principle. The coefficient H_{pqrs} is called a Volterra kernel.

The principle of Volterra theory can be applied to the case where the input phasors are the $A_{s,n}$ and where the output phasors are the $B_{r,m}$. This case is different from the above two-tone case for three reasons. First, there is only one fundamental frequency present, and as such the frequency of the phasors $A_{s,n}$ and $B_{r,m}$ is indicated by their second integer index. Secondly there is a plurality of signal ports, where the port is indicated by the first index of the phasors $A_{s,n}$ and $B_{r,m}$. The port association of a kernel will be denoted by its first two indices. The third difference is that the description is limited to terms that are at most of order 1 in any of the $A_{s,n}$ and the conjugates of the $A_{s,n}$ for all indices (s,n) different from (1,1). The result is the following:

$$B_{r,m} = \sum_{sn}\left(\sum_{\substack{\forall p,q \geq 0 \\ p-q+n-m=0}} H_{rs,mn,pq10} A_{1,1}^{p} A_{1,1}^{*q} A_{s,n} + \sum_{\substack{\forall p,q \geq 0 \\ p-q-n-m=0}} H_{rs,mn,pq01} A_{1,1}^{p} A_{1,1}^{*q} A_{s,n}^{*}\right). \tag{B.4}$$

In (B.4) the summation is not only constrained by the equations $p - q + n - m = 0$ and $p - q - n - m = 0$, respectively, but also to indices p and q that are both non-negative integers.

B.4 Derivation of the McLaurin series

Next, (B.4) is rewritten by substituting $A_{1,1}$ by $|A_{1,1}|P$:

$$B_{r,m} = \sum_{sn}\left(\sum_{\substack{\forall p,q \geq 0 \\ p-q+n-m=0}} H_{rs,mn,pq10} |A_{1,1}|^{p+q} P^{p-q} A_{s,n} + \sum_{\substack{\forall p,q \geq 0 \\ p-q-n-m=0}} H_{rs,mn,pq01} |A_{1,1}|^{p+q} P^{p-q} A_{s,n}^{*}\right). \tag{B.5}$$

Because of the constraint on the p's and q's the exponent $(p - q)$ can be substituted by $(m - n)$ and $(m + n)$, respectively:

$$B_{r,m} = \sum_{sn}\left(\sum_{\substack{\forall p,q \geq 0 \\ p-q+n-m=0}} H_{rs,mn,pq10} |A_{1,1}|^{p+q} P^{m-n} A_{s,n} + \sum_{\substack{\forall p,q \geq 0 \\ p-q-n-m=0}} H_{rs,mn,pq01} |A_{1,1}|^{p+q} P^{m+n} A_{s,n}^{*}\right). \tag{B.6}$$

Next, the above expression will be simplified further by rewriting the exponent $(p + q)$.

Consider the first term in (B.6). If $(n - m)$ is positive, q can be substituted by $(p + n - m)$ whereby the summation in p ranges from 0 to infinity. If $(n - m)$ is negative, p can be substituted by $(q - n + m)$ whereby the summation in q ranges from 0 to infinity. This is illustrated by the following equations:

$$\text{if } n - m \geq 0, \quad \sum_{\substack{\forall p,q \geq 0 \\ p-q+n-m=0}} H_{rs,mn,pq10} |A_{1,1}|^{p+q} = \sum_{p=0}^{\infty} H_{rs,mn,p(p+n-m)10} |A_{1,1}|^{2p+n-m}; \tag{B.7}$$

$$\text{if } n-m<0, \quad \sum_{\substack{\forall p,q\geq 0\\ p-q+n-m=0}} H_{rs,mn,pq10}|A_{1,1}|^{p+q} = \sum_{q=0}^{\infty} H_{rs,mn,(q-n+m)q10}|A_{1,1}|^{2q-n+m}. \tag{B.8}$$

Consider the second term in (B.6). The variable p can be substituted by $(q + n + m)$ whereby the summation in q ranges from 0 to infinity:

$$\sum_{\substack{\forall p,q\geq 0\\ p-q-n-m=0}} H_{rs,mn,pq10}|A_{1,1}|^{p+q} = \sum_{q=0}^{\infty} H_{rs,mn,(q+n+m)q10}|A_{1,1}|^{2q+n+m}. \tag{B.9}$$

To simplify the mathematical notation, the symbol q is replaced by the symbol p as the iterator, and a new set of kernels is defined as follows:

$$\alpha_{rs,pmn} = H_{rs,mn,p(p+n-m)10} \quad \text{for} \quad n-m \geq 0; \tag{B.10}$$

$$\alpha_{rs,pmn} = H_{rs,mn,p(p+n-m)10} \text{ for } \quad n-m < 0; \tag{B.11}$$

$$\beta_{rs,pmn} = H_{rs,mn,(p+n+m)p10}. \tag{B.12}$$

Using (B.7) to (B.12) and substituting the results in (B.6) results in the following simplified expression:

$$B_{r,m} = \sum_{sn}\left(\sum_{p=0}^{\infty} \alpha_{rs,pmn}|A_{1,1}|^{2p+|m-n|}P^{m-n}A_{s,n} + \sum_{p=0}^{\infty}\beta_{rs,pmn}|A_{1,1}|^{2p+m+n}P^{m+n}A^*_{s,n}\right). \tag{B.13}$$

Comparing (B.13) with (B.2), the McLaurin series are identified for the functions $X^{(S)}_{r,m;s,n}(.)$ and $X^{(T)}_{r,m;s,n}(.)$ as follows:

$$X^{(S)}_{r,m;s,n} = \sum_{p=0}^{\infty} \alpha_{rs,pmn}|A_{1,1}|^{2p+|m-n|}; \tag{B.14}$$

$$X^{(T)}_{r,m;s,n} = \sum_{p=0}^{\infty} \beta_{rs,pmn}|A_{1,1}|^{2p+m+n}. \tag{B.15}$$

The McLaurin series for the function $X^{(FB)}_{r,m}(.)$ is found by looking at the special case of the summation in (B.13) where $(s,n) = (1,1)$:

$$X^{(FB)}_{r,m}(|A_{1,1}|)P^m = \sum_{p=0}^{\infty}\alpha_{r1,pm1}|A_{1,1}|^{2p+|m-1|}P^{m-1}A_{1,1} + \sum_{p=0}^{\infty}\beta_{r1,pm1}|A_{1,1}|^{2p+m+1}P^{m+1}A^*_{1,1}. \tag{B.16}$$

Eliminating P^m on both sides of (B.16) and substituting $A_{1,1}$ by $|A_{1,1}|P$ results in the following:

$$X^{(FB)}_{r,m}\left(|A_{1,1}|\right) = \sum_{p=0}^{\infty} \alpha_{r1,pm1} |A_{1,1}|^{2p+|m-1|+1} + \sum_{p=0}^{\infty} \beta_{r1,pm1} |A_{1,1}|^{2p+m+2}. \tag{B.17}$$

Excluding the DC output, which corresponds to $m = 0$, all m are greater than or equal to 1. In that case, $|m - 1|$ equals $m - 1$ and (B.17) can be rewritten as follows:

$$X^{(FB)}_{r,m}\left(|A_{1,1}|\right) = \sum_{p=0}^{\infty} \alpha_{r1,pm1} |A_{1,1}|^{2p+m} + \sum_{p=0}^{\infty} \beta_{r1,pm1} |A_{1,1}|^{2p+m+2}. \tag{B.18}$$

Finally, (B.18) can be written as a McLaurin series as follows:

$$X^{(FB)}_{r,m}\left(|A_{1,1}|\right) = \alpha_{r1,0m1} |A_{1,1}|^{m} + \sum_{p=0}^{\infty} \left(\alpha_{r1,(p+1)m1} + \beta_{r1,pm1}\right) |A_{1,1}|^{2p+m+2}. \tag{B.19}$$

B.5 McLaurin series for the DC output

The properties of the McLaurin series for the DC output equations can be found by performing the above analysis for $m = 0$. All equations up to (B.17) remain valid. The result is the following:

$$X^{(Y)}_{r;s,n}\left(|A_{1,1}|\right) = \sum_{p=0}^{\infty} \left(\alpha_{rs,p0n} + \beta_{rs,p0n}\right) |A_{1,1}|^{2p+n}; \tag{B.20}$$

$$X^{(Z)}_{r;s,n}\left(|A_{1,1}|\right) = \sum_{p=0}^{\infty} \left(\alpha_{rs,p0n} + \beta_{rs,p0n}\right) |A_{1,1}|^{2p+n}; \tag{B.21}$$

$$X^{(FI)}_{r}\left(|A_{1,1}|\right) = \alpha_{r1,001} + \sum_{p=0}^{\infty} \left(\alpha_{r1,(p+1)01} + \beta_{r1,p01}\right) |A_{1,1}|^{2p+2}. \tag{B.22}$$

B.6 Conclusions

Several terms are missing in the McLaurin series for the $X^{(FB)}_{r,m}(.)$, $X^{(S)}_{r,m;s,n}(.)$, and $X^{(T)}_{r,m;s,n}(.)$ functions. The results can be summarized as follows.

The lowest order of the $|A_{1,1}|$ term in the McLaurin series for $X^{(FB)}_{r,m}(|A_{1,1}|)$ is m.
The lowest order of the $|A_{1,1}|$ term in the McLaurin series for $X^{(S)}_{r,m;s,n}(|A_{1,1}|)$ is $|m - n|$.
The lowest order of the $|A_{1,1}|$ term in the McLaurin series for $X^{(T)}_{r,m;s,n}(|A_{1,1}|)$ is $m + n$.
For each of the functions the order of any consecutive $|A_{1,1}|$ term increases by 2.

The above conclusions also hold for the DC output functions ($m = 0$).

References

[1] J. C. Peyton Jones and S. A. Billings, "Describing functions, Volterra series, and the analysis of non-linear systems in the frequency domain," *Int. J. Control*, vol. **53**, no. 4, pp. 871–887, 1991.

[2] J. Verspecht, "Describing functions can better model hard nonlinearities in the frequency domain than the Volterra theory," Ph.D. thesis annex, Vrije Univ. Brussel, Belgium, Nov. 1995; available at http://www.janverspecht.com/skynet/Work/annex.pdf.

Appendix C: Parallel Hammerstein symmetry

According to [1], among the best performing behavioral models are the parallel Hammerstein (PH) models. This appendix shows that a PH model introduces a symmetry not consistent with realistic behavior in so-called complementary large-signal step responses.

Consider the response, denoted $B_{LS}(A_1, A_2, t)$, of an amplifier to a large-signal step from an initial amplitude of A_1 to an amplitude A_2. The complementary large-signal step response is just the response to a step input from amplitude A_2 to an amplitude A_1, namely $B_{LS}(A_2, A_1, t)$. It will be shown in the following that for PH models the sum of $B_{LS}(A_1, A_2, t)$ and $B_{LS}(A_2, A_1, t)$ is a constant, independent of time, that depends only on A_1 and A_2.

In a general PH model the relationship between the input $A(t)$ and the output $B(t)$ is given by

$$B(t) = \sum_{i=1}^{N} \int_{0}^{\infty} F_i\left(\left|A(t-u)\right|\right) e^{j\phi\left(A(t-u)\right)} h_i(u)du. \tag{C.1}$$

Expressed in words, the input signal goes through a set of N parallel static nonlinear operators $F_i(.)$, whereby each output is convolved with a specific linear filter impulse response $h_i(u)$. The output is then equal to the sum of the results of the convolutions. It is important to note that the PH model of (C.1) is actually an extension of the PH models described in [1] and [2] in the sense that the nonlinear operators $F_i(.)$ as described in [1] and [2] are restricted to simple monomials rather than general nonlinear functions, namely $F_i(x) = x^{2i-1}$.

The complementary large-signal step responses are calculated as follows:

$$B_{LS}(A_1, A_2, t) = \sum_{i=1}^{N} \left[\int_{0}^{t} F_i(A_2) h_i(u)du + \int_{t}^{\infty} F_i(A_1) h_i(u)du \right]; \tag{C.2}$$

$$B_{LS}(A_2, A_1, t) = \sum_{i=1}^{N} \left[\int_{0}^{t} F_i(A_1) h_i(u)du + \int_{t}^{\infty} F_i(A_2) h_i(u)du \right]. \tag{C.3}$$

The sum of the complementary large-signal step responses is given by

$$B_{LS}(A_1, A_2, t) + B_{LS}(A_2, A_1, t) = \sum_{i=1}^{N} \left[\int_0^\infty F_i(A_1) h_i(u) du + \int_0^\infty F_i(A_2) h_i(u) du \right], \quad \text{(C.4)}$$

which can be written as

$$B_{LS}(A_1, A_2, t) + B_{LS}(A_2, A_1, t) = \sum_{i=1}^{N} (F_i(A_1) + F_i(A_2)) \int_0^\infty h_i(u) du. \quad \text{(C.5)}$$

The right-hand side of (C.5) is independent of time, proving that the sum of complementary large-signal step responses generated by any PH model is constant in time, as claimed.

A simple example demonstrates that this is not consistent with realistic behavior. Consider the case whereby one switches from an amplitude equal to zero to a large amplitude, and this for a system with significant modulation-induced baseband memory. The large-signal step response will show a transient with a duration that corresponds to the time constants associated with the baseband memory effects. For any PH model, the complementary large-signal step response, the response of the amplifier when switching from a high input amplitude to zero, will have the same transient response, but with an opposite sign. As soon as the input amplitude is switched off, however, one would expect the output signal to go rapidly to zero, in a time span that corresponds to the very short time constants associated with the carrier memory effects, not to the long time constants associated with the modulation-induced baseband effects. This nonsensical relationship between complementary large-signal step responses is not present in the dynamical X-parameter model.

References

[1] J. C. Pedro and S. A. Maas, "A comparative overview of microwave and wireless power-amplifier behavioral modeling approaches," *IEEE Trans. Microw. Theory Tech.*, vol. **53**, no.4, Apr. 2005.

[2] M. Isaksson, D. Wisell and Daniel Rönnow, "A comparative analysis of behavioral models for RF power amplifiers," *IEEE Trans. Microw. Theory Tech.*, vol. **54**, no. 1, pp. 348–359, Jan. 2006.

Appendix D: Wide-band memory approximation

In this appendix it is proved that

$$\int_0^\infty G(|A(t)|, |A(t-\tau)|, \tau)d\tau \approx \int_0^\infty \left(\int_0^\infty G(|A(t)|, a, \tau)d\tau \right) \rho(a)da$$

under the following conditions:

(1) the input envelope is varying quickly compared to the time constants associated with the function $G(x,y,t)$;
(2) the statistical properties of the input envelope amplitude are stationary and are described by the PDF $\rho(a)$.

The proof begins by splitting the time axis into small intervals, each of duration T. The result is

$$\int_0^\infty G\Big(|A(t)|, |A(t-\tau)|, \tau\Big)d\tau = \sum_{i=0}^{\infty} \int_{iT}^{(i+1)T} G\Big(|A(t)|, |A(t-u)|, u\Big)du. \tag{D.1}$$

For any T that is small enough, the value of the function $G(.,.,.)$ will not significantly depend on the time variable over each individual interval. This results in

$$\int_0^\infty G\Big(|A(t)|, |A(t-\tau)|, \tau\Big)d\tau \approx \sum_{i=0}^{\infty} \int_{iT}^{(i+1)T} G\Big(|A(t)|, |A(t-u)|, iT\Big)du. \tag{D.2}$$

Note that the value of $|A(t-u)|$ may still vary significantly over each interval in the case of a fast varying envelope, which corresponds to a wideband modulation. The remaining integrals on the right-hand side can be written as the limit of a Riemann sum:

$$\int_0^\infty G\Big(|A(t)|, |A(t-\tau)|, \tau\Big)d\tau \approx \sum_{i=0}^{\infty} \lim_{M\to\infty} \sum_{j=0}^{M-1} G\left(|A(t)|, \left|A\left(t - iT - j\frac{T}{M}\right)\right|, iT \right) \frac{T}{M}, \tag{D.3}$$

which can be written as

$$\int_0^\infty G\Big(|A(t)|,|A(t-\tau)|,\tau\Big)d\tau \approx \sum_{i=0}^{\infty} T \lim_{M\to\infty}\left\{\frac{\sum_{j=0}^{M-1} G\left(|A(t)|,\left|A\left(t-iT-j\frac{T}{M}\right)\right|,iT\right)}{M}\right\}. \tag{D.4}$$

Consider now the expression denoted by the curly brackets {}. This expression simply represents the sample average of a large number of sampled values. The statistical law of large numbers states that the value of this average will converge to the expected value as the number of samples increases to infinity. This is mathematically expressed by

$$\lim_{M\to\infty}\left\{\frac{\sum_{j=0}^{M-1} G\left(|A(t)|,\left|A\left(t-iT-j\frac{T}{M}\right)\right|,iT\right)}{M}\right\} = \int_0^\infty G\Big(|A(t)|,a,iT\Big)\rho(a)da, \tag{D.5}$$

whereby $\rho(a)$ represents the probability density function of $A(t)$. Note that, in deriving the above equation, one uses the fact that $A(t)$ has stationary statistical properties. Substitution of (D.5) in (D.4) results in

$$\int_0^\infty G\Big(|A(t)|,|A(t-\tau)|,\tau\Big)d\tau \approx \sum_{i=0}^{\infty} T\int_0^\infty G\Big(|A(t)|,a,iT\Big)\rho(a)da. \tag{D.6}$$

The remaining sum corresponds to a Riemann sum. Taking the limit of this expression for ever smaller time intervals T leads to another Riemann integral:

$$\int_0^\infty G\Big(|A(t)|,|A(t-\tau)|,\tau\Big)d\tau \approx \lim_{T\to 0}\sum_{i=0}^{\infty} T\int_0^\infty G\Big(|A(t)|,a,iT\Big)\rho(a)da \approx \int_0^\infty\left(\int_0^\infty G\Big(|A(t)|,a,\tau\Big)d\tau\right)\rho(a)da. \tag{D.7}$$

Q.E.D.

Appendix E: Solutions to exercises

Exercise 1.1 Referring to Figure 1.9, Eq. (1.5) is used to write KCL at the internal node according to (E.1):

$$\begin{aligned} i(t) &= i_2^{(1)}(t) + i_1^{(2)}(t) = 0 \\ &= a_2^{(1)}(t) - b_2^{(1)}(t) + a_1^{(2)}(t) - b_1^{(2)}(t). \end{aligned} \tag{E.1}$$

The voltage at the internal node can be written using Eq. (1.5) according to (E.2):

$$v(t) = a_2^{(1)}(t) + b_2^{(1)}(t) = a_1^{(2)}(t) + b_1^{(2)}(t). \tag{E.2}$$

Solving (E.2) for $a_2^{(1)}(t)$ yields

$$a_2^{(1)}(t) = -b_2^{(1)}(t) + a_1^{(2)}(t) + b_1^{(2)}(t). \tag{E.3}$$

Substituting for $a_2^{(1)}(t)$ using (E.3) into (E.1) results in (E.4):

$$b_2^{(1)}(t) = a_1^{(2)}(t). \tag{E.4}$$

Substituting (E.4) into (E.1) results in

$$a_2^{(1)}(t) = b_1^{(2)}(t). \tag{E.5}$$

Transforming (E.4) and (E.5) into the frequency domain results in Eqs (1.17) and (1.18) for the relationships between the complex amplitudes of the waves, reproduced here as (E.6):

$$\begin{aligned} B_2^{(1)} &= A_1^{(2)}, \\ B_1^{(2)} &= A_2^{(1)}. \end{aligned} \tag{E.6}$$

Exercise 1.2 The objective is to reduce the problem to a 2×2 matrix defining the linear dependence of $B_1^{(1)}$ and $B_2^{(2)}$ on the incident waves, $A_1^{(1)}$ and $A_2^{(2)}$, since these are the reflected and incident waves of the composite structure.

Using the definitions of the S-parameter matrices of the two constituent systems, and the solution of Exercise 1.1, the set of equations governing the composite system is given by the sets of equations (E.7) and (E.6):

$$\begin{bmatrix} B_1^{(1)} \\ B_2^{(1)} \\ B_1^{(2)} \\ B_2^{(2)} \end{bmatrix} = \begin{bmatrix} S_{11}^{(1)} & S_{12}^{(1)} & 0 & 0 \\ S_{21}^{(1)} & S_{22}^{(1)} & 0 & 0 \\ 0 & 0 & S_{11}^{(2)} & S_{12}^{(2)} \\ 0 & 0 & S_{21}^{(2)} & S_{22}^{(2)} \end{bmatrix} \begin{bmatrix} A_1^{(1)} \\ A_2^{(1)} \\ A_1^{(2)} \\ A_2^{(2)} \end{bmatrix}. \tag{E.7}$$

Substituting for the incident waves $A_2^{(1)}$ and $A_1^{(2)}$ in terms of the scattered waves $B_1^{(2)}$ and $B_2^{(1)}$, respectively, using (E.6), (E.7) can now be written as (E.8):

$$\begin{bmatrix} B_1^{(1)} \\ B_2^{(1)} \\ B_1^{(2)} \\ B_2^{(2)} \end{bmatrix} = \begin{bmatrix} S_{11}^{(1)} & S_{12}^{(1)} & 0 & 0 \\ S_{21}^{(1)} & S_{22}^{(1)} & 0 & 0 \\ 0 & 0 & S_{11}^{(2)} & S_{12}^{(2)} \\ 0 & 0 & S_{21}^{(2)} & S_{22}^{(2)} \end{bmatrix} \begin{bmatrix} A_1^{(1)} \\ B_1^{(2)} \\ B_2^{(1)} \\ A_2^{(2)} \end{bmatrix}. \tag{E.8}$$

The equations given by the second and third rows of (E.8) can be used to solve for $B_1^{(2)}$ and $B_2^{(1)}$ in terms of $A_1^{(1)}$ and $A_2^{(2)}$ and four of the eight S-parameters of the original systems. Solving for $B_1^{(2)}$ and $B_2^{(1)}$ results in (E.9) and (E.10):

$$B_1^{(2)} = \frac{S_{21}^{(1)} S_{11}^{(2)}}{1 - S_{11}^{(2)} S_{22}^{(1)}} A_1^{(1)} + \frac{S_{12}^{(2)}}{1 - S_{11}^{(2)} S_{22}^{(1)}} A_2^{(2)}, \tag{E.9}$$

$$B_2^{(1)} = \frac{S_{21}^{(1)}}{1 - S_{11}^{(2)} S_{22}^{(1)}} A_1^{(1)} + \frac{S_{12}^{(2)} S_{22}^{(1)}}{1 - S_{11}^{(2)} S_{22}^{(1)}} A_2^{(2)}. \tag{E.10}$$

Equations (E.9) and (E.10) are used to substitute for $B_1^{(2)}$ and $B_2^{(1)}$ when evaluating the first and fourth rows of (E.8) for $B_1^{(1)}$ and $B_2^{(2)}$, respectively. After collecting terms in $A_1^{(1)}$ and $A_2^{(2)}$, the resulting 2×2 matrix expressing the linear relationships of $B_1^{(1)}$ and $B_2^{(2)}$ to $A_1^{(1)}$ and $A_2^{(2)}$ is the S-parameter matrix of the composite system. This is simply expression (1.19), reproduced here as (E.11):

$$S^{composite} = \begin{pmatrix} S_{11}^{(1)} + S_{11}^{(2)} \dfrac{S_{12}^{(1)} S_{21}^{(1)}}{1 - S_{22}^{(1)} S_{11}^{(2)}} & \dfrac{S_{12}^{(1)} S_{12}^{(2)}}{1 - S_{22}^{(1)} S_{11}^{(2)}} \\ \dfrac{S_{21}^{(1)} S_{21}^{(2)}}{1 - S_{22}^{(1)} S_{11}^{(2)}} & S_{22}^{(1)} \dfrac{S_{12}^{(2)} S_{21}^{(2)}}{1 - S_{22}^{(1)} S_{11}^{(2)}} + S_{22}^{(2)} \end{pmatrix}. \tag{E.11}$$

Exercise 1.3 Start from the definition of the S-parameter matrix, given in terms of the matrix relationship between the vector of scattered and incident waves, namely Eq. (A.30):

$$[B] = [S] \cdot [A]. \tag{E.12}$$

Substitute the vector relationships for the incident and scattered waves in terms of the currents and voltages using Eq. (1.1). The result is given by (E.13):

$$[I - Z_0 Y][V] = [S][I + Z_0 Y][V]. \tag{E.13}$$

Here I is the unit matrix and Y is the admittance matrix defined in Appendix A. This must hold for all vectors, $[V]$, so the matrix relationship (E.14) follows:

$$[I - Z_0 Y] = [S][I + Z_0 Y]. \tag{E.14}$$

Assuming the matrix $[I + Z_0 Y]$ has an inverse, (E.14) is post-multiplied by $[I + Z_0 Y]^{-1}$ to obtain (E.15), which is equivalent to Eq. (1.31):

$$[I - Z_0 Y][I + Z_0 Y]^{-1} = [S] \tag{E.15}$$

Exercise 1.4 Start from Eq. (1.30) for the model admittance matrix calculated in Eq. (1.10). The relevant matrices to multiply, using (E.15), are given in (E.16) and (E.17), where the bias-dependent arguments of the element values are not shown:

$$[I - Z_0 Y] = \begin{bmatrix} 1 - j\omega c_{GS} Z_0 & 0 \\ g_m Z_0 & 1 - g_{DS} Z_0 \end{bmatrix}, \tag{E.16}$$

$$[I + Z_0 Y]^{-1} = \frac{1}{1 + (g_{DS} + j\omega c_{GS}) Z_0 + j\omega c_{GS} g_{DS} Z_0^2} \begin{bmatrix} 1 + g_{DS} Z_0 & 0 \\ -g_m Z_0 & 1 + j\omega c_{GS} Z_0 \end{bmatrix}. \tag{E.17}$$

Matrix multiplying (E.16) and (E.17) produces, after algebraic simplification, Eq. (1.32), reproduced here as (E.18):

$$S = \begin{pmatrix} \dfrac{1 - j\omega c_{GS} Z_0}{1 + j\omega c_{GS} Z_0} & 0 \\ \dfrac{-2 g_m Z_0}{(1 + g_{DS} Z_0)(1 + j\omega c_{GS} Z_0)} & \dfrac{1 - g_{DS} Z_0}{1 + g_{DS} Z_0} \end{pmatrix}. \tag{E.18}$$

Exercise 2.1 Equation (2.51) is a time-invariant map provided that whenever $A_{1,1} \to A_{1,1} e^{j\phi}$ then $B_{2,1} \to B_{2,1} e^{j\phi}$ for any angle ϕ.

Evaluating the proposed map (2.51) for $A_{1,1} e^{j\phi}$ yields (E.19):

$$\begin{aligned} G A_{1,1} e^{j\phi} + \gamma (A_{1,1} e^{j\phi})^3 &= G A_{1,1} e^{j\phi} + \gamma A_{1,1}^3 e^{3j\phi} \\ &\neq \left(G A_{1,1} + \gamma A_{1,1}^3 \right) e^{j\phi} = B_{2,1} e^{j\phi}. \end{aligned} \tag{E.19}$$

Therefore, Eq. (2.51) is not a time-invariant map and cannot be a correct model of a time-invariant component.

Exercise 2.2 Evaluating the proposed map (2.52) for $A_{1,1}e^{j\phi}$, (E.20) is obtained as follows:

$$\begin{aligned} GA_{1,1}e^{j\phi} + \gamma\left|A_{1,1}e^{j\phi}\right|^2 A_{1,1}e^{j\phi} &= GA_{1,1}e^{j\phi} + \gamma\left|A_{1,1}\right|^2 A_{1,1}e^{j\phi} \\ &= \left(GA_{1,1} + \gamma\left|A_{1,1}\right|^2 A_{1,1}\right)e^{j\phi} = B_{2,1}e^{j\phi}. \end{aligned} \tag{E.20}$$

Therefore, Eq. (2.52) defines a map that is time invariant.

Exercise 2.3 Equation (2.52) can be written in the following form:

$$B_{2,1} = \left(G - \gamma|A_{1,1}|^2\right)A_{1,1} = \left(G - \gamma|A_{1,1}|^2\right)|A_{1,1}|e^{j\phi(A_{1,1})}. \tag{E.21}$$

Equation (E.21) can be written as (E.22):

$$B_{2,1}(A_{1,1}) = X_{2,1}(|A_{1,1}|)P, \tag{E.22}$$

where

$$X_{2,1}(|A_{1,1}|) = \left(G - \gamma|A_{1,1}|^2\right)|A_{1,1}| \tag{E.23}$$

and

$$P = e^{j\phi(A_{1,1})}. \tag{E.24}$$

Exercise 3.1 The response is determined by the X-parameter model:

$$B_{p,k} \cong X^{(FB)}_{p.k}(refLSOPS)P^k + X^{(S)}_{p,k;q,l}(refLSOPS)A_{q,l}P^{k-l} + X^{(T)}_{p,k;q,l}(refLSOPS)A^*_{q,l}P^{k+l}. \tag{E.25}$$

A simpler notation is used in this exercise:

$$\begin{aligned} X^{(S)}_{p,k;q,l}(refLSOPS) &= Se^{j\phi^{(XS)}}, \\ X^{(T)}_{p,k;q,l}(refLSOPS) &= Te^{j\phi^{(XT)}}, \\ A_{q,l} &= ae^{j\phi^{(a)}}. \end{aligned} \tag{E.26}$$

The first term in $B_{p,k}$, $X^{(FB)}_{p.k}(refLSOPS)P^k$, does not depend on the small-signal perturbation. It represents the response at the LSOP. It is thus sufficient to prove that the difference $B_{p,k} - X^{(FB)}_{p.k}(refLSOPS)P^k$ describes an ellipse when the phase of the small-signal perturbation $A_{q,l}$ varies.

The difference $B_{p,k} - X^{(FB)}_{p.k}(refLSOPS)P^k$ formally shifts the origin of the orthogonal system of coordinates to the tip of the vector $X^{(FB)}_{p.k}(refLSOPS)P^k$.

An additional notation is used for convenience in this exercise, as shown in (E.27):

$$
\begin{aligned}
&X^{(S)}_{p,k;q,l}(refLSOPS)P^{k-l} = \vec{S} = Se^{j\phi^{(S)}},\\
&X^{(T)}_{p,k;q,l}(refLSOPS)P^{k+l} = \vec{T} = Te^{j\phi^{(T)}},\\
&A_{q,l} = \vec{a} = ae^{j\phi^{(a)}},\\
&B_{p,k} - X^{(FB)}_{p.k}(refLSOPS)P^{k} = \vec{z}.
\end{aligned}
\tag{E.27}
$$

The notation P is related to the phase of the large-signal stimulus, $A_{1,1} = |A_{1,1}|e^{j\phi(A_{1,1})}$, as shown in its definition (E.28):

$$P = e^{j\phi(A_{1,1})}. \tag{E.28}$$

Using the definition (E.28), the phases $\phi^{(S)}$ and $\phi^{(T)}$ have the form shown in (E.29):

$$
\begin{aligned}
\phi^{(S)} &= \phi^{(XS)} + (k-l)\phi^{(A_{1,1})},\\
\phi^{(T)} &= \phi^{(XT)} + (k+l)\phi^{(A_{1,1})}.
\end{aligned}
\tag{E.29}
$$

Using the above notation, (E.25) transforms into (E.30):

$$\vec{z} = \vec{S}\vec{a} + \vec{T}\,\vec{a}^{*}, \tag{E.30}$$

$$\vec{z} = Sae^{j\left(\phi^{(S)}+\phi^{(a)}\right)} + Tae^{j\left(\phi^{(T)}-\phi^{(a)}\right)}. \tag{E.31}$$

The vector $\vec{z}$ has maximum amplitude when its two components are aligned. The notation $\phi^{(M)}$ designates the phase of $\vec{z}$ when it has maximum amplitude and $\phi^{(a_M)}$ denotes the phase of $\vec{a}$ when this condition exists; $\phi^{(M)}$ is determined by condition (E.32):

$$\phi^{(M)} = \phi^{(S)} + \phi^{(a_M)} = \phi^{(T)} - \phi^{(a_M)} + 2k\pi, \quad k \in \mathbb{Z}. \tag{E.32}$$

Equation (E.32) also determines the value of $\phi^{(a_M)}$, as shown in (E.33):

$$\phi^{(a_M)} = \frac{\phi^{(T)} - \phi^{(S)}}{2} + k\pi. \tag{E.33}$$

The direction of the maximum is shown in (E.34):

$$\phi^{(M)} = \phi^{(S)} + \phi^{(a_M)} = \phi^{(S)} + \frac{\phi^{(T)} - \phi^{(S)}}{2} + k\pi = \frac{\phi^{(S)} + \phi^{(T)}}{2} + k\pi. \tag{E.34}$$

Considering the periodicity of the phase, only two values represent distinct positions: the ones obtained for $k = 0$ and for $k = 1$, as shown in (E.35):

$$
\begin{aligned}
\phi_1^{(M)} &= \frac{\phi^{(S)} + \phi^{(T)}}{2} = \phi^{(M)},\\
\phi_2^{(M)} &= \frac{\phi^{(S)} + \phi^{(T)}}{2} + \pi = \phi^{(M)} + \pi.
\end{aligned}
\tag{E.35}
$$

The phase $\phi^{(M)}$ can be expressed as a function of the X-parameter phases and the phase of the original large-signal stimulus, as shown in (E.36), where (E.29) is used:

$$\phi^{(M)} = \frac{\phi^{(XS)} + \phi^{(XT)}}{2} + k\phi^{(A_{1,1})}. \tag{E.36}$$

The two positions of the vectors $\vec{z}$ with maximum amplitude reside on the same line, and the vectors extend into opposite directions from their origin. The distance between the tips of these two vectors has a value of $2a(S + T)$, as shown in (E.37):

$$\left| a(S+T)e^{j\phi_1^{(M)}} - a(S+T)e^{j\left(\phi_1^{(M)}+\pi\right)} \right| = 2a(S+T). \tag{E.37}$$

In a similar manner, it can be proven that $\vec{z}$ has minimum amplitude when its two components are counter-aligned:

$$\phi^{(m)} = \phi^{(S)} + \phi^{(a_m)} = \phi^{(T)} - \phi^{(a_m)} + 2k\pi + \pi, \quad k \in \mathbb{Z}. \tag{E.38}$$

The notation $\phi^{(m)}$ designates the phase of $\vec{z}$ when it has minimum amplitude and the notation $\phi^{(a_m)}$ designates the phase of $\vec{a}$ when this condition exists; $\phi^{(m)}$ is determined by condition (E.39):

$$\phi^{(a_m)} = \frac{\phi^{(T)} - \phi^{(S)}}{2} + (2k+1)\frac{\pi}{2}. \tag{E.39}$$

The direction of the minimum is shown in (E.40):

$$\phi^{(m)} = \phi^{(S)} + \phi^{(a_m)} = \frac{\phi^{(T)} + \phi^{(S)}}{2} + (2k+1)\frac{\pi}{2}. \tag{E.40}$$

Considering the periodicity of the phase, only two values represent distinct positions: the ones obtained for $k = 0$ and for $k = 1$, as shown in (E.41):

$$\begin{aligned} \phi_1^{(m)} &= \frac{\phi^{(S)} + \phi^{(T)}}{2} + \frac{\pi}{2} = \phi^{(m)}, \\ \phi_2^{(m)} &= \frac{\phi^{(S)} + \phi^{(T)}}{2} + 3\frac{\pi}{2} = \phi^{(m)} + \pi. \end{aligned} \tag{E.41}$$

The phase $\phi^{(m)}$ can be expressed as a function of the X-parameter phases and the phase of the original large-signal stimulus, as shown in (E.42), where (E.29) is used:

$$\phi^{(m)} = \frac{\phi^{(XS)} + \phi^{(XT)}}{2} + k\phi^{(A_{1,1})} + \frac{\pi}{2}. \tag{E.42}$$

The two positions of the vectors $\vec{z}$ with minimum amplitude reside on the same line, and the vectors extend into opposite directions from their origin. The distance between the tips of these two vectors has a value of $2a(S - T)$, as shown in (E.43):

$$\left| a(S-T)e^{j\phi_1^{(m)}} - a(S-T)e^{j\left(\phi_1^{(m)}+\pi\right)} \right| = 2a(S-T). \tag{E.43}$$

It is also important to emphasize that the two directions determined so far (of the maximum and minimum amplitudes) are perpendicular to each other, as shown in (E.44):

$$\phi^{(m)} - \phi^{(M)} = \frac{\pi}{2}. \tag{E.44}$$

It is thus conceivable that these two segments represent the two axes of an ellipse: the major axis with a value of $2a(S + T)$ and the minor axis with a value of $2a(S - T)$.

The foci need to be located on the major axis, symmetrically from the origin of $\vec{z}$. Let us consider the two points identified as $\vec{f}_1$ and $\vec{f}_2$, located on the major axis, symmetrically from the center, at distances x and $-x$, $x \geq 0$. The sum of the distances from any of the two ends of the major axis to these two points is shown in (E.45):

$$\left[(S+T)a - x\right] + \left[(S+T)a + x\right] = 2(S+T)a. \tag{E.45}$$

The sum of the distances from any of the two ends of the minor axis to these two points is shown in (E.46):

$$2\sqrt{(S-T)^2 a^2 + x^2}. \tag{E.46}$$

The two sums expressed in (E.45) and (E.46) must be the same if $\vec{f}_1$ and $\vec{f}_2$ are the foci of an ellipse. The value of x determined from this condition is shown in (E.47):

$$x = 2a\sqrt{ST}. \tag{E.47}$$

The two potential foci are thus identified by the following vectors:

$$\begin{aligned} \vec{f}_1 &= 2a\sqrt{ST}e^{j\phi^{(M)}}, \\ \vec{f}_2 &= 2a\sqrt{ST}e^{j\left(\phi^{(M)}+\pi\right)}. \end{aligned} \tag{E.48}$$

A rotation of the system of coordinates with $\phi^{(M)}$ is needed for a simpler mathematical treatment. After this rotation, the x-axis is along the major axis. All vectors maintain their magnitude, but their phases change with $-\phi^{(M)}$. After the rotation of the coordinate system, all vectors are identified with the letter "r" in front of them, as shown in (E.49):

$$\begin{aligned} \vec{rS} &= \vec{S}\, e^{-j\phi^{(M)}} = Se^{j\left(\phi^{(S)}-\phi^{(M)}\right)}, \\ \vec{rT} &= \vec{T}\, e^{-j\phi^{(M)}} = Te^{j\left(\phi^{(T)}-\phi^{(M)}\right)}, \\ \vec{ra} &= \vec{a}\, e^{-j\phi^{(M)}} = ae^{j\left(\phi^{(a)}-\phi^{(M)}\right)}; \end{aligned} \tag{E.49}$$

$$\begin{aligned} \vec{rf}_1 &= \vec{f}_1 e^{-j\phi^{(M)}} = 2a\sqrt{ST}e^{j\left(\phi^{(M)}-\phi^{(M)}\right)} = 2a\sqrt{ST}, \\ \vec{rf}_2 &= \vec{f}_2 e^{-j\phi^{(M)}} = 2a\sqrt{ST}e^{j\left(\phi^{(M)}-\phi^{(M)}+\pi\right)} = 2a\sqrt{ST}e^{j\pi} = -2a\sqrt{ST}. \end{aligned} \tag{E.50}$$

After the coordinate system rotation, the vector $\vec{z}$ can be expressed as in (E.51):

$$\begin{aligned} \vec{rz} &= \vec{rS}\,\vec{ra} + \vec{rT}\,\vec{ra}^* \\ &= Se^{j\left(\phi^{(S)}-\phi^{(M)}\right)}ae^{j\left(\phi^{(a)}-\phi^{(M)}\right)} + Te^{j\left(\phi^{(T)}-\phi^{(M)}\right)}ae^{-j\left(\phi^{(a)}-\phi^{(M)}\right)} \\ &= Sae^{j\left(\phi^{(S)}+\phi^{(a)}-2\phi^{(M)}\right)} + Tae^{j\left(\phi^{(T)}-\phi^{(a)}\right)}. \end{aligned} \tag{E.51}$$

Considering the value of $\phi^{(M)}$ from (E.35), $\vec{rz}$ can be expressed as in (E.52):

$$\vec{rz} = Sae^{j\left(\phi^{(a)}-\phi^{(T)}\right)} + Tae^{-j\left(\phi^{(a)}-\phi^{(T)}\right)}. \tag{E.52}$$

Using the notation in (E.53),

$$\alpha = \phi^{(a)} - \phi^{(T)}, \tag{E.53}$$

$\vec{rz}$ is reformulated as shown in (E.54):

$$\vec{rz} = Sae^{j\alpha} + Tae^{-j\alpha}. \tag{E.54}$$

The distance from the tip of $\vec{rz}$ to the tip of $\vec{rf}_1$ is shown in (E.55):

$$d_1 = \sqrt{\left[a(S+T)\cos(\alpha) - 2a\sqrt{ST}\right]^2 + [a(S-T)\sin(\alpha)]^2}. \tag{E.55}$$

After some algebraic transformations, d_1 can be expressed as shown in (E.56):

$$d_1 = a(S+T) - 2a\sqrt{ST}\cos(\alpha). \tag{E.56}$$

In a similar fashion, the distance from the tip of $\vec{rz}$ to the tip of $\vec{rf}_2$ is shown in (E.57):

$$d_2 = \sqrt{\left[a(S+T)\cos(\alpha) + 2a\sqrt{ST}\right]^2 + [a(S-T)\sin(\alpha)]^2}. \tag{E.57}$$

After some algebraic transformations, d_2 can be expressed as shown in (E.58):

$$d_2 = a(S+T) + 2a\sqrt{ST}\cos(\alpha). \tag{E.58}$$

The sum of the two distances is shown in (E.59); it is constant with respect to the angle α, and hence independent of the phase of the small-signal perturbation:

$$d = d_1 + d_2 = 2a(S+T). \tag{E.59}$$

This proves that the contour described by the tip of the vector $B_{p,k}$ is an ellipse.

Exercise 5.1 The signals can be written, according to Appendix A, as (E.60) and (E.61):

$$a_1(t) = \mathrm{Re}\{A_{1,1}e^{j\omega t}\} = \mathrm{Re}\left\{|A_{1,1}|e^{j(\phi(A_{1,1})-\omega t)}\right\}, \tag{E.60}$$

$$a_2(t) = \mathrm{Re}\{A_{2,1}e^{j\omega t}\} = \mathrm{Re}\left\{|A_{2,1}|e^{j(\phi(A_{2,1})-\omega t)}\right\}. \tag{E.61}$$

At any instant in time, the phase difference between the signals, $\phi_{2,1}$, is simply given by (E.62):

$$\phi_{2,1} \equiv \phi(A_{2,1}) - \omega t - (\phi(A_{1,1}) - \omega t) = \phi(A_{2,1}) - \phi(A_{1,1}). \tag{E.62}$$

Clearly, from (E.62), $\phi_{2,1}$ is independent of time (constant).

Equation (E.63) then follows trivially from (E.62), for $P = e^{j\phi(A_{1,1})}$:

$$e^{j\phi_{2,1}} = e^{j\left[\phi(A_{2,1}) - \phi(A_{1,1})\right]} = e^{j\phi(A_{2,1})}P^{-1}. \tag{E.63}$$

This is just Eq. (5.3).

Exercise 5.2 The number of ideal X-parameter measurements is the sum of the number of measurements with one large incident tone and the number of measurements with both large and small probe tones applied simultaneously.

In the former case, for each power and frequency, there are just N_{Z_1} measurements, equal to the number of specified complex loads at the fundamental frequency.

For the cases where there are perturbation tones in addition to the large incident signal, there are two measurements (one for each of the two distinct phases needed to identify the $X^{(S)}$ and $X^{(T)}$ terms) at each of the second and third harmonics considered in this exercise.

The sum of these numbers is given by (E.64), which is equivalent to Eq. (5.14):

$$N_{X\text{-}par} = N_{Z_1} + 2{\cdot}2{\cdot}N_{Z_1} = (1 + 2{\cdot}2)N_{Z_1} = 5N_{Z_1}. \tag{E.64}$$

For independent control of the fundamental, second, and third harmonic impedances, the total number of measurements, per power per frequency, is just the product of the number of complex loads independently presented at each harmonic, given by (E.65), which is the result presented in Eq. (5.15):

$$N_{harm\ l\text{-}p} = N_{Z_1}{\cdot}N_{Z_2}{\cdot}N_{Z_3}. \tag{E.65}$$

Exercise 5.3 Start from Eq. (5.27b), written in the manifestly time-invariant form (E.66):

$$B_{q,[n,m]} = B_{q,[n,m]}\left(DCS,\ |A_{p_1,[1,0]}|,\ |A_{p_2,[0,1]}|,\ \ldots, A_{p',[n,m]}P_{[1,0]}^{-n}P_{[0,1]}^{-m}\right)P_{[1,0]}^{n}P_{[0,1]}^{m}. \tag{E.66}$$

Expanding the non-analytic function (E.66) for small values of the arguments of all non-trivial intermodulation products of the fundamental tones, Eq. (E.67) is obtained:

$$\begin{aligned} B_{q,[n,m]} &= X^{(F)}_{q,[n,m]}(refLSOPS)\cdot P_{[1,0]}^{n}P_{[0,1]}^{m} \\ &\quad + \sum_{n',m',q'}\left[\left.\frac{\partial B_{q,[n,m]}}{\partial\left(A_{q',[n'm']}P_{[1,0]}^{-n'}P_{[0,1]}^{-m'}\right)}\right|_{refLSOPS} A_{q',[n'm']}P_{[1,0]}^{-n'}P_{[0,1]}^{-m'} + \left.\frac{\partial B_{q,[n,m]}}{\partial\left(A^{*}_{q',[n'm']}P_{[1,0]}^{n'}P_{[0,1]}^{m'}\right)}\right|_{refLSOPS} A^{*}_{q',[n'm']}P_{[1,0]}^{n'}P_{[0,1]}^{m'}\right]P_{[1,0]}^{n}P_{[0,1]}^{m} \\ &= X^{(FB)}_{q,[n,m]}(refLSOPS)\cdot P_{[1,0]}^{n}P_{[0,1]}^{m} \\ &\quad + \sum_{n',m',q'}\left[X^{(S)}_{q,[n,m];q',[n',m']}(refLSOPS)A_{q',[n'm']}P_{[1,0]}^{n-n'}P_{[0,1]}^{m-m'} + X^{(T)}_{q,[n,m];q',[n',m']}(refLSOPS)A^{*}_{q',[n'm']}P_{[1,0]}^{n+n'}P_{[0,1]}^{m+m'}\right]. \end{aligned} \tag{E.67}$$

In (E.67)

$$X^{(S)}_{q,[n,m];q',[n',m']} = \left.\frac{\partial B_{q,[n,m]}}{\partial\left(A_{q',[n'm']}P^{-n'}_{[1,0]}P^{-m'}_{[0,1]}\right)}\right|_{refLSOPS} = \left.\frac{\partial B_{q,[n,m]}}{\partial A_{q',[n',m']}}\right|_{refLSOPS} P^{n'}_{[1,0]}P^{m'}_{[0,1]} \tag{E.68}$$

and

$$X^{(T)}_{q,[n,m];q',[n',m']} = \left.\frac{\partial B_{q,[n,m]}}{\partial\left(A^{*}_{q',[n'm']}P^{n'}_{[1,0]}P^{m'}_{[0,1]}\right)}\right|_{refLSOPS} = \left.\frac{\partial B_{q,[n,m]}}{\partial A^{*}_{q',[n',m']}}\right|_{refLSOPS} P^{-n'}_{[1,0]}P^{-m'}_{[0,1]}. \tag{E.69}$$

Therefore (E.68) and (E.69) are equivalent to Eq. (5.32), and (E.67) is equivalent to (5.29), for the refLSOPS given by (5.28), reproduced here as (E.70):

$$refLSOPS = \left(DCS_p,\ |A_{p_1,[1,0]}|,\ |A_{p_2,[0,1]}|, 0,\ \ldots, 0\right); \tag{E.70}$$

(E.71) has also been used:

$$X^{(F)}_{q,[n,m]}(refLSOPS) = B_{q,[n,m]}\left(DCS,\ |A_{p_1,[1,0]}|,\ |A_{p_2,[0,1]}|, 0, 0,\ \ldots, 0\right). \tag{E.71}$$

Index

Printed in the United States
By Bookmasters